An Introduction to Central Simple Algebras and Their Applications to Wireless Communication

Mathematical
Surveys
and
Monographs

Volume 191

An Introduction to Central Simple Algebras and Their Applications to Wireless Communication

Grégory Berhuy
Frédérique Oggier

American Mathematical Society
Providence, Rhode Island

2010 *Mathematics Subject Classification.* Primary 12E15; Secondary 11T71, 16W10.

For additional information and updates on this book, visit
www.ams.org/bookpages/surv-191

Library of Congress Cataloging-in-Publication Data

Berhuy, Grégory.
   An introduction to central simple algebras and their applications to wireless communications
/ Grégory Berhuy, Frédérique Oggier.
     pages cm. – (Mathematical surveys and monographs ; volume 191)
   Includes bibliographical references and index.
   ISBN 978-0-8218-4937-8 (alk. paper)
   1. Division algebras. 2. Skew fields. I. Oggier, Frédérique. II. Title.

QA247.45.B47   2013
512′.3–dc23
                                2013009629

Contents

Foreword

Mathematics continually surprises and delights us with how useful its most abstract branches turn out to be in the real world. Indeed, physicist Eugene Wigner's memorable phrase[1] "The unreasonable effectiveness of mathematics" captures a critical aspect of this utility. Abstract mathematical ideas often prove to be useful in rather "unreasonable" situations: places where one, a priori, would not expect them at all! For instance, no one who was not actually following the theoretical explorations in multi-antenna wireless communication of the late 1990s would have predicted that division algebras would turn out to be vital in the deployment of multi-antenna communication. Yet, once performance criteria for space-time codes (as coding schemes for multi-antenna environments are called) were developed and phrased as a problem of design of matrices, it was completely natural that division algebras should arise as a solution of the design problem. The fundamental performance criterion ask for $n \times n$ matrices M_i such that the difference of any two of the M_i is of full rank. To anyone who has worked with division algebras, the solution simply leaps out: any division algebra of index n embeds into the $n \times n$ matrices over a suitable field, and the matrices arising from the embedding naturally satisfy this criterion.

But there is more. Not only did division algebras turn out to be the most natural context in which to solve the fundamental design problem above, they also proved to be the correct objects to satisfy various other performance criteria that were developed. For instance, a second, and critical, performance criterion called the coding gain criterion turned out to be naturally satisfied by considering division algebras over number fields and using natural $\mathbb{Z}$-orders within them that arise from rings of integers of maximal subfields. Other criteria (for instance "DMG optimality," "good shaping," "information-losslessness" to name just a few) all turned out to be satisfied by considering suitable orders inside suitable division algebras over number fields. Indeed, this exemplifies another phenomenon Wigner describes: after saying that "mathematical concepts turn up in entirely unexpected connections," he goes on to say that "they often permit an unexpectedly close and accurate description of the phenomena in these connections." The match between division algebras and the requirement of space-time codes is simply uncanny.

The subject of multi-antenna communication has several unsolved mathematical problems still, for instance, in the area of decoding for large numbers of antennas. Nevertheless, division algebras are already being deployed for practical two-antenna

---

[1]Eugene P. Wigner, The unreasonable effectiveness of mathematics in the natural sciences, Comm. Pure Appl. Math., **13** Feb. 1960, 1–14

systems, and codes based on them are now part of various standards of the Institute of Electrical and Electronics Engineers (IEEE). It would behoove a student of mathematics, therefore, to know something about the applicability of division algebras while studying their theory; in parallel, it is vital for a communications engineer working in coding for multiple-antenna wireless to know something about division algebras.

Berhuy and Oggier have written a *charming* text on division algebras and their application to multiple-antenna wireless communication. There is a wealth of examples here, particularly over number fields and local fields, with explicit calculations, that one does not see in other texts on the subject. By pairing almost every chapter with a discussion of issues from wireless communication, the authors have given a very concrete flavor to the subject of division algebras. The book can be studied profitably not just by a graduate student in mathematics, but also by a mathematically sophisticated coding theorist. I suspect therefore that this book will find wide acceptability in both the mathematics and the space-time coding community and will help cross-communication between the two. I applaud the authors' efforts behind this very enjoyable book.

B.A. Sethuraman

Northridge, California

Introduction

A central simple algebra over a field k is a finite dimensional k-algebra with center k which does not have any proper two-sided ideals. The most elementary example is the Hamilton quaternion algebra. More generally, a division ring with center k can be viewed as a central simple k-algebra, where the algebra structure is induced by the multiplication law. Central simple algebras and division rings have been extensively studied, and have appeared in many other areas of mathematics, such as ring theory, number theory, representation theory of finite groups, algebraic geometry or classification theory of quadratic forms. Surprisingly, they have recently been proven useful in coding theory.

The ambition of this book is to provide an introduction to the theory of central simple algebras accessible at a graduate level, starting from scratch and including fundamental concepts such as splitting fields, Brauer group, crossed product algebras, index and exponent, as well as algebras with involution. Even though most of our exposition is rather classical, we have tried to focus on explicit techniques and examples, most of them coming from coding theory. The codes presented in this book are there to illustrate the theory of central simple algebras, and do not give an exhaustive view of the work done on the theory of algebraic space-time coding.

The use of division algebras for space-time coding is usually attributed to the seminal work by B. A. Sethuraman et al. [48]. Number fields and cyclic algebras were discussed, which have been a favourite tool for space-time design (see for example [12, 6, 40, 13, 32, 55]). Other algebras have been explored, such as Clifford algebras [27], or crossed product algebras (e.g. [57]).

Alternative studies considered the use of maximal orders (e.g. [56]) or non-associative algebras (e.g. [42]).

Some surveys on coherent space-time coding [36, 45] and one survey on non-coherent space-time coding [35] are now available. These works are just representing a few of the different approaches studied so far in the area of space-time coding, which is still an active field of research. These are just pointers for the interested reader, and by no mean provide a complete list.

In Chapter I, we introduce the concept of a central simple k-algebra and give the first examples of such algebras, including quaternion algebras. We then explain how they can be embedded into matrix algebras, and how this result may be used in coding theory. In Chapter II, we have a closer look at the properties of quaternion algebras. We also prove that the only finite dimensional division $\mathbb{R}$-algebras are, up to isomorphism, $\mathbb{R}, \mathbb{C}$ or the Hamilton quaternion algebra $\mathbb{H}$. We then provide examples of quaternion based codes. The results presented in Chapter III are the core of the theory. We first study the stability of central simple k-algebras under algebraic operations such as tensor product or base field extension. We then prove that any central simple k-algebra is isomorphic to a matrix algebra over a central division k-algebra, and establish that every k-automorphism of such an algebra is inner. We also focus on the structure of the centralizer of a simple subalgebra, which is a crucial tool in the study of maximal subfields and splitting fields of central simple algebras, which will be developed in Chapter IV. As an application of this theory, we define the reduced characteristic polynomial of an element of a central simple algebra, and introduce the concept of the reduced norm, which generalizes the determinant of a matrix. The latter can in turn be used to reinterpret code parameters. In Chapter V, we define the Brauer group $\mathrm{Br}(k)$ of a field k, which allows us to study globally all central simple k-algebras. We show that this group is an abelian torsion group, and use this result to define the exponent of a central simple k-algebra. We end this chapter by establishing the existence of a primary decomposition of a central simple k-algebra. In Chapter VI, we characterize central simple algebras which have a Galois maximal subfield. This leads to the notion of a crossed product algebra. We then present the standard results on these particular algebras. At the end of this chapter, crossed product algebras are used to construct families of codes. Chapter VII is devoted to cyclic algebras, that is, the case where the Galois maximal subfield is cyclic. At this occasion, an overview of the theory of central simple algebras over local and number fields is given without proofs. Explicit criteria to decide whether a given central simple algebra over a global field is division are established. Finally, these criteria are used to design codes based on cyclic division algebras. Chapter VIII focuses on central simple k-algebras of degree 4. We show that these algebras are crossed products over a biquadratic extension L/k, and a full description by generators and relations is given. We also provide a criterion to check if such an algebra is division in terms of the parameters defining the algebra when k is a number field, and applications to code constructions are given. In Chapter IX, the concept of a unitary involution on a central simple algebra is defined. The existence of unitary involutions is then investigated. We particularly focus on the case of crossed product algebras. We then explain how central simple algebras with a unitary involution may be used in coding theory via the construction of unitary codes, and we give various examples.

We would like to sincerely thank N. Markyn, S. Pumpluen, A. Quéguiner-Mathieu, B.A. Sethuraman, J.-P. Tignol, T. Unger and R. Vehkalahti for their careful reading of substantial parts of this book. Their pertinent comments enabled us to dramatically improve the quality of the exposition.

CHAPTER I

Central simple algebras

This chapter contains the necessary definitions and background on central simple algebras. After some preliminaries on k-algebras and tensor products, we introduce central simple algebras and give some examples. We then show how to identify central simple algebras with matrix subalgebras. As a first illustration, we explain how central simple algebras may be used in coding theory, and examples of code constructions are presented.

I.1. Preliminaries on k-algebras

In the sequel, k will denote an arbitrary field.

DEFINITION I.1.1. A k-**algebra** is a pair (A, μ), where A is a k-vector space and $\mu : A \times A \longrightarrow A$ is a k-bilinear map, called the **product law** of A. We write aa' for $\mu(a, a')$, and call it the **product** of the elements a and a'.

A k-algebra A is called **associative** (resp. **commutative**, resp. **unital**) if the product law is associative (resp. commutative, resp. has a unit element 1_A).

EXAMPLES I.1.2.

(1) The ring of polynomials $k[X]$ is a commutative, associative and unital k-algebra.

(2) If L/k is a field extension, then L is a commutative, associative and unital k-algebra.

$\square$

DEFINITION I.1.3. A k-**algebra morphism** is a k-linear map $f : A \longrightarrow B$ satisfying

$$f(aa') = f(a)f(a') \text{ for all } a, a' \in A.$$

If A and B are unital, we require in addition that $f(1_A) = 1_B$. A k-**algebra isomorphism** is a k-algebra morphism which is bijective. In this case, the inverse map f^{-1} is also a k-algebra morphism.

DEFINITION I.1.4. A **subalgebra** of a k-algebra A is a linear subspace B of A which is closed under the product. If A is unital, we require in addition that $1_A \in B$. It is unital, (resp. associative, resp. commutative) whenever A is.

EXAMPLES I.1.5.

(1) The intersection of an arbitrary family of subalgebras of a k-algebra A is again a subalgebra of A.

(2) The image of any k-algebra morphism $f : A \longrightarrow B$ is a subalgebra of B.

$\square$

DEFINITION I.1.6. The **center** of a k-algebra A is by definition the set

$$Z(A) = \{z \in A \mid az = za \text{ for all } a \in A\}.$$

It is a commutative subalgebra of A whenever A is associative.

EXAMPLE I.1.7. The matrix algebra $\mathrm{M}_n(k)$, consisting of $n \times n$ matrices with entries from k, is a unital k-algebra with center k (we identify k with the set of scalar matrices). $\square$

REMARK I.1.8. If A is an associative unital k-algebra, then addition and product naturally endow A with a ring structure. In particular, every subalgebra of A is also a subring, and every k-algebra morphism is also a ring morphism. Moreover if $1_A \neq 0_A$ (i.e. A is not zero), k identifies with a subalgebra of $Z(A)$ (hence a subalgebra of A).

Indeed the k-bilinearity of the product law and the properties of 1_A imply that we have

$$(\lambda \cdot 1_A)a = 1_A(\lambda \cdot a) = (\lambda \cdot a)1_A = a(\lambda \cdot 1_A)$$

for all $a \in A$ and $\lambda \in k$, so $k \cdot 1_A \subset Z(A)$. One may verify that $k \cdot 1_A$ is a k-subalgebra of $Z(A)$. Hence the map

$$k \longrightarrow Z(A)$$
$$\lambda \longmapsto \lambda \cdot 1_A$$

is a non-trivial k-algebra morphism, which is injective since k is a field. $\square$

In this book, all k-algebras will implicitly be assumed to be unital, associative, and finite-dimensional over k. Moreover, we will systematically identify k and $k \cdot 1_A$.

DEFINITION I.1.9. A **division** k-algebra is a k-algebra which is also a division ring (that is, every non-zero element is invertible).

At this stage, it may be worth making a few remarks on subalgebras of finite dimensional division algebras generated by a single element.

Let D be a finite dimensional division k-algebra, and let $d \in D$. We denote by $k[d]$ the smallest subalgebra of D containing d, and by $k(d)$ the smallest division subalgebra of D containing d. Clearly, we have

$$k[d] = \{P(d) \mid P \in k[X]\}.$$

Since D is finite dimensional over k, so is $k[d]$. Therefore, the successive powers of d cannot be linearly independent, and the evaluation morphism

$$ev_d : \begin{array}{c} k[X] \longrightarrow D \\ P \longmapsto P(d) \end{array}$$

cannot be injective. Hence, its kernel is generated by a unique monic polynomial $\mu_{d,k} \in k[X]$, and we have an isomorphism of k-algebras

$$k[X]/(\mu_{d,k}) \cong_k k[d].$$

Since D has no zero divisors, $k[d]$ is an integral domain and $(\mu_{d,k})$ is a prime ideal, hence maximal. Thus $k[d]$ is a field, $k[d] = k(d)$ and we have

$$[k(d) : k] = \deg(\mu_{d,k}).$$

Moreover, $\mu_{d,k}$ is irreducible since it generates a maximal ideal of $k[X]$, and $\mu_{d,k}(d) = 0$.

We will use these facts without further reference from now on.

DEFINITION I.1.10. Let D be a division k-algebra, and let $d \in D$. The polynomial $\mu_{d,k}$ is called **the minimal polynomial** of $d \in D$ over k.

We now recall the main properties of the tensor product of k-algebras.

If A and B are k-algebras, their **tensor product** $A \otimes_k B$ may be viewed as the k-vector space spanned by the symbols $a \otimes b, a \in A, b \in B$ subject to the relations:

$$(a + a') \otimes b = a \otimes b + a' \otimes b$$
$$a \otimes (b + b') = a \otimes b + a \otimes b'$$
$$(\lambda a) \otimes b = a \otimes (\lambda b) = \lambda(a \otimes b)$$

for all $a, a' \in A, b, b' \in B, \lambda \in k$. The symbols $a \otimes b$ are called **elementary tensors**.

The product on $A \otimes_k B$ is the unique product law satisfying

$$(a \otimes b)(a' \otimes b') = aa' \otimes bb' \text{ for all } a, a' \in A, b, b' \in B.$$

If $(e_i)_{i \in I}$ and $(e'_j)_{j \in J}$ are k-bases of A and B as k-vector spaces, then $(e_i \otimes e'_j)_{(i,j) \in I \times J}$ is a k-basis of $A \otimes_k B$. In particular $A \otimes_k B$ is finite-dimensional as a k-vector space if and only A and B are, and in this case we have

$$\dim_k(A \otimes_k B) = \dim_k(A) \dim_k(B).$$

Moreover, if $\varphi : A \longrightarrow C$ and $\psi : B \longrightarrow C$ are two morphisms of unital k-algebras satisfying

$$\varphi(a)\psi(b) = \psi(b)\varphi(a) \text{ for all } a \in A, b \in B,$$

there exists a unique morphism $h : A \otimes_k B \longrightarrow C$ of unital k-algebras satisfying

$$h(a \otimes 1_B) = \varphi(a) \text{ and } h(1_A \otimes b) = \psi(b) \text{ for all } a \in A, b \in B.$$

In particular, if $f : A \longrightarrow A'$ and $g : B \longrightarrow B'$ are two morphisms of unital k-algebras, there exists a unique k-algebra morphism

$$f \otimes g : A \otimes_k B \longrightarrow A' \otimes_k B'$$

satisfying

$$(f \otimes g)(a \otimes b) = f(a) \otimes g(b) \text{ for all } a \in A, b \in B.$$

If f and g are isomorphisms, so is $f \otimes g$.

Finally, if A and B are unital, the k-algebra morphisms

$$\begin{array}{ccc} A \longrightarrow A \otimes_k B & & B \longrightarrow A \otimes_k B \\ a \longmapsto a \otimes 1_B & \text{and} & b \longmapsto 1_A \otimes b \end{array}$$

are injective.

Now let L/k be an arbitrary field extension. If A is a k-algebra and B is an L-algebra, then $A \otimes_k B$ has a natural structure of L-algebra, where the structure of L-vector space is defined on elementary tensors by

$$\lambda \cdot (a \otimes b) = a \otimes \lambda b \text{ for all } \lambda \in L, a \in A, b \in B.$$

In particular, $A \otimes_k L$ has a natural structure of an L-algebra. Moreover, $A \otimes_k L$ is finite dimensional over L if and only if A is finite dimensional over k. In this case, we have

$$\dim_L(A \otimes_k L) = \dim_k(A).$$

If A and B are unital, we have a natural isomorphism of L-algebras

$$(A \otimes_k L) \otimes_L B \cong_L A \otimes_k B.$$

Similarly, $B \otimes_k A$ and $L \otimes_k A$ have a natural structure of L-algebras, and we have an isomorphism of L-algebras

$$B \otimes_L (L \otimes_k A) \cong_L B \otimes_k A.$$

If now A and B are two unital k-algebras, we have a natural L-algebra isomorphism

$$(A \otimes_k B) \otimes_k L \cong_L (A \otimes_k L) \otimes_L (B \otimes_k L).$$

Finally, if $k \subset K \subset L$ is a tower of field extensions, we have

$$(A \otimes_k K) \otimes_K L \cong_L A \otimes_k L.$$

The justification of the tensor product properties described above is quite lengthy, so we leave the details for now. For the sake of completeness, the reader may find full constructions and proofs in Appendix A.

We end this section with an elementary lemma.

LEMMA I.1.11. *Let A be a k-algebra, let $n \geq 1$ be an integer and let L/k be a field extension. Then the following properties hold:*

(1) *we have a natural k-algebra isomorphism $\mathrm{M}_n(k) \otimes_k A \cong_k \mathrm{M}_n(A)$;*

(2) *we have a natural L-algebra isomorphism $\mathrm{M}_n(k) \otimes_k L \cong_L \mathrm{M}_n(L)$.*

Proof.

(1) The k-algebra morphisms

$$\begin{array}{cc} \mathrm{M}_n(k) \longrightarrow \mathrm{M}_n(A) & A \longrightarrow \mathrm{M}_n(A) \\ M \longmapsto M & \text{and} \quad a \longmapsto aI_n \end{array}$$

have commuting images, and therefore there is a unique k-algebra morphism $\varphi : \mathrm{M}_n(k) \otimes_k A \longrightarrow \mathrm{M}_n(A)$ satisfying

$$\varphi(M \otimes a) = aM, \text{ for all } M \in \mathrm{M}_n(k), a \in A.$$

Since $\mathrm{M}_n(k) \otimes_k A$ and $\mathrm{M}_n(A)$ have the same dimension over k, it suffices to prove that φ is surjective. Let E_{ij} be the matrix with coefficient 1 at row i and column j and coefficients 0 elsewhere. For any matrix $M' = (m'_{ij}) \in \mathrm{M}_n(A)$, we have

$$\varphi\Big(\sum_{i,j} E_{ij} \otimes m'_{ij} \Big) = M',$$

which proves the surjectivity of φ.

(2) By (1), we have an isomorphism of k-algebras $M_n(k) \otimes_k L \cong_k M_n(L)$. One may check that this isomorphism is also L-linear. □

REMARK I.1.12. In particular, we have a natural isomorphism

$$M_m(k) \otimes_k M_n(k) \cong_k M_{mn}(k)$$

which maps $M \otimes N$ onto the Kronecker product of M and N. □

I.2. Central simple algebras: the basics

We now define the main object of this book.

DEFINITION I.2.1. Let k be a field. A k-algebra A is **simple** if it has no non-trivial two-sided ideals.

The next lemma gives an elementary but very useful property of simple algebras.

LEMMA I.2.2. *Let k be a field, and let $\phi : A \longrightarrow B$ be a k-algebra morphism. If A is simple, then ϕ is injective. If moreover A and B are finite dimensional over k and $\dim_k(A) = \dim_k(B)$, then ϕ is an isomorphism.*

Proof. Assume that A is simple. Since $\ker(\phi)$ is a two-sided ideal of A, we have $\ker(\phi) = (0)$ or A. The latter case cannot happen since $\phi(1) = 1$. Hence ϕ is injective; the last part is clear. □

We now give examples of simple algebras.

EXAMPLES I.2.3.

(1) Any division ring D is a simple $Z(D)$-algebra.

(2) Let k be an arbitrary field. Then $M_n(k)$ is a simple k-algebra.

Indeed, let J be a non-zero ideal of $M_n(k)$, and let $M = (m_{ij})_{i,j}$ be a non-zero element of J. Fix two integers r, s such that $m_{rs} \neq 0$. For all $i = 1, \ldots, n$, we have

$$m_{rs}^{-1} E_{ir} M E_{si} = E_{ii},$$

and therefore,

$$I_n = \sum_i E_{ii} = \sum_i m_{rs}^{-1} E_{ir} M E_{si} \in J$$

since J is a two-sided ideal. Hence J contains a unit, so $J = M_n(k)$.

(3) Similar arguments show that if D is a division k-algebra, then $M_r(D)$ is a simple k-algebra for all $r \geq 1$.

□

We now give our first concrete example of a simple k-algebra. Let k be a field of characteristic different from 2.

Let $a, b \in k^\times$, and consider the k-linear subspace $(a, b)_k$ of $M_4(k)$ generated by the matrices

$$I_4 = \begin{pmatrix} 1 & 0 & 0 & 0 \\ 0 & 1 & 0 & 0 \\ 0 & 0 & 1 & 0 \\ 0 & 0 & 0 & 1 \end{pmatrix}, i = \begin{pmatrix} 0 & a & 0 & 0 \\ 1 & 0 & 0 & 0 \\ 0 & 0 & 0 & a \\ 0 & 0 & 1 & 0 \end{pmatrix},$$

$$j = \begin{pmatrix} 0 & 0 & b & 0 \\ 0 & 0 & 0 & -b \\ 1 & 0 & 0 & 0 \\ 0 & -1 & 0 & 0 \end{pmatrix}, ij = \begin{pmatrix} 0 & 0 & 0 & -ab \\ 0 & 0 & b & 0 \\ 0 & -a & 0 & 0 \\ 1 & 0 & 0 & 0 \end{pmatrix}.$$

Straightforward computations show that these matrices are linearly independent over k, and that we have

$$i^2 = a, j^2 = b, (ij)^2 = -ab \text{ and } ji = -ij.$$

It easily follows that $(a, b)_k$ is a k-subalgebra of $M_4(k)$ of dimension 4 over k.

DEFINITION I.2.4. Let k be a field of characteristic different from 2. The k-algebra $(a, b)_k$ is called a **quaternion k-algebra**.

PROPOSITION I.2.5. *Let k be a field of characteristic different from 2. For every $a, b \in k^{\times}$, the k-algebra $(a, b)_k$ is a simple k-algebra, with center isomorphic to k.*

Proof. Let us first determine the center of $(a, b)_k$.

Let $q_1 = x + yi + zj + tij \in (a, b)_k$ and assume that $q_1 \in Z((a, b)_k)$. Then we have

$$iq_1 = i(x + yi + zj + tij) = xi + ay + zij + taj$$

and

$$q_1 i = (x + yi + zj + tij)i = xi + ay - zij - taj.$$

Since by assumption $iq_1 = q_1 i$, we have therefore $z = t = 0$ and thus $q_1 = x + yi$. Since we have $jq_1 = q_1 j$, we get $xj - yij = xj + yij$ in a similar way, so $y = 0$ and $q_1 = x \in k$. Hence $Z((a, b)_k) = k$.

Let us prove now that $(a, b)_k$ is simple. For, let I be a non-zero two-sided ideal of $(a, b)_k$, and let $q_1 = x + yi + zj + tij \in I, q_1 \neq 0$. We then have

$$\frac{1}{2}(iq_1 - q_1 i) = zij + taj \in I \text{ and } \frac{1}{2}(iq_1 + q_1 i) = xi + ay \in I.$$

Since by assumption x, y, z or t is non-zero, it follows that $zij + taj$ or $xi + ay$ is non-zero. Assume for example that $q_2 = zij + taj$ is not zero, that is $z \neq 0$ or $t \neq 0$. We have

$$\frac{1}{2}(jq_2 - q_2 j) = -bzi \in I \text{ and } \frac{1}{2}(jq_2 + q_2 j) = tab \in I.$$

If $t \neq 0$, then $tab \in k^{\times}$ is a unit of $(a, b)_k$; if $z \neq 0$, then $-bzi \in k^{\times}$ is a unit of $(a, b)_k$ (with inverse $-(abz)^{-1}i$). In both cases, I contains a unit, so $I = (a, b)_k$. The case $xi + ay \neq 0$ may be dealt with in a similar way and is left to the reader. $\square$

REMARK I.2.6. Later on, we will see a criterion to decide whether or not $(a, b)_k$ is a division algebra. For the moment, let us just point out that it can actually be a division algebra for some well-chosen values of a and b. For example, if $k = \mathbb{R}$ and $a = b = -1$, we obtain the Hamilton quaternion algebra $\mathbb{H}$, which is known to be a division ring. We will recover this fact in the next chapter. $\square$

DEFINITION I.2.7. A k-algebra A is called **central** if $Z(A) = k$. A **central simple** k-algebra is a k-algebra which is central and simple.

EXAMPLES I.2.8.

(1) The k-algebra $\mathrm{M}_n(k)$ is central simple.

(2) If D is a division ring, then its center $Z(D)$ is a field and D is a central $Z(D)$-algebra.

(3) If D is a central division k-algebra, then $\mathrm{M}_r(D)$ is a central simple k-algebra for all $r \geq 1$.

Indeed, the fact that $\mathrm{M}_r(D)$ is a simple k-algebra was already pointed out in Example I.2.3 (3). Now if $M \in Z(\mathrm{M}_r(D))$, the equality

$$E_{ij}M = ME_{ij} \text{ for all } i, j$$

shows that M is diagonal and that $m_{11} = \cdots = m_{rr}$. Let us denote this common value by d. The fact that $M \in Z(\mathrm{M}_r(D))$ then easily implies that $d \in Z(D) = k$. Hence $M \in Z(D)$, so $Z(\mathrm{M}_r(D)) = k$ (where k is identified with the set of scalar matrices).

(4) If L/k is a field extension, then L is a simple k-algebra which is not central.

(5) By Proposition I.2.5, any quaternion k-algebra is a central simple k-algebra.

$\square$

DEFINITION I.2.9. We say that a central simple k-algebra is **split** if it is isomorphic to a matrix algebra.

Matrix algebras are in some sense the simplest examples of central simple algebras. Even if not all simple algebras are split, they can be naturally viewed as subalgebras of matrix algebras, as we proceed to show now. This property is particularly interesting for explicit computations. Let us give a definition first.

DEFINITION I.2.10. Let A be a k-algebra. A **subfield** of A is a commutative subalgebra L of A which is also a field. In particular, A is a right L-vector space. Moreover, L contains k since it is a k-algebra. However, notice that A may not be an L-algebra (unless $L = k$), since L does not necessarily commute with all the elements of A.

We may now state the next result.

LEMMA I.2.11. *Let A be a k-algebra, and let L be a subfield of A. For all $a \in A$, the map*

$$\ell_a : \begin{array}{c} A \longrightarrow A \\ z \longmapsto az \end{array}$$

is an endomorphism of the right L-vector space A, and the map

$$\phi : \begin{array}{c} A \longrightarrow \mathrm{End}_L(A) \\ a \longmapsto \ell_a \end{array}$$

is a k-algebra morphism. In particular, if A is simple, ϕ is injective.

Proof. Recall that the structure of L-vector space on $\mathrm{End}_L(A)$ is defined by

$$\mathrm{End}_L(A) \times L \longrightarrow \mathrm{End}_L(A)$$

$$(u, \lambda) \longmapsto u\lambda,$$

where

$$(u\lambda)(z) = u(z)\lambda \text{ for all } z \in A.$$

Let us check that ℓ_a is an endomorphism of the right L-vector space A and that the map

$$\phi: \begin{array}{c} A \longrightarrow \mathrm{End}_L(A) \\ a \longmapsto \ell_a \end{array}$$

is a k-algebra morphism. We have

$$\ell_a(z + z') = a(z + z') = az + az' = \ell_a(z) + \ell_a(z'),$$

for all $z, z' \in A$. Moreover, for all $\lambda \in L$, we have

$$\ell_a(z\lambda) = a(z\lambda) = (az)\lambda = \ell_a(z)\lambda = (\ell_a\lambda)(z).$$

Hence ℓ_a is an endomorphism of the right L-vector space A.

Clearly, we have $\ell_1 = \mathrm{Id}_A$. Moreover, for every $a, a', z \in A$ and $\xi \in k$, we have

$$\ell_{a+a'}(z) = (a + a')z = az + a'z = \ell_a(z) + \ell_{a'}(z),$$

$$\ell_{aa'}(z) = aa'z = a(a'z) = \ell_a(a'z) = (\ell_a \circ \ell_{a'})(z),$$

$$\ell_{a\xi}(z) = (a\xi)z = a(\xi z) = a(z\xi) = (az)\xi = (\ell_a\xi)(z).$$

We then get

$$\ell_{a+a'} = \ell_a + \ell_{a'}, \ell_{aa'} = \ell_a \circ \ell_{a'}, \ell_{a\xi} = \ell_a\xi,$$

and the result follows. The last part comes from Lemma I.2.2. □

REMARK I.2.12. Let A be a **simple** k-algebra. Let us choose a basis of the right L-vector space A, and set

$$m = \dim_L(A) = \frac{\dim_k(A)}{[L : k]}.$$

Then composing the injective k-algebra morphism ϕ defined in the previous lemma with the isomorphism $\mathrm{End}_L(A) \cong_k \mathrm{M}_m(L)$ gives rise to an **injective** k-algebra morphism

$$\varphi_{A,L}: \begin{array}{c} A \hookrightarrow \mathrm{M}_m(L) \\ a \longmapsto M_a, \end{array}$$

where M_a is the matrix of left multiplication by a in the chosen L-basis of A.

For example, if $L = k$, we obtain an injection $\varphi_{A,k} : A \hookrightarrow \mathrm{M}_d(k)$, where $d = \dim_k(A)$. □

The reader will find plenty of examples of computations of such an injection $\varphi_{A,L}$ in the next section and the other chapters, as we will use it to provide all code constructions presented in this book.

I.3. Introducing space-time coding

Coding theory deals with the problem of transmitting data reliably over a communication channel which is noisy. The coding problem addressed depends on the characteristics of the channel. In classical coding theory, the channel involves a transmitter and a receiver, with between the two of them a discrete channel given by

$$y = x + v,$$

where x, y, v are n-dimensional vectors over a finite field F (typically of characteristic 2). The vector x is the transmitted signal, the vector v is the noise, and y is the noisy received vector. Transmission takes place during n time slots. A linear code over F is a subspace of F^n, where n is called the length of the code. Encoding consists of mapping a string of data of length k into a redundant coded version of length n. We call k the rank or dimension of the code. Vectors in F^n are called codewords. The minimum distance of a code is the smallest Hamming distance between two distinct codewords, where the Hamming distance counts the number of entries in which the two codewords differ. Since the minimum distance is a performance parameter of the code, a fundamental problem in the design of such codes is to find codes with large rank and minimum distance with respect to the block length.

The above channel model was introduced for wired discrete channels. There exists an analogous model for continuous channels, given by

$$y = x + v,$$

where x, y, v are now n-dimensional vectors over the complex field $\mathbb{C}$, and the noise vector v has random independent and identically distributed (i.i.d.) Gaussian entries with zero mean and unit variance. This is called a Gaussian channel, and it has a corresponding coding theory of its own. A further generalization appeared with the introduction of wireless communication. Now transmitter and receiver are both equipped with one antenna (see Fig. 1). Let $x = (x_1, \ldots, x_n) \in \mathbb{C}^n$ be the signal to be transmitted. At time t, $t = 1, \ldots, n$ the transmit antenna sends x_t, which will reach the receive antenna via different paths, due to the nature of the wireless environment. This is taken into account in the channel model, given by

$$y_t = x_t h_t + v_t, \ t = 1, \ldots, n,$$

where the coefficients h_t and v_t model respectively fading (coming from the signal propagation through multipaths) and noise. The wireless channel from the transmitter to the receiver during n time slots can thus be modeled as follows:

$$y = xH + v,$$

where $y \in \mathbb{C}^n$ is the received vector, and H is a diagonal $n \times n$ matrix called the **fading matrix** or **channel matrix**. All noise and fading coefficients are assumed to be i.i.d. complex Gaussian random variables with zero mean and unit variance.

In order to transmit more and more data in wireless environments, systems with multiple antennas at both transmitter and receiver have been introduced. They are commonly called Multiple Input Multiple Output (MIMO) systems or channels. Let us first consider a channel with two transmit and two receive antennas. At time t, the first and second antennas respectively send x_{1t} and x_{2t}. Both signals will be received by the two receive antennas, and will follow different paths to reach each of them. The signals y_{1t}, y_{2t} sensed by each receive antenna are

$$y_{1t} = h_{11}x_{1t} + h_{12}x_{2t} + v_{1t}$$
$$y_{2t} = h_{21}x_{1t} + h_{22}x_{2t} + v_{2t}$$

where h_{ji} denotes the fading from the ith transmit antenna to the jth receive antenna, and v_{jt} denotes the noise at the jth receive antenna at time t. Note that in the above equations, the fading coefficients h_{ji} should depend on t. However, it is reasonable to assume that the environment does not change so fast, and that there is a period of time T during which the channel (that is h_{ji}) remains constant. This period T is called a **coherence interval**, and the length of T depends on the channel considered.

For example, let us assume that the channel stays approximately constant over a period of length $T = 2$, and the transmission starts at time $t = 1$. The first antenna transmits at time $t = 1$ and $t + 1 = 2$ the signals x_{11} and x_{12} respectively. Similarly, the second antenna transmits at time t and $t + 1$ the signals x_{21} and x_{22} respectively. The first antenna receives consecutively a signal which is the sum of the two transmitted signals with fading and some noise, that is

$$y_{11} = h_{11}x_{11} + h_{12}x_{21} + v_{11}$$
$$y_{12} = h_{11}x_{12} + h_{12}x_{22} + v_{12}.$$

Similarly, the second antenna gets

$$y_{21} = h_{21}x_{11} + h_{22}x_{21} + v_{21}$$
$$y_{22} = h_{21}x_{12} + h_{22}x_{22} + v_{22}.$$

This can be written in a matrix equation as

$$\begin{pmatrix} y_{11} & y_{12} \\ y_{21} & y_{22} \end{pmatrix} = \begin{pmatrix} h_{11} & h_{12} \\ h_{21} & h_{22} \end{pmatrix} \begin{pmatrix} x_{11} & x_{12} \\ x_{21} & x_{22} \end{pmatrix} + \begin{pmatrix} v_{11} & v_{12} \\ v_{21} & v_{22} \end{pmatrix}.$$

This model can be generalized to the case where we have M transmit antennas and N receive antennas. At time t, the M antennas each send one signal. Those M signals can be collected and written as a vector $x_t = (x_{1t}, \ldots, x_{Mt})^t$. Each x_{it} will be received by all the N antennas. Thus x_{it} follows N different paths, each corresponding to a given fading denoted by h_{ji}, $j = 1, \ldots, N$ to reach its N destinations. Now, each receive antenna will sense a signal, which is the sum of noisy and faded copies of the signals transmitted by all antennas. Let us now consider T instances of the transmission, where we recall that T is the coherence time interval, during which the channel is assumed to be constant. The model for transmission with multiple antennas over a coherence time T can be summarized as follows:

$$\mathbf{Y}_{N \times T} = \mathbf{H}_{N \times M} \mathbf{X}_{M \times T} + \mathbf{V}_{N \times T},$$

where all matrices have coefficients in $\mathbb{C}$, and their dimensions are written as subscript. The tth column of the matrix $\mathbf{X}$ contains the vector x_t, sent at time t. The matrices $\mathbf{H}$ and $\mathbf{V}$ are random matrices whose coefficients are complex i.i.d. Gaussian random variables.

Coding for the above MIMO channel consists of designing the codewords $\mathbf{X}$ as a function of the data (or **information symbols**), which typically adds redundancy, similarly to the classical case. The set of codewords is called a **codebook** or simply a code. We will typically consider linear codes, that is, the encoding map from the information symbols to a codeword $\mathbf{X}$ will be linear. Since the data is encoded during **time** (we consider a time interval of T slots) and **space** (since we have M antennas), codes for multiple antennas systems are often called **space-time codes**.

We are now left with discussing how to design good space-time codes. In what follows, we denote by $\mathbf{X}^*$ the Hermitian conjugate of $\mathbf{X}$. Coding should be done so as to help the receiver to recover the transmitted signal $\mathbf{X}$ from the received signal $\mathbf{Y}$, despite the fading and noise. When there is no fading, a transmitted signal $\mathbf{X}$ will only be affected by noise. Geometrically, $\mathbf{X}$ can be seen as a point in an MT-dimensional space, and the received signal $\mathbf{Y}$ lies within a ball centered in $\mathbf{X}$ of radius given by the variance of the noise. In this case, the decoder which knows all the possible codewords can compute $\|\mathbf{X} - \mathbf{Y}\|^2$ for all possible $\mathbf{X}$ in the codebook, where the norm is the Frobenius norm: $\|A\|^2 = \mathrm{Tr}(AA^*)$, and Tr denotes the trace. It then decides that its estimate $\hat{\mathbf{X}}$ of $\mathbf{Y}$ is given by the matrix which minimizes $\|\mathbf{X} - \mathbf{Y}\|^2$. If the codewords are designed such that there is only one codeword in a ball of radius the variance of the noise, then the decoder will with high probability get the right estimate. The situation is different in case of fading.

Let us for now assume that the receiver has the knowledge of the channel $\mathbf{H}$. This is called the **coherent** case. The **non-coherent** case considers the scenario when the receiver does not know the channel, and will be discussed later on. A decoding rule is obtained as follows. Let $\mathcal{C}$ denote the codebook. The receiver knows $\mathbf{Y} = \mathbf{H}\mathbf{X} + \mathbf{V}$, the codebook, and an estimate of $\mathbf{H}$. It thus computes the "faded" codebook $\{\mathbf{H}\mathbf{X} \mid \mathbf{X} \in \mathcal{C}\}$ by multiplying every codeword by $\mathbf{H}$. It then chooses as decoded codeword the one which minimizes the distance between $\mathbf{H}\mathbf{X}$ and $\mathbf{Y}$. We thus have that the decoded codeword $\hat{\mathbf{X}}$ is given by

$$\hat{\mathbf{X}} = \min_{\mathbf{X} \in \mathcal{C}} \|\mathbf{H}\mathbf{X} - \mathbf{Y}\|^2.$$

An error will occur if the decoded codeword $\hat{\mathbf{X}}$ is different from the transmitted codeword $\mathbf{X}$. A way of formalizing the reliability of a channel is thus to compute its **pairwise probability of error**, namely, the probability of sending $\mathbf{X}$ and decoding erroneously $\hat{\mathbf{X}} \neq \mathbf{X}$. We write such probability $\mathbb{P}(\mathbf{X} \to \hat{\mathbf{X}})$. In [**52**], the following upper bound on this probability of error has been computed:

$$\mathbb{P}(\mathbf{X} \to \hat{\mathbf{X}}) \leq \left((\prod_{i=1}^{r} \lambda_i)^{1/r} c(\rho) \right)^{-rN}$$

where N is the number of receive antennas, r is the rank of the matrix $A(\mathbf{X}, \hat{\mathbf{X}}) := (\mathbf{X} - \hat{\mathbf{X}})(\mathbf{X} - \hat{\mathbf{X}})^*$, and λ_i, $i = 1, \ldots, r$ are the non-zero eigenvalues of $A(\mathbf{X}, \hat{\mathbf{X}})$. Furthermore, $c(\rho)$ is a constant that depends on channel parameters, and importantly on the **signal-to-noise ratio (SNR)** of the channel at the receiver, denoted by ρ, and defined by

$$\rho = \frac{\mathbb{E}[\|\mathbf{H}\mathbf{X}\|^2]}{\mathbb{E}[\|\mathbf{V}\|^2]}.$$

We indeed expect the probability of error to depend on how strong the signal is compared to the noise occurring over the channel.

We call the negative exponent of $c(\rho)$ given here by rN in the above expression the **diversity order** of the pairwise error probability. The higher the diversity order is, the smaller the upper bound will be. Since $0 \leq r \leq M$, the best diversity order is MN, which is obtained when the matrix $A(\mathbf{X}, \hat{\mathbf{X}})$ is full rank.

Design criteria to build the codebook $\mathcal{C}$ are derived from the above upper bound. Codes that have parameters that minimize the bound will give the best performance. The design criteria are summarized as follows:

(1) **The rank criterion**: in order to achieve the maximum diversity MN, the matrix $A(\mathbf{X}, \hat{\mathbf{X}}) = (\mathbf{X} - \hat{\mathbf{X}})(\mathbf{X} - \hat{\mathbf{X}})^*$ has to be full rank for any pair of codewords $\mathbf{X} \neq \hat{\mathbf{X}}$. Codes that achieve the maximal diversity are called **fully diverse**.

(2) **The determinant criterion**: once a given diversity is obtained, the minimum of the determinant of $A(\mathbf{X}, \hat{\mathbf{X}})$ taken over all pairs of distinct codewords must be maximized. If the code is not fully diverse, then the determinant is understood as the product of the non-zero eigenvalues $\lambda_1, \ldots, \lambda_r$. The product $\left(\prod_{i=1}^{r} \lambda_i \right)^{1/r}$ is called the **coding gain**. Note that the obvious solution which consists in scaling a whole codebook by multiplying it by a constant does not work: it only increases the signal-to-noise ratio ρ.

Since the coefficients of the codewords are complex, we may see them as included in a subfield L of $\mathbb{C}$. Furthermore, since rectangular codewords can be obtained from square ones by removing the appropriate number of rows or columns, we can suppose that $T = M$. Moreover, in order to obtain a decoding process with optimal performances, we need $M = N$ (see [**36**] for more details). From now on, we will therefore assume that $T = M = N = n$.

Looking at the above design criteria and remarks, we can summarize the coding problem as follows: find a family $\mathcal{C}$ of matrices in $\mathrm{M}_n(L)$ such that the matrix $(\mathbf{X}' - \mathbf{X}'')(\mathbf{X}' - \mathbf{X}'')^*$ is full rank, for all $\mathbf{X}' \neq \mathbf{X}'' \in \mathcal{C}$, or equivalently, such that

$$\det(\mathbf{X}' - \mathbf{X}'') \neq 0, \text{ for all } \mathbf{X}' \neq \mathbf{X}'' \in \mathcal{C}.$$

In this case, the previous estimation of the probability error may be rewritten as

$$\mathbb{P}(\mathbf{X}' \to \mathbf{X}'') \leq \frac{c(\rho)^{-n^2}}{|\det(\mathbf{X}' - \mathbf{X}'')|^{2n}}.$$

Let us note here that the cardinality of $\mathcal{C}$ plays a role, since higher coding gain can be obtained with a smaller cardinality. The cardinality of the code is often normalized and expressed in terms of its **rate**.

DEFINITION I.3.1. The **rate** R of the code $\mathcal{C}$ is defined by

$$R = \frac{1}{M} \log_2 |\mathcal{C}|.$$

The difficulty in building fully-diverse matrices clearly comes from the nonlinearity of the determinant. Not much can be said about the determinant of the difference of two matrices. In order to overcome this obstacle, one natural solution is to look for a linear codebook $\mathcal{C}$, namely one that satisfies

$$\mathbf{X}', \mathbf{X}'' \in \mathcal{C} \Rightarrow \mathbf{X}' \pm \mathbf{X}'' \in \mathcal{C}.$$

This indeed simplifies the design criterion to

$$\det(\mathbf{X}) \neq 0, \ \mathbf{0} \neq \mathbf{X} \in \mathcal{C}.$$

We thus restrict our attention to additive subgroups D of $\mathrm{M}_n(L)$. The condition that $\det(\mathbf{X}) \neq 0$ now reads that we need all the non-zero matrices in D to be invertible.

Using the work presented in Section I.2, we now illustrate how fully diverse space-time codes can be obtained by taking D to be a simple algebra. Let A be a simple k-algebra, and let L be a subfield of A. We will restrict ourselves to k-algebras A such that $k \subset \mathbb{C}$ since $L \subset \mathbb{C}$. By Remark I.2.12, we have an injective k-algebra morphism

$$\varphi_{A,L} : \begin{array}{l} A \hookrightarrow \mathrm{M}_m(L) \\ a \longmapsto M_a, \end{array}$$

where $m = \dim_L(A)$ (and can thus chosen to be n, the number of transmit antennas).

DEFINITION I.3.2. We will call a **code** or an **algebra based code**, any set $\mathcal{C} \subset \mathrm{M}_n(\mathbb{C})$ of matrices satisfying

$$\mathcal{C} \subset \mathcal{C}_{A,L} = \{M_a \mid a \in A\},$$

where A is a simple k-algebra and L is a subfield of A.

We could define a slightly more general notion of algebra based code by replacing $\varphi_{A,L}$ by any injective k-algebra morphism $\varphi : A \longrightarrow \mathrm{M}_m(L)$. However, in the following (as well as in the existing literature), all the examples considered will make use of the map $\varphi_{A,L}$.

REMARK I.3.3. It is clear that if A is a k-division algebra (thus simple), then any code $\mathcal{C}_{A,L} = \{\mathbf{X} = M_a \mid a \in A\}$ is fully diverse, and therefore so is any algebra based code $\mathcal{C}$.

Indeed, if $\mathbf{X}', \mathbf{X}'' \in \mathcal{C}_{A,L}, \mathbf{X}' \neq \mathbf{X}''$, then there exist $a', a'' \in A, a' \neq a''$ such that

$$\mathbf{X}' = M_{a'} \text{ and } \mathbf{X}'' = M_{a''}.$$

Now the map

$$\varphi_{A,L} : \begin{array}{l} A \hookrightarrow \mathrm{M}_m(L) \\ a \longmapsto M_a, \end{array}$$

is a k-algebra morphism, hence in particular a group morphism. We then have

$$\mathbf{X}' - \mathbf{X}'' = M_{a'} - M_{a''} = M_{a'-a''}.$$

Now since A is a k-division algebra and $a' - a'' \neq 0$, $a' - a''$ is a unit of A. Since $\varphi_{A,L}$ is a ring morphism, it maps units to units, and thus $\mathbf{X}' - \mathbf{X}'' = \varphi_{A,L}(a' - a'')$ is an invertible matrix, implying that $\mathcal{C}_{A,L}$ (and therefore $\mathcal{C}$) is fully diverse. $\square$

EXAMPLE I.3.4. Let us start with the easy case where we consider a field extension F/k of degree m, where $F = k(\theta)$ and θ has minimal polynomial

$$\mu_{\theta,k}(X) = \mu_0 + \mu_1 X + \cdots + \mu_{m-1} X^{m-1} + X^m.$$

Now take $A = F$, and $L = k$. We thus have

$$\varphi_{F,k}\colon \begin{array}{c} F \hookrightarrow \mathrm{M}_m(k) \\ a \longmapsto M_a. \end{array}$$

Then a k-basis for F is given by $\{1, \theta, \ldots, \theta^{m-1}\}$. Let $a \in F$, that is, $a = a_0 + a_1\theta + \cdots + a_{m-1}\theta^{m-1}$. One can compute that

$$M_a = \begin{pmatrix} a_0 & -\mu_0 a_{m-1} & \cdots \\ a_1 & a_0 - \mu_1 a_{m-1} & \cdots \\ a_2 & a_1 - \mu_2 a_{m-1} & \cdots \\ \vdots & \vdots & \cdots \\ a_{m-1} & a_{m-2} - \mu_{m-1} a_{m-1} & \cdots \end{pmatrix}.$$

By Remark I.3.3, the code

$$\mathcal{C}_{F,k} = \{M_a \mid a \in F\}$$

is fully diverse, since F is a field. Since F/k is finite, and thus F is an algebraic extension of k, this can also be seen as follows. We have that

$$\det(\mathbf{X}) = \det(\varphi_{F,k}(a)) = N_{F/k}(a),$$

where $N_{F/k}(a)$ denotes the norm map. Since F is a field, we have

$$N_{F/k}(a) = 0 \iff a = 0.$$

In the particular case where $\mu_{\theta,k}(X) = X^m - \lambda$, for some $\lambda \in k^\times$, we can write explicitly

$$M_a = \begin{pmatrix} a_0 & \lambda a_{m-1} & \lambda a_{m-2} & \cdots & \lambda a_1 \\ a_1 & a_0 & \lambda a_{m-1} & \cdots & \lambda a_2 \\ a_2 & a_1 & a_0 & \cdots & \lambda a_3 \\ \vdots & \vdots & \vdots & \cdots & \vdots \\ a_{m-2} & a_{m-3} & a_{m-4} & \cdots & \lambda a_{m-1} \\ a_{m-1} & a_{m-2} & a_{m-3} & \cdots & a_0 \end{pmatrix}.$$

We then obtain the following explicit description of the code $\mathcal{C}_{F,k}$:

$$\mathcal{C}_{F,k} = \left\{ \begin{pmatrix} a_0 & \lambda a_{m-1} & \lambda a_{m-2} & \cdots & \lambda a_1 \\ a_1 & a_0 & \lambda a_{m-1} & \cdots & \lambda a_2 \\ a_2 & a_1 & a_0 & \cdots & \lambda a_3 \\ \vdots & \vdots & \vdots & \cdots & \vdots \\ a_{m-2} & a_{m-3} & a_{m-4} & \cdots & \lambda a_{m-1} \\ a_{m-1} & a_{m-2} & a_{m-3} & \cdots & a_0 \end{pmatrix}, a_0, \ldots, a_{m-1} \in k \right\}.$$

Note that the elements $a_0, a_1, \ldots, a_{m-1}$ are the information symbols to be sent over the channel. This code is valid if there are m transmit antennas at the transmitter end. Transmission takes place over m periods of time. At time $t = 1$, each of the m transmit antennas sends one information symbol a_i, that is, the first column is sent. During the $m - 1$ other time slots, the $m - 1$ other columns are sent. The first column contains the data. The other columns contain the redundancy that protects the data. Such a code has been proposed in [**47**]. $\qquad\square$

Note that in the above code, we are sending m^2 coefficients for communicating only m information symbols. We can now define another notion of **rate**, similar this time to the one found in classical coding theory.

DEFINITION I.3.5. We call **rate** (or sometimes **throughput**) the ratio of information symbols per coefficients sent.

The rate of the above code is thus $m/m^2 = 1/m$.

EXAMPLE I.3.6. Let $\mathbb{H} = (-1, -1)_{\mathbb{R}}$ be the Hamilton quaternion algebra. Notice that $\mathbb{H}$ contains a subfield isomorphic to $\mathbb{C}$, namely $\mathbb{R}(i)$, where i denotes one of the generators of $\mathbb{H}$. In particular, one may consider the element $\zeta_8 = \dfrac{1 + i}{\sqrt{2}}$, which is a primitive 8-th root of unity. Let

$$A = \mathbb{Q}(\zeta_8) \oplus j\mathbb{Q}(\zeta_8).$$

Then one may check easily that A is a division $\mathbb{Q}(i)$-algebra of dimension 4. We thus have

$$\varphi_{A, \mathbb{Q}(i)} : \begin{aligned} A &\hookrightarrow \mathrm{M}_4(\mathbb{Q}(i)) \\ a &\longmapsto M_a. \end{aligned}$$

A $\mathbb{Q}(i)$-basis for A is given by $(1, \zeta_8, j, j\zeta_8)$.

Let $a = a_1 + \zeta_8 a_2 + j a_3 + j\zeta_8 a_4 \in A$. Using the fact that we have

$$z\zeta_8 = \zeta_8 \bar{z} \text{ and } zj = j\bar{z} \text{ for all } z \in \mathbb{R}(i),$$

one can compute that

$$M_a = \begin{pmatrix} a_1 & ia_2 & -\bar{a}_3 & -\bar{a}_4 \\ a_2 & a_1 & i\bar{a}_4 & -\bar{a}_3 \\ a_3 & ia_4 & \bar{a}_1 & \bar{a}_2 \\ a_4 & a_3 & -i\bar{a}_2 & \bar{a}_1 \end{pmatrix}.$$

Similarly as in the previous example, matrices M_a, $a \in A$, can be used to define a code $\mathcal{C}$ as follows:

$$\mathcal{C} = \left\{ \begin{pmatrix} a_1 & ia_2 & -\bar{a}_3 & -\bar{a}_4 \\ a_2 & a_1 & i\bar{a}_4 & -\bar{a}_3 \\ a_3 & ia_4 & \bar{a}_1 & \bar{a}_2 \\ a_4 & a_3 & -i\bar{a}_2 & \bar{a}_1 \end{pmatrix}, \ a_1, a_2, a_3, a_4 \in \mathbb{Q}(i) \right\}.$$

This construction (together with further improvements) has been proposed in [20]. It is a codebook designed for 4 transmit antennas and has rate $4/16 = 1/4$. It is fully diverse by Remark I.3.3, since $A = \mathbb{Q}(\zeta_8) \oplus j\mathbb{Q}(\zeta_8)$ is a division algebra, as already noticed above.   □

As it can be seen from these two examples, code constructions require an explicit presentation of the algebra considered. We will thus see further code examples once more simple algebras have been studied.

Let us consider now the general case:

EXAMPLE I.3.7. Let $\mathcal{C}$ be an algebra based code

$$\mathcal{C} \subset \mathcal{C}_{A, L} = \{\mathbf{X} = \varphi_{A, L}(a), \ a \in \mathcal{A}\} \subset \mathrm{M}_n(\mathbb{C}),$$

where A is a simple k-algebra, and L is a subfield of A.

Let us compute the rate of $\mathcal{C}$, that we will denote by $r(\mathcal{C})$.

The information symbols that we would like to transmit are elements of k, which may be used to define elements of A, after the choice of a k-basis of A. Each element of a may then carry $\dim_k(A)$ information symbols. However, an $m \times m$ matrix may contain m^2 information symbols, so we have

$$r(\mathcal{C}) = \frac{\dim_k(A)}{m^2},$$

where $m = \dim_L(A)$. Since $\dim_k(A) = \dim_L(A)[L : K] = m[L : K]$, this may be rewritten as

$$r(\mathcal{C}) = \frac{[L : k]^2}{\dim_k(A)},$$

so we should choose L/k such that $[L : k]$ is as large as possible. If A is a central division k-algebra, there exists a subfield L of A containing k such that $[L : k] = \deg(A)$ (as it will be shown in Chapter IV).

In particular, we obtain a code with a rate equal to 1 in this case, which is the best possible value. $\square$

Exercises

1. Show that a k-subalgebra of a simple k-algebra is not necessarily simple.

2. Let k be a field. Let $n \geq 2$ be an integer. Assume that $\mathrm{char}(k)$ is prime to n, and that $\mu_n \subset k$. Let $\zeta_n \in k^\times$ be a primitive n^{th}-root of 1, and let $a, b \in k^\times$. Let $e, f \in \mathrm{M}_n(\bar{k})$ be the matrices defined by

$$
e = \begin{pmatrix} 0 & & & a \\ 1 & \ddots & & \\ & \ddots & \ddots & \\ & & 1 & 0 \end{pmatrix}, f = \begin{pmatrix} \beta & & & \\ & \zeta_n^{-1}\beta & & \\ & & \ddots & \\ & & & \zeta_n^{-(n-1)}\beta \end{pmatrix},
$$

where $\beta \in \bar{k}$ satisfies $\beta^n = b$.

(a) Show that we have $e^n = a$, $f^n = b$ and $ef = \zeta_n fe$.

(b) Let $\{a, b\}_{n,\zeta_n}$ be the k-subalgebra of $\mathrm{M}_n(\bar{k})$ generated by e and f. Show that $\{a, b\}_n$ is a central simple k-algebra of dimension n^2.

(c) Assume that $L = k(f) \cong_k k(\sqrt[n]{b})$ has degree n. Compute the left multiplication matrix of an element of $\{a, b\}_{n,\zeta_n}$ with respect to the L-basis $(1, e, \ldots, e^{n-1})$.

3. Let p be a prime number, and let k be a field of characteristic p. Let $a \in k^{\times}$ and $b \in k$. Let $e, f \in M_p(\overline{k})$ be the matrices defined by

$$e = \begin{pmatrix} 0 & & & a \\ 1 & \ddots & & \\ & \ddots & \ddots & \\ & & 1 & 0 \end{pmatrix}, f = \begin{pmatrix} \beta & & & \\ & \beta - 1 & & \\ & & \ddots & \\ & & & \beta - p + 1 \end{pmatrix},$$

where $\beta \in \overline{k}$ satisfies $\beta^p - \beta = b$.

(a) Show that we have $e^p = a$, $f^p - f = b$ and $ef = fe + e$.

(b) Let $(a, b]_p$ be the k-subalgebra of $M_n(\overline{k})$ generated by e and f. Show that $(a, b]_p$ is a central simple k-algebra of dimension p^2.

(c) Assume that $L = k(f) \cong_k k(\wp^{-1}(b))$ has degree p, where

$$\wp : \begin{aligned} \overline{k} &\longrightarrow \overline{k} \\ x &\longmapsto x^p - x \end{aligned}$$

is the Weierstrass function. Compute the left multiplication matrix of an element of $(a, b]_p$ with respect to the L-basis $(1, e, \ldots, e^{n-1})$.

CHAPTER II

Quaternion algebras

Let k be a field of characteristic different from 2. Let us first recall the definition of a quaternion k-algebra, already introduced in Definition I.2.4. Let $a, b \in k^{\times}$. Then the quaternion algebra $(a, b)_k$ is the k-algebra generated by two elements i and j, and subject to the relations

$$i^2 = a, \ j^2 = b, \ ij = -ji.$$

This is a central simple k-algebra by Proposition I.2.5. As a k-vector space, $(a, b)_k$ is spanned by $1, i, j$ and ij. Hence $(a, b)_k$ is a central simple k-algebra of dimension 4. If $k = \mathbb{R}$ and $a = b = -1$, $(-1, -1)_{\mathbb{R}}$ is called the Hamilton quaternion algebra, and is denoted by $\mathbb{H}$.

II.1. Properties of quaternion algebras

We start with three easy properties of quaternion algebras.

LEMMA II.1.1. *For $a, b, \lambda, \mu \in k^{\times}$, we have*

(1) $(a, b)_k \cong_k (b, a)_k$;

(2) $(a\lambda^2, b\mu^2) \cong_k (a, b)_k$;

(3) $(1, b)_k \cong_k \mathrm{M}_2(k)$.

Proof.

(1) Let $1, i, j, ij$ be the standard basis of $(a, b)_k$, and let $1, i', j', i'j'$ be the standard basis of $(b, a)_k$. Let $f : (a, b)_k \longrightarrow (b, a)_k$ be the unique k-linear map satisfying

$$f(1) = 1, f(i) = j', f(j) = i' \text{ and } f(ij) = -i'j'.$$

In other words, we have

$$f(x + yi + zj + tij) = x + yj' + zi' - ti'j' \text{ for all } x, y, z, t \in k.$$

We claim that f is a k-algebra morphism. Since f is k-linear, a distributivity argument shows that it is enough to check that

$$f(u_1 u_2) = f(u_1)f(u_2) \text{ for } u_1, u_2 \in \{1, i, j, ij\},$$

which follows from direct computations. For example, we have

$$f(ij) = -i'j' = j'i' = f(i)f(j).$$

It remains to prove that f is bijective. Since $(a, b)_k$ and $(b, a)_k$ have the same dimension over k, Lemma I.2.2 leads to the conclusion.

(2) Let $1, e, f, ef$ be the standard basis of $(a\lambda^2, b\mu^2)_k$, and let $1, i, j, ij$ be the standard basis of $(a, b)_k$. Arguing as before, one can show that the k-linear map $g : (a\lambda^2, b\mu^2)_k \longrightarrow (a, b)_k$ defined by

$$g(1) = 1, g(e) = \lambda i, g(f) = \mu j, g(ef) = \lambda\mu ij$$

is a k-algebra isomorphism.

(3) As in the previous cases, the map $h : (1, b)_k \longrightarrow M_2(k)$ defined by

$$h(x + yi + zj + tij) = \begin{pmatrix} x + y & b(z + t) \\ z - t & x - y \end{pmatrix}$$

is easily seen to be a k-algebra isomorphism. $\square$

Before we continue, let us introduce some definitions.

DEFINITION II.1.2. Let $(a, b)_k$ be a quaternion k-algebra.

For $q = x + yi + zj + tij \in (a, b)_k$, we define the **conjugate** of q to be

$$\gamma(q) = x - (yi + zj + tij).$$

A direct computation shows that the map $\gamma : (a, b)_k \longrightarrow (a, b)_k$ is k-linear and satisfies

$$\gamma(q_1 q_2) = \gamma(q_2)\gamma(q_1) \text{ for all } q_1, q_2 \in (a, b)_k.$$

DEFINITION II.1.3. Let $Q = (a, b)_k$ be a quaternion k-algebra. The **reduced norm** of $q = x + yi + zj + tij \in Q$ is defined by

$$\mathrm{Nrd}_Q(q) = x^2 - ay^2 - bz^2 + abt^2 \in k.$$

One can easily check that

$$\mathrm{Nrd}_Q(q) = q\gamma(q) = \gamma(q)q.$$

In particular, for all $q_1, q_2 \in (a, b)_k$, we have

$$\mathrm{Nrd}_Q(q_1 q_2) = \mathrm{Nrd}_Q(q_1)\mathrm{Nrd}_Q(q_2).$$

We now give an explicit criterion which permits to decide whether or not $(a, b)_k$ is a division algebra.

PROPOSITION II.1.4. *The quaternion algebra $Q = (a, b)_k$ is either split or a division k-algebra. It is a division k-algebra if and only if the equation*

$$\mathrm{Nrd}_Q(q) = 0$$

admits only the trivial solution $q = 0$, that is

$$x^2 - ay^2 - bz^2 + abt^2 = 0 \Rightarrow x = y = z = t = 0 \text{ for all } x, y, z, t \in k.$$

Proof. For all $q \in Q$, we have

$$q\gamma(q) = \gamma(q)q = \mathrm{Nrd}_Q(q).$$

If $\mathrm{Nrd}_Q(q) \neq 0$ whenever $q \neq 0$, this implies that every quaternion $q \neq 0$ is invertible in Q with inverse $\dfrac{1}{\mathrm{Nrd}_Q(q)}\gamma(q)$. Therefore, Q is a division k-algebra.

Assume now that the equation $\mathrm{Nrd}_Q(q) = 0$ has a non-trivial solution. In order to show that Q is split, we are going to construct two elements e and f of Q satisfying

$$e^2 = 1, f^2 = c \text{ and } ef = -fe,$$

for some suitable $c \in k^{\times}$. This will lead to the desired conclusion as follows: the k-vector space spanned by $1, e, f$ and ef is easily seen to be a subalgebra of Q isomorphic to $(1, c)_k$. Since $(1, c)_k$ and Q have the same dimension over k, we have

$$(a, b)_k \cong_k (1, c)_k \cong_k M_2(k),$$

the last isomorphism coming from Lemma II.1.1. It thus remains to construct e and f.

Let us first assume that $a \in k^{\times 2}$. In this case, we may set $e = u^{-1}i$, where $u \in k^{\times}$ satisfies $u^2 = a$, and $f = j$ with $f^2 = b$. Notice that we did not use our assumption here, but in this case the equation $\mathrm{Nrd}_Q(q) = 0$ has automatically non-trivial solutions, such as $u - i$ (that is $x = u, y = -1, z = t = 0$) for example.

We now suppose for the rest of the proof that a is not a square. Let $q = x + yi + zj + tij \neq 0$ be such that $\mathrm{Nrd}_Q(q) = 0$. One may assume that $x \neq 0$ without loss of generality. Indeed, if $x = 0$, then we have

$$\mathrm{Nrd}_Q(q) = -ay^2 - bz^2 + abt^2 = 0.$$

Since $q \neq 0$, one of the scalars y, z, t is non-zero, say y. Then $ay \neq 0$ by assumption, and multiplying the equation above by $-a$ yields

$$(ay)^2 - b(at)^2 + abz^2 = 0.$$

The quaternion $q' = ay + atj + zij$ then satisfies $\mathrm{Nrd}_Q(q') = 0$ and has a non-zero constant term.

Assuming now that $x \neq 0$, set $e = \dfrac{1}{x}(yi + zj + tij)$. We then have

$$e^2 = \frac{1}{x^2}(yi + zj + tij)^2 = \frac{1}{x^2}(ay^2 + bz^2 - abt^2) = 1,$$

since $\mathrm{Nrd}_Q(q) = 0$. One may verify that we have

$$ie - ei = \frac{2}{x}(atj + zij) \text{ and } ej - je = \frac{2}{x}(bti + yij).$$

One of the scalars y, z, t is non-zero (since the equality $\mathrm{Nrd}_Q(q) = 0$ would imply that $x = 0$ otherwise), so we have $ei - ie \neq 0$ or $ej - je \neq 0$. Assume for example that $ei - ie \neq 0$, that is $(z, t) \neq (0, 0)$, and set $f = x(ie - ei) = 2(atj + zij)$. We then have

$$ef = -fe \text{ and } f^2 = 4ab(at^2 - z^2).$$

Since $a \notin k^{\times 2}$, we have $c = f^2 = 4ab(at^2 - z^2) \in k^{\times}$, since $(z, t) \neq (0, 0)$. The case $je - ej \neq 0$ may be dealt with in a similar way, and this concludes the proof. $\square$

COROLLARY II.1.5. *Let $Q = (a, b)_k$ be a quaternion k-algebra. Then the following properties are equivalent:*

(1) *Q is split;*

(2) *$b \in N_{k(\sqrt{a})/k}(k(\sqrt{a})^{\times})$;*

(3) *$a \in N_{k(\sqrt{b})/k}(k(\sqrt{b})^{\times})$.*

Proof. Let us prove (1) $\iff$ (2). Assume that $b \in N_{k(\sqrt{a})/k}(k(\sqrt{a})^{\times})$. Then $b = x^2 - ay^2$ for some $x, y \in k^{\times}$ not both equal to zero. Then $\mathrm{Nrd}_Q(x + yi + j) = 0$, and $(a, b)_k$ is split by the previous proposition. Conversely, assume that $(a, b)_k$ is split. By the previous proposition, there exist $x, y, z, t \in k$, not all zero, such that

$$x^2 - ay^2 - bz^2 + abt^2 = 0.$$

If $a = \lambda^2 \in k^{\times 2}$, then $k(\sqrt{a}) = k$, and property (2) is clearly satisfied.

Thus one can assume that $a \notin k^{\times 2}$. We now construct an element $\alpha \in k(\sqrt{a})^{\times}$ satisfying $N_{k(\sqrt{a})/k}(\alpha) = b$.

If $z = t = 0$, we have $x^2 - ay^2 = 0$, and since x or y is not zero, we get that a is a square in k, which is a contradiction. Hence z or t is not zero, and therefore $z^2 - at^2 \neq 0$, since a is not a square. Thus we may write

$$b = \frac{x^2 - ay^2}{z^2 - at^2} = \frac{(x^2 - ay^2)(z^2 - at^2)}{(z^2 - at^2)^2} = N_{k(\sqrt{a})/k}(\alpha),$$

with $\alpha = \dfrac{(x + y\sqrt{a})(z + t\sqrt{a})}{z^2 - at^2}$. Since $(b, a)_k \cong_k (a, b)_k$ by Lemma II.1.1 (1), the previous point shows that we also have (1) $\iff$ (3). This concludes the proof. $\qquad\square$

EXAMPLE II.1.6. Using the previous result, one can see immediately that the quaternion algebras

$$(\lambda^2, c)_k, (a, -a)_k \text{ and } (b, 1 - b)_k$$

are split for all $\lambda, a, b \in k^{\times}$, $b \neq 1$. $\qquad\square$

EXAMPLE II.1.7. Let $p > 2$ be a prime number, and let $\varepsilon \in \mathbb{Z}$ be such that $p \nmid \varepsilon$. Assume that ε is not a square modulo p.

Let us show that the quaternion algebra $Q = (p, \varepsilon)_{\mathbb{Q}}$ is a division algebra.

Assume to the contrary that Q is split. By Corollary II.1.5, we have

$$p = N_{\mathbb{Q}(\sqrt{\varepsilon})/\mathbb{Q}}(\xi) \text{ for some } \xi \in \mathbb{Q}(\sqrt{\varepsilon})^{\times}.$$

Write $\xi = \dfrac{x + y\sqrt{\varepsilon}}{z}$, for some $x, y, z \in \mathbb{Z}$, $z \neq 0$. We then have

$$pz^2 = x^2 - \varepsilon y^2.$$

Dividing x, y, z by their greatest common divisor if necessary, we may assume without any loss of generality that x, y, z are coprime.

Assume first that $p \mid y$. Then $p \mid x^2$ and therefore $p \mid x$. Consequently, $p^2 \mid pz^2$, which implies that $p \mid z$, contradicting the fact that x, y, z are coprime. Hence $p \nmid y$, and we deduce that ε is a square modulo p, which is another contradiction.

Thus Q is a division algebra. $\qquad\square$

EXAMPLE II.1.8. Let us show that $Q = (i, 1 + 2i)_{\mathbb{Q}(i)}$ is a division algebra.

Assume to the contrary that Q is split. By Corollary II.1.5, we have

$$1 + 2i = N_{\mathbb{Q}(i)(\zeta_8)/\mathbb{Q}(i)}(\xi) \text{ for some } \xi \in \mathbb{Q}(i)(\zeta_8)^{\times},$$

where ζ_8 denotes a primitive 8-th root of 1. As in the previous example, this implies that

$$(1 + 2i)z^2 = x^2 - iy^2, \text{ for some } x, y, z \in \mathbb{Z}[i].$$

Since $\mathbb{Z}[i]$ is a unique factorization domain, one may assume that x, y, z are coprime, and show as before that $1 + 2i \nmid y$, using that $1 + 2i$ is an irreducible element of $\mathbb{Z}[i]$.

Now recall that the map

$$\mathbb{Z}[i]/(1 + 2i) \longrightarrow \mathbb{F}_5$$
$$a + bi \longmapsto \overline{a + 2b}$$

is a well-defined ring isomorphism. Since this isomorphism maps the class of i onto $\overline{2}$, reducing modulo $1 + 2i$ the previous equation yields that 2 is a square modulo 5, which is a contradiction.

Hence Q is a division algebra. $\qquad\qquad\qquad\qquad\qquad\qquad\qquad\qquad\square$

We continue with a very useful property of quaternion algebras.

PROPOSITION II.1.9. *For $a, b, c \in k^\times$, we have*

$$(a, b)_k \otimes (a, c)_k \cong_k \mathrm{M}_2((a, bc)_k) \ \text{and} \ (a, c)_k \otimes (b, c)_k \cong_k \mathrm{M}_2((ab, c)_k).$$

Proof. In view of Lemma II.1.1 (1), it is enough to prove the first assertion. Notice that by Lemma I.1.11 (1), we have

$$\mathrm{M}_2((a, bc)_k) \cong_k \mathrm{M}_2(k) \otimes_k (a, bc)_k.$$

Now using Corollary II.1.5, we have $\mathrm{M}_2(k) \cong_k (c, -a^2c)_k$. Hence we have to prove that

$$(a, b)_k \otimes_k (a, c)_k \cong_k (c, -a^2c)_k \otimes_k (a, bc)_k.$$

Our first goal is to construct a k-algebra morphism

$$\rho : (a, b)_k \otimes_k (a, c)_k \longrightarrow (c, -a^2c)_k \otimes_k (a, bc)_k.$$

Let $1, i_1, j_1, i_1j_1$ be the standard basis of $(a, b)_k$ and let $1, i_2, j_2, i_2j_2$ be the standard basis of $(a, c)_k$. Notice that the 16 elementary tensors

$$1 \otimes 1, 1 \otimes i_2, \ldots, i_1j_1 \otimes i_2j_2$$

form a k-basis of $(a, b)_k \otimes_k (a, c)_k$. Now let A be the k-linear subspace with basis elements

$$1 \otimes 1, e_1 = 1 \otimes j_2, f_1 = i_1 \otimes i_2j_2 \ \text{and} \ e_1f_1 = -ci_1 \otimes i_2.$$

It is easy to check that A is a k-subalgebra of $(a, b)_k \otimes_k (a, c)_k$ which is isomorphic to $(c, -a^2c)_k$. Similarly, the k-linear subspace B with basis elements

$$1 \otimes 1, e_2 = i_1 \otimes 1, f_2 = j_1 \otimes j_2 \ \text{and} \ e_2f_2 = i_1j_1 \otimes j_2$$

is a k-subalgebra of $(a, b)_k \otimes_k (a, c)_k$ isomorphic to $(a, bc)_k$. Hence we get two injective k-algebra morphisms

$$(c, -a^2c)_k \longrightarrow (a, b)_k \otimes_k (a, c)_k \ \text{and} \ (a, bc)_k \longrightarrow (a, b)_k \otimes_k (a, c)_k,$$

whose images are respectively equal to A and B. Since elements of A and B commute (as may be seen on the generators), these two morphisms induce in turn a k-algebra morphism

$$\rho : (a, bc)_k \otimes_k (c, -a^2c)_k \longrightarrow (a, b)_k \otimes_k (a, c)_k,$$

whose image contains A and B.

We are now going to prove that ρ is an isomorphism. A dimension argument shows that it is enough to prove the surjectivity of this map. As a k-algebra, $(a,b)_k \otimes_k (a,c)_k$ is generated by the 4 elements

$$i_1 \otimes 1, j_1 \otimes 1, 1 \otimes i_2 \text{ and } 1 \otimes j_2.$$

Therefore, since the image of ρ is a subalgebra of $(a,b)_k \otimes_k (a,c)_k$, it is enough to check that it contains these 4 elements. Since $\text{Im}(\rho)$ contains A and B, it contains $e_1 = 1 \otimes j_2$ and $e_2 = i_1 \otimes 1$, so it remains to check that it contains $j_1 \otimes 1$ and $1 \otimes i_2$. But we have

$$-\frac{1}{ac}e_1 f_1 e_2 = \frac{1}{a}(i_1 \otimes i_2)(i_1 \otimes 1) = 1 \otimes i_2$$

and

$$\frac{1}{c}e_1 f_2 = \frac{1}{c}(1 \otimes j_2)(j_1 \otimes j_2) = j_1 \otimes 1,$$

and this concludes the proof. $\square$

Notice that the quaternion k-algebra $Q = (a,b)_k$ contains a subfield of degree 2 isomorphic to $k(\sqrt{b})$, namely $k(j)$, as well as a subfield of degree 2 isomorphic to $k(\sqrt{a})$, namely $k(i)$. The lemma below describe the corresponding injection given by Remark I.2.12 for these two subfields.

LEMMA II.1.10. *Let $Q = (a,b)_k$ be a quaternion k-algebra, and let $q = x + yi + zj + tij \in (a,b)_k,$.*

(1) *If $L = k(j)$ is identified to $k(\sqrt{b})$, the matrix of left multiplication by q with respect to the L-basis $(1,i)$ is*

$$M_q = \begin{pmatrix} x + z\sqrt{b} & a(y - t\sqrt{b}) \\ y + t\sqrt{b} & x - z\sqrt{b} \end{pmatrix}.$$

(2) *If $L = k(i)$ is identified to $k(\sqrt{a})$, the matrix of left multiplication by q with respect to the L-basis $(1,j)$ is*

$$M_q = \begin{pmatrix} x + y\sqrt{a} & b(z + t\sqrt{a}) \\ z - t\sqrt{a} & x - y\sqrt{a} \end{pmatrix}.$$

In particular, in both cases, the map

$$\varphi_{Q,L} : \begin{array}{c} Q \longrightarrow \text{M}_2(L) \\ q \longmapsto M_q \end{array}$$

is an injective k-algebra morphism.

Proof. We only prove the first point of the lemma, the proof of the second one being similar. Clearly, $1, i$ is a $k(j)$-basis of the right $k(j)$-vector space Q. For $q \in Q$, let us denote by M_q the matrix of left multiplication by q with respect to this particular basis. By Remark I.2.12, the map

$$\varphi_{Q,k(j)} : \begin{array}{c} Q \longrightarrow \text{M}_2(k(j)) \\ q \longmapsto M_q \end{array}$$

is then an injective k-algebra morphism. Since we have

$$i \cdot 1 = i, i^2 = a = 1 \cdot a,$$

the matrix M_i of left multiplication by i in the right $k(j)$-vector space $(a, b)_k$ is

$$M_i = \begin{pmatrix} 0 & a \\ 1 & 0 \end{pmatrix}.$$

Since we have

$$j \cdot 1 = j = 1 \cdot j, ji = -ij = i \cdot (-j),$$

the matrix M_j of left multiplication by j in the right L-vector space $(a, b)_k$ is

$$M_j = \begin{pmatrix} j & 0 \\ 0 & -j \end{pmatrix}.$$

We then get

$$\varphi_{Q,k(j)}(i) = \begin{pmatrix} 0 & a \\ 1 & 0 \end{pmatrix}, \varphi_{Q,k(j)}(j) = \begin{pmatrix} j & 0 \\ 0 & -j \end{pmatrix}.$$

Since $\varphi_{Q,k(j)}$ is a k-algebra morphism, we also have $\varphi_{Q,k(j)}(1) = I_2$ and

$$\varphi_{Q,k(j)}(ij) = \varphi_{Q,k(j)}(i)\varphi_{Q,k(j)}(j) = \begin{pmatrix} 0 & -aj \\ j & 0 \end{pmatrix}.$$

Putting things together, and after identifying $k(j)$ with L, we get the desired result.

$\square$

REMARK II.1.11. An immediate computation shows that we have

$$\det(M_q) = \mathrm{Nrd}_Q(q) \text{ for all } q \in (a, b)_k.$$

$\square$

II.2. Hamilton quaternions

We now have a closer look at the Hamilton quaternion algebra

$$\mathbb{H} = (-1, -1)_{\mathbb{R}}.$$

An immediate application of Proposition II.1.4 shows that $\mathbb{H}$ is a division algebra. The Hamilton quaternion algebra was actually the first example of a finite dimensional non-commutative division $\mathbb{R}$-algebra appearing in the literature, and was discovered by Hamilton. The following result, due to Frobenius, shows that it is the only one, up to isomorphism.

THEOREM II.2.1 (Frobenius). *Let D be a finite dimensional division $\mathbb{R}$-algebra. Then D is isomorphic to $\mathbb{R}, \mathbb{C}$ or $\mathbb{H}$. In particular, $\mathbb{R}$ and $\mathbb{H}$ are the unique central simple division $\mathbb{R}$-algebras, up to isomorphism.*

Proof. Recall that $\mathbb{R}$ is a subalgebra of $Z(D)$ by Remark I.1.8, so D contains $\mathbb{R}$. Notice also that $Z(D)$ is a field containing $\mathbb{R}$, so $Z(D)$ is isomorphic to either $\mathbb{R}$ or $\mathbb{C}$, since $Z(D)/\mathbb{R}$ has finite degree by assumption. If D is commutative, then D is a field extension of $\mathbb{R}$, hence is isomorphic to $\mathbb{R}$ or $\mathbb{C}$.

Assume now that D is not commutative, and let $k = Z(D)$. Then k cannot be isomorphic to $\mathbb{C}$. Otherwise, for all $d \in D$, the minimal polynomial $\mu_{d,k}$ of d would be of degree 1, meaning that $d \in k$. We then would have that $D = k$, contradicting the fact that D is not commutative. Hence k is isomorphic to $\mathbb{R}$.

Since D is not commutative, there exists an element $d \in D, d \notin k = \mathbb{R}$. In particular, $k(d)$ is a proper field extension of $\mathbb{R}$, hence isomorphic to $\mathbb{C}$. Therefore, there exists $i \in k(d)$ such that $i^2 = -1$. In particular, $i \notin k = \mathbb{R}$, and there exists $y \in D$ which does not commute with i. Then $z = yi - iy$ is not zero and satisfies $zi = -iz$. In particular, $z \notin k$.

Let us prove that $z^2 \in k$ and that $z^2 < 0$. Since $z \notin k$, $k(z)/k$ has degree 2. Since $k(z^2)$ is a subfield of $k(z)$ containing $k = \mathbb{R}$, this implies that the field extensions $k(z^2)/k$ and $k(z)/k(z^2)$ have degree at most 2. Now notice that $z^2 i = iz^2$, and thus any element of $k(z^2)$ commutes with i. In particular, $k(z) \neq k(z^2)$ and $k(z)/k(z^2)$ has degree 2. This implies that $[k(z^2) : k] = 1$, that is $z^2 \in k$. Since $k(z)/k$ has degree 2, necessarily $z^2 < 0$ (otherwise $z = \sqrt{z^2}$ would lie in k).

Hence we have shown that there exists $\lambda \in k^\times$ such that $z^2 = -\lambda^2$. Now $j = \lambda^{-1} z$ satisfies $j^2 = -1$ and $ij = -ij$. The linear subspace spanned by $1, i, j$ and ij is then a subalgebra of D isomorphic to $\mathbb{H}$, that we will still denote by $\mathbb{H}$ for simplicity.

We now prove that $D = \mathbb{H}$. Assume to the contrary that $D \neq \mathbb{H}$, and let $u \in D, u \notin \mathbb{H}$. Then $v = ui - iu$ satisfies $vi = -iv$, and we can show that $v^2 = -\mu^2$ for some $\mu \in k^\times$ as before. Then $w = \mu^{-1} v$ satisfies $wi = -iw$ and $w^2 = -1$. We then have $jwi = ijw$. Then $k(i, jw)$ is a field containing $k(i)$. Since $k(i)$ is isomorphic to $\mathbb{C}$, we get that $jw \in k(i) \subset \mathbb{H}$, and therefore $w \in \mathbb{H}$. It follows that we also have $v \in \mathbb{H}$. Similarly, $w' = ui + iu$ commutes with i and therefore $w' \in k(i) \subset \mathbb{H}$. Then $u = i^{-1} \dfrac{w + w'}{2} \in \mathbb{H}$, a contradiction. This concludes the proof. $\qquad\square$

The description of the injection described in Lemma II.1.10 immediately yields the following lemma.

LEMMA II.2.2. *The map* $\varphi_{\mathbb{H},\mathbb{C}} : \mathbb{H} \longrightarrow M_2(\mathbb{C})$ *given by*

$$q = x + yi + zj + tij = x_0 + x_1 j \longmapsto M_q = \begin{pmatrix} x_0 & -\overline{x}_1 \\ x_1 & \overline{x}_0 \end{pmatrix}$$

is an injective k-algebra morphism.

In particular, the division $\mathbb{R}$-algebra $\mathbb{H}$ is isomorphic to the real subalgebra of $M_2(\mathbb{C})$ consisting of all the matrices

$$\begin{pmatrix} x_0 & -\overline{x}_1 \\ x_1 & \overline{x}_0 \end{pmatrix}, \ x_0, x_1 \in \mathbb{C}.$$

II.3. Quaternion algebras based codes

In this section, we present two code constructions, one based on Hamilton quaternions and the other on generalized quaternion algebras. We refer the reader to Section I.3 for the coding motivation and related definitions.

In 1998, S.M. Alamouti [1] published a simple code construction for a wireless system with 2 transmit antennas, given as follows. Let x_0, x_1 be two complex numbers that represent the information symbols to be sent, and let the codebook $\mathcal{C}$ be given by the following set of matrices

$$\mathcal{C} = \left\{ \begin{pmatrix} x_0 & -\overline{x}_1 \\ x_1 & \overline{x}_0 \end{pmatrix} \mid x_0, \ x_1 \in \mathbb{C} \right\}.$$

In order to get an efficient code, Alamouti designed his code to be fully-diverse . Indeed, let $\mathbf{X}$ and $\mathbf{X}'$ be two codewords in $\mathcal{C}$. We have that

$$\det(\mathbf{X} - \mathbf{X}') = |x_0 - x_0'|^2 + |x_1 - x_1'|^2 \geq 0,$$

with equality if and only if $x_0 = x_0'$ and $x_1 = x_1'$.

Good performance combined with simplicity made the Alamouti code very attractive. Attempts have been made to understand it better in order to generalize it, and in [46], Sethuraman et al. understood that codewords from the Alamouti code can be seen as left multiplication matrices by elements of $\mathbb{H}$ as described in Lemma II.2.2:

$$\varphi_{\mathbb{H},\mathbb{C}}: \quad \begin{aligned} \mathbb{H} &\longrightarrow M_2(\mathbb{C}) \\ x_0 + x_1 j &\longmapsto \begin{pmatrix} x_0 & -\overline{x}_1 \\ x_1 & \overline{x}_0 \end{pmatrix}. \end{aligned}$$

From the quaternion point of view, full-diversity is immediate, since if we take a codeword $\mathbf{X} \in \mathcal{C}$, then $\mathbf{X} = M_q$ for some quaternion $q \in \mathbb{H}$, and by Remark II.1.11, we have

$$\det(M_q) = \mathrm{Nrd}(q) = 0 \iff q = 0.$$

Equivalently, full-diversity comes from the fact that $\mathbb{H}$ is a division algebra.

Similarly, one can consider a generalized quaternion algebra $Q = (a, b)_k$. Codes over generalized quaternion algebras were introduced in [5], with k a number field. Recall from Remark I.2.12 that we have an injection

$$\varphi_{Q,k(\sqrt{b})}: \quad \begin{aligned} Q &\hookrightarrow M_2(k(\sqrt{b})) \\ q &\longmapsto M_q \end{aligned}$$

given by

$$\varphi_{Q,k(\sqrt{b})}(q) = M_q = \begin{pmatrix} x + z\sqrt{b} & a(y - t\sqrt{b}) \\ y + t\sqrt{b} & x - z\sqrt{b} \end{pmatrix},$$

by Lemma II.1.10.

We thus have a codebook $\mathcal{C}_{Q,k(\sqrt{b})}$ built on Q of the form

$$\mathcal{C}_{Q,k(\sqrt{b})} = \left\{ \begin{pmatrix} x + z\sqrt{b} & a(y - t\sqrt{b}) \\ y + t\sqrt{b} & x - z\sqrt{b} \end{pmatrix}, \; x, y, z, t \in k \right\}.$$

Similarly, we have a codebook $\mathcal{C}_{Q,k(\sqrt{a})}$ built on Q of the form

$$\mathcal{C}_{Q,k(\sqrt{a})} = \left\{ \begin{pmatrix} x + y\sqrt{a} & b(z + t\sqrt{a}) \\ z - t\sqrt{a} & x - y\sqrt{a} \end{pmatrix}, \; x, y, z, t \in k \right\}.$$

For these codes to be fully diverse, we need Q to be a division algebra by Remark I.3.3. Division quaternion algebras are characterized in Proposition II.1.4.

REMARK II.3.1. The definition of codes based on Hamilton and general quaternions is consistent with the definition of code introduced in Definition I.3.2 since $\varphi_{\mathbb{H},\mathbb{C}}$ and $\varphi_{Q,L}$ ($L = k(\sqrt{a})$ or $k(\sqrt{b})$) are indeed suitable injective $\mathbb{R}$-algebra and k-algebra morphisms respectively. $\qquad\square$

Notice that each coefficient of a codeword in $\mathcal{C}_{Q,k(\sqrt{b})}$ is an element of $k(\sqrt{b})$, which is a vector space of dimension 2 over k. Thus while an element of $k(\sqrt{b})$ can be seen as one signal sent, if information symbols to be sent are chosen in k, we have that one signal actually contains two information symbols. Note here the important difference between a code based on general quaternion algebras and one based on Hamilton quaternions. A codeword in $\mathcal{C}_{Q,k(\sqrt{b})}$ can transmit up to 4 information symbols, $x, y, z, t \in k$. However, we need the information symbols to be in $\mathbb{C}$. Since $\mathbb{H}$ is an $\mathbb{R}$-vector space of dimension 4, and thus a $\mathbb{C}$-vector space of dimension 2, one coefficient of the codeword (say $x + z\sqrt{b}$) contains only one element in $\mathbb{C}$ (namely $x + z\sqrt{b}$ itself). By contrast, if Q is chosen with $k \not\subset \mathbb{R}$, then one coefficient (say $x + z\sqrt{b}$) contains two elements in $\mathbb{C}$ (x and y). Thus if we consider the amount of information transmitted, we have 2 information symbols ($x_0 = x + z\sqrt{b}$ and $x_1 = y + t\sqrt{b}$) using 4 signals with the Alamouti code (which means a rate of $2/4 = 1/2$) and 4 information symbols (x, y, z, t) using 4 signals with a code based on a generalized quaternion algebra, assuming $k \not\subset \mathbb{R}$ (which is thus a rate of $4/4 = 1$).

There is now a natural question to address, which is: how to generalize the quaternion codes in higher dimensions ? Since we know by Theorem II.2.1 that the only finite-dimensional division $\mathbb{R}$-algebras are $\mathbb{R}$, $\mathbb{C}$ and $\mathbb{H}$, it is thus clear that we need to look for other base fields than $\mathbb{R}$.

Exercises

1. Let $a, b \in \mathbb{Q}^\times$. Assume that $a < 0$ and $b < 0$. Show that the quaternion $\mathbb{Q}$-algebra $(a, b)_\mathbb{Q}$ is a division algebra.

2. Let k be a field of characteristic different from 2, and let t_1, t_2 be two independent indeterminates over k. Show that the quaternion algebra $(t_1, t_2)_{k(t_1, t_2)}$ is a division algebra.

3. Solve the equation $q^2 = -1, q \in \mathbb{H}$.

4. Using a suitable quaternion algebra, show that the set
$$\{a^2 + b^2 + c^2 + d^2 \mid a, b, c, d \in \mathbb{Z}\}$$
is closed under multiplication.

5. Let k be a field of characteristic different from 2, and let $Q = (a, b)_k$. Let Q^0 be the k-linear subspace of Q spanned by i, j and ij.

   (a) Let $q \in Q$. Show that $q \in Q^0$ if and only if $q^2 \in k$.

   (b) Let $e \in Q^0 \cap Q^\times$. Show that there exists $f \in Q^0 \cap Q^\times$ such that $ef = -fe$, then show that $(1, e, f, ef)$ is a k-basis of Q.

CHAPTER III

Fundamental results on central simple algebras

In this chapter, we study the stability of central simple algebras under the classical operations, and we prove three important theorems in the theory of central simple algebras, namely Skolem-Noether's Theorem, Wedderburn's Theorem and the Centralizer Theorem, after collecting some definitions and results on R-modules.

Recall that in this book, all the k-algebras are unital, associative and finite dimensional over k, unless specified otherwise.

III.1. Operations on central simple algebras

In this section, we study the stability of central simple algebras under classical operations (tensor product, scalar extension). We start with a very simple lemma on tensor products:

LEMMA III.1.1. *Let A and B be two k-algebras. If $b_1, \ldots, b_m \in B$ are linearly independent over k, then for all $x_1, \ldots, x_m \in A$, we have*

$$x_1 \otimes b_1 + \cdots + x_m \otimes b_m = 0 \Rightarrow x_1 = \cdots = x_m = 0.$$

Similarly, if $a_1, \ldots, a_n \in A$ are linearly independent over k, then for all $y_1, \ldots, y_n \in B$, we have

$$a_1 \otimes y_1 + \cdots + a_n \otimes y_n = 0 \Rightarrow y_1 = \cdots = y_n = 0.$$

Proof. We only prove the first part, since the proof of the second part is completely similar. We may extend $b_1, \ldots, b_m$ into a k-basis $(b_1, \ldots, b_s)$ of B. Now let $(a_1, \ldots, a_r)$ be a k-basis of A, and write $x_j = \sum_i x_{ij} a_i$, for $x_{ij} \in k$ and $j = 1, \ldots, m$.

By assumption, we then have

$$\sum_{i,j} x_{ij} a_i \otimes b_j = 0.$$

Since $(a_i \otimes b_j)_{i,j}$ is a k-basis of $A \otimes_k B$, we get that $x_{ij} = 0$ for all i, j, and thus $x_1 = \cdots = x_m = 0$. $\square$

REMARK III.1.2. The result is true more generally if A and B are arbitrary k-vector spaces. $\square$

We now define the centralizer of a subset of a k-algebra.

DEFINITION III.1.3. Let A be a k-algebra, and let $B \subset A$ be a subset of A. The **centralizer** of B in A is the set $C_A(B)$ defined by

$$C_A(B) = \{a \in A \mid ab = ba \text{ for all } b \in B\}.$$

Clearly, this is a subalgebra of A and $C_A(A) = Z(A)$.

PROPOSITION III.1.4. *Let A and B be two k-algebras. Assume that A' and B' are subalgebras of A and B respectively. Then we have*

$$C_{A \otimes_k B}(A' \otimes_k B') = C_A(A') \otimes_k C_B(B').$$

Proof. Assume that $z = \sum_i a_i \otimes b_i \in C_A(A') \otimes_k C_B(B')$. To prove that $z \in C_{A \otimes_k B}(A' \otimes_k B')$, it is enough to show that z commutes with elementary tensors of $A' \otimes_k B'$. But this follows directly from the definitions in this case. Hence $C_{A \otimes_k B}(A' \otimes_k B') \supset C_A(A') \otimes_k C_B(B')$. To prove the missing inclusion, let $(b_1, \ldots, b_m)$ be a k-basis of B, and let x be an element of $C_{A \otimes_k B}(A' \otimes_k B')$. Writing elements of B as linear combinations of the b_j's, one can see that $x = x_1 \otimes b_1 + \cdots + x_m \otimes b_m$, for some $x_i \in A$. By assumption, we have $(a' \otimes 1)x = x(a' \otimes 1)$ for all $a' \in A'$. It follows that we have

$$(a'x_1 - x_1 a') \otimes b_1 + \cdots + (a'x_m - x_m a') \otimes b_m = 0 \text{ for all } a' \in A'.$$

By the previous lemma, for all j we get

$$a'x_j = x_j a' \text{ for all } a' \in A'.$$

Hence $x_1, \ldots, x_m \in C_A(A')$. Now let $a_1, \ldots, a_n$ be a k-basis of $C_A(A')$. Writing the x_j's in this basis shows that we have

$$x = a_1 \otimes y_1 + \cdots + a_n \otimes y_n$$

for some $y_i \in B$. Reasoning as previously shows that $y_1, \ldots, y_n$ lie in $C_B(B')$. Therefore $x \in C_A(A') \otimes_k C_B(B')$ and we are done. $\qquad\square$

COROLLARY III.1.5. *Let A and B be two k-algebras, and let L/k be a field extension. Then the following properties hold:*

(1) *$A \otimes_k B$ is central over k if and only if A and B are central over k;*

(2) *$A \otimes_k L$ is central over L if and only if A is central over k.*

Proof.

(1) Proposition III.1.4 shows that $Z(A \otimes_k B) = Z(A) \otimes_k Z(B)$. Therefore, $\dim_k(Z(A \otimes_k B)) = \dim_k(Z(A)) \dim_k(Z(B))$. Now $A \otimes_k B$ is central over k if and only if $\dim_k(Z(A \otimes_k B)) = 1$. The previous equality shows that it is equivalent to $\dim_k(Z(A)) = \dim_k(Z(B)) = 1$, that is A and B are central over k.

(2) By Proposition III.1.4, we have $Z(A \otimes_k L) = Z(A) \otimes_k L$. Hence $\dim_L(Z(A \otimes_k L)) = \dim_k(Z(A))$, and we conclude as in (1). $\qquad\square$

PROPOSITION III.1.6. *If A is a central simple k-algebra and B is a simple k-algebra, then $A \otimes_k B$ is simple.*

Proof. Let I be a non-trivial two-sided ideal of $A \otimes_k B$, and let

$$x = a_1 \otimes b_1 + \cdots + a_m \otimes b_m \in I, \ x \neq 0$$

such that $m \geq 1$ is minimal. In particular, $b_1, \ldots, b_m$ are k-linearly independent.

Now since $a_m \neq 0$ (by minimality of m) and A is simple, the two-sided ideal of A generated by a_m is A. Hence there exist $x_i, x_i' \in A$ such that $\sum_i x_i a_m x_i' = 1$. We then have

$$\sum_i (x_i \otimes 1) x (x_i' \otimes 1) = \Big(\sum_i x_i a_1 x_i'\Big) \otimes b_1 + \cdots + \Big(\sum_i x_i a_m x_i'\Big) \otimes b_m \in I.$$

Hence we may assume without loss of generality that $a_m = 1$.

We now prove that $m = 1$. Suppose that $m > 1$. By minimality of m, a_{m-1} and $a_m = 1$ are k-linearly independent, so $a_{m-1} \notin k$. Since A is central, it means that $a_{m-1} \notin Z(A)$. Hence, there exists $a \in A$ such that $a a_{m-1} - a_{m-1} a \neq 0$. Since $a_m = 1$, we have

$$(a \otimes 1) x - x (a \otimes 1) = (a a_1 - a_1 a) \otimes b_1 + \cdots + (a a_{m-1} - a_{m-1} a) \otimes b_{m-1}.$$

Since $a a_{m-1} - a_{m-1} a \neq 0$ and $b_1, \ldots, b_{m-1}$ are linearly independent, this element is a non-zero element of I by Lemma III.1.1. This contradicts the minimality of m. Hence $m = 1$, so I contains an element of the form $1 \otimes b$. Since B is simple, arguing as at the beginning of the proof shows that I contains $1 \otimes 1$, so $I = A \otimes_k B$ and we are done. $\square$

REMARK III.1.7. This result is not true if A is not central. For example, $\mathbb{C}$ is a simple $\mathbb{R}$-algebra. However, we have

$$\mathbb{C} \otimes_{\mathbb{R}} \mathbb{C} \cong_{\mathbb{R}} \mathbb{C} \times \mathbb{C}.$$

Since $\mathbb{C} \times \{0\}$ is a non-trivial ideal of $\mathbb{C} \times \mathbb{C}$, it follows that $\mathbb{C} \otimes_{\mathbb{R}} \mathbb{C}$ is not simple. $\square$

COROLLARY III.1.8. *Let A and B be two k-algebras, and let L/k be a field extension. Then the following properties hold:*

(1) *if A and B are central simple, so is $A \otimes_k B$;*
(2) *A is central simple over k if and only if $A \otimes_k L$ is central simple over L.*

Proof. Point (1) and the direct implication of (2) readily follow from Corollary III.1.5 and the previous proposition. Assume that $A \otimes_k L$ is central simple over L. Then A is central over k by Corollary III.1.5. Now let I be a non-zero two-sided ideal of A. Then $I \otimes_k L$ is a non-zero two-sided ideal of $A \otimes_k L$, and therefore $I \otimes_k L = A \otimes_k L$. Thus we have

$$\dim_k(I) = \dim_L(I \otimes_k L) = \dim_L(A \otimes_k L) = \dim_k(A),$$

and so $I = A$. Hence, A is simple. This concludes the proof. $\square$

We will go back to the study of the centralizer at the end of this chapter. For the moment, we would like to end this section by introducing the opposite algebra of a central simple k-algebra A.

DEFINITION III.1.9. Let A be k-algebra, and let A^{op} be the set

$$A^{op} = \{a^{op} \mid a \in A\}.$$

The operations

$$k \times A^{op} \longrightarrow A^{op} \qquad A^{op} \times A^{op} \longrightarrow A^{op}$$
$$(\lambda, a^{op}) \longmapsto (\lambda \cdot a)^{op}, \qquad (a_1^{op}, a_2^{op}) \longmapsto (a_1 + a_2)^{op}$$

and
$$A^{op} \times A^{op} \longrightarrow A^{op}$$
$$(a_1^{op}, a_2^{op}) \longmapsto (a_2 a_1)^{op}$$

endow A^{op} with the structure of a k-algebra, called the **opposite algebra**.

REMARK III.1.10. It is easy to see that $Z(A^{op}) = Z(A)$ and that the left (resp. right, resp. two-sided) ideals of A are in one-to-one correspondence with the right (resp. left, resp. two-sided) ideals of A^{op}. In particular, A^{op} is a central simple (resp. division) k-algebra if and only if A is. ☐

LEMMA III.1.11. *Let A, B be k-algebras, let L/k be a field extension, and let $n \geq 1$ be an integer. Then we have:*

(1) $\mathrm{M}_n(k)^{op} \cong_k \mathrm{M}_n(k)$;

(2) $(A^{op})^{op} \cong_k A$;

(3) $(A \otimes_k B)^{op} \cong_k A^{op} \otimes_k B^{op}$;

(4) $(A \otimes_k L)^{op} \cong_L A^{op} \otimes_k L$.

Proof.

(1) It is easy to check that
$$t: \quad \begin{array}{c} \mathrm{M}_n(k)^{op} \longrightarrow \mathrm{M}_n(k) \\ M^{op} \longmapsto M^t \end{array}$$

is an isomorphism of k-algebras.

(2) The desired isomorphism is given by
$$(A^{op})^{op} \xrightarrow{\sim} A$$
$$(a^{op})^{op} \longmapsto a.$$

(3) Notice that the maps
$$f: \quad \begin{array}{c} A^{op} \longrightarrow (A \otimes_k B)^{op} \\ a^{op} \longmapsto (a \otimes 1)^{op} \end{array}$$

and
$$g: \quad \begin{array}{c} B^{op} \longrightarrow (A \otimes_k B)^{op} \\ b^{op} \longmapsto (1 \otimes b)^{op} \end{array}$$

are k-algebra morphisms with commuting images. Then there exists a unique k-algebra morphism
$$\varphi : A^{op} \otimes_k B^{op} \longrightarrow (A \otimes_k B)^{op}$$

satisfying

$\varphi(a^{op} \otimes 1) = (a \otimes 1)^{op}$ and $\varphi(1 \otimes b^{op}) = (1 \otimes b)^{op}$ for all $a \in A, b \in B$.

Then we have easily

$$\varphi(a^{op} \otimes b^{op}) = (a \otimes b)^{op} \text{ for all } a \in A, b \in B,$$

and since the elements of the form $(a \otimes b)^{op}$ span $(A \otimes_k B)^{op}$, it follows that φ is therefore surjective. Since $(A \otimes_k B)^{op}$ and $A^{op} \otimes_k B^{op}$ have the same dimension over k, φ is bijective.

(4) Observe that the obvious map induces an isomorphism of k-algebras

$$L \cong_k L^{op},$$

since L is commutative. Using the previous point, we get an isomorphism of k-algebras

$$\rho : A^{op} \otimes_k L \longrightarrow (A \otimes_k L)^{op}$$

satisfying

$$\rho(a^{op} \otimes \lambda) = (a \otimes \lambda)^{op} \text{ for all } a \in A, \lambda \in L.$$

One may verify that ρ is also L-linear. $\qquad\square$

REMARK III.1.12. The reader may show as an exercise that the isomorphism $(A \otimes_k B)^{op} \cong_k A^{op} \otimes_k B^{op}$ still hold in the infinite-dimensional case. $\qquad\square$

Let A be a central simple k-algebra. For $a \in A$, we denote by ℓ_a and r_a the elements of $\mathrm{End}_k(A)$ (where A is a considered as a k-vector space) defined by

$$\ell_a : \begin{array}{c} A \longrightarrow A \\ z \longmapsto az \end{array} \quad \text{and } r_a : \begin{array}{c} A \longrightarrow A \\ z \longmapsto za. \end{array}$$

The maps

$$\begin{array}{c} A \longrightarrow \mathrm{End}_k(A) \\ a \longmapsto \ell_a \end{array} \quad \text{and} \quad \begin{array}{c} A^{op} \longrightarrow \mathrm{End}_k(A) \\ a^{op} \longmapsto r_a \end{array}$$

are easily seen to be k-algebra morphisms, with commuting images. Therefore, there exists a unique k-algebra morphism

$$\mathrm{Sand} : A \otimes_k A^{op} \longrightarrow \mathrm{End}_k(A)$$

satisfying

$$\mathrm{Sand}(a \otimes b^{op})(z) = azb \text{ for all } a, b, z \in A,$$

called the **Sandwich morphism**.

LEMMA III.1.13. *For every central simple k-algebra A, Sand induces a k-algebra isomorphism $A \otimes_k A^{op} \cong_k \mathrm{End}_k(A)$.*

Proof. Since A and A^{op} are central simple k-algebras, so is $A \otimes_k A^{op}$ by Corollary III.1.8. Since $A \otimes_k A^{op}$ and $\mathrm{End}_k(A)$ have the same dimension over k, Lemma I.2.2 implies that Sand is an isomorphism. $\qquad\square$

III.2. Simple modules

DEFINITION III.2.1. Let R be a ring (not necessarily commutative). A **left R-module** is an abelian group $(M, +)$ endowed with a scalar multiplication

$$\begin{array}{c} R \times M \longrightarrow M \\ (a, x) \longmapsto a{\cdot}x \end{array}$$

satisfying:

(1) $1_R{\cdot}x = x$ for all $x \in M$;
(2) $(aa'){\cdot}x = a{\cdot}(a'{\cdot}x)$ for all $a, a' \in R, x \in M$;
(3) $a{\cdot}(x + y) = a{\cdot}x + a{\cdot}y$ for all $a \in R, x, y \in M$;
(4) $(a + a'){\cdot}x = a{\cdot}x + a'{\cdot}x$ for $a, a' \in R, x \in M$.

A **right** R**-module** is an abelian group $(M, +)$ endowed with a scalar multiplication

$$M \times R \longrightarrow M$$

$$(x, a) \longmapsto x \cdot a$$

satisfying:

(1) $x \cdot 1_R = x$ for all $x \in M$;

(2) $x \cdot (aa') = (x \cdot a) \cdot a'$ for all $a, a' \in R, x \in M$;

(3) $(x + y) \cdot a = x \cdot a + y \cdot a$ for all $a \in R, x, y \in M$;

(4) $x \cdot (a + a') = x \cdot a + x \cdot a'$ for $a, a' \in R, x \in M$.

A **submodule** of a left (resp. right) R-module M is a non-empty subset N of M which is closed under addition and scalar multiplication. In this case, N is an R-module for the addition and scalar multiplication of M.

A left (resp. right) R-module M is said to be **finitely generated** if there exist $x_1, \ldots, x_r \in M, r \geq 0$ such that every element of M is a left (resp. right) A-linear combination of $x_1, \ldots, x_r$.

EXAMPLES III.2.2.

(1) A ring R has a natural left (resp. right) R-module structure, the scalar multiplication being given by multiplication in R. In this setting, a submodule is just a left (resp. right) ideal of R.

(2) If $R = k$ is a field, then an R-module is nothing but a k-vector space.

(3) If M is a left (resp. right) R-module, so is M^n for any $n \geq 1$, an element $a \in R$ acting componentwise. In particular, R^n is a left (resp. right) R-module for any $n \geq 1$.

(4) If I is a left ideal of R, the abelian group R/I has a natural left R-module structure, where the scalar multiplication is given by

$$R \times R/I \longrightarrow R/I$$

$$(a, \overline{x}) \longmapsto \overline{a \cdot x}.$$

The reader will check that this map is indeed well-defined. Of course, if I is a right ideal, we may endow R/I with a structure of a right R-module in a similar way.

$\square$

Now that we have defined the objects, we need to define the maps between them. This is provided by the following definition.

DEFINITION III.2.3. Let R be a ring, and let M and N be two left R-modules. An R**-module morphism** $f : M \longrightarrow N$ is a group morphism satisfying

$$f(a \cdot x) = a \cdot f(x) \text{ for all } a \in R, x \in M.$$

Similarly, if M and N are right R-modules, an R**-module morphism** $f : M \longrightarrow N$ is a group morphism satisfying

$$f(x \cdot a) = f(x) \cdot a \text{ for all } a \in R, x \in M.$$

We also say that f is an R**-linear** map. If $M = N$, we say that f is an **endomorphism**. We will denote by $\mathrm{End}_R(M)$ the ring of endomorphisms of M. An

isomorphism of R-modules is a R-linear map which is bijective. In this case, one may verify that f^{-1} is also R-linear.

We continue this list of definitions by introducing the notion of a free R-module.

DEFINITION III.2.4. A finitely generated R-module M is **free** if it is isomorphic to R^n for some $n \geq 1$.

We would like now to give a couple of technical results concerning the endomorphism ring of some R-modules. We start with an easy observation: let R be a ring and let $e \in R$ be a non-zero idempotent, that is $e^2 = e$. Then eRe is a ring for the addition and multiplication of R, with unit e (we have $e \in eRe$ since $e = e^2 = e1_Re$).

We are ready to state and prove the following lemma.

LEMMA III.2.5. *Let R be a ring, and let e be a non-zero idempotent of R. Then we have a ring isomorphism*

$$eRe \cong \mathrm{End}_R(eR)$$

where eR is considered as a right R-module.

Proof. For $a \in R$, let ℓ_a be the map defined by

$$\ell_a \colon \begin{array}{c} R \longrightarrow R \\ x \longmapsto ax. \end{array}$$

Clearly, ℓ_a is a group morphism. Moreover for all $a, a', x \in R$, we have

$$\ell_a(x{\cdot}a') = \ell_a(xa') = a(xa') = (ax){\cdot}a' = \ell_a(x){\cdot}a'.$$

Therefore, $\ell_a \in \mathrm{End}_R(R)$. Moreover, if $a \in eRe$, then clearly ℓ_a restricts to an endomorphism of eR. We claim that the map

$$\varphi \colon \begin{array}{c} eRe \longrightarrow \mathrm{End}_R(eR) \\ a \longmapsto \ell_a \end{array}$$

is a ring isomorphism.

Let us prove first that φ is a ring morphism. First, we have

$$\varphi(e)(ex) = e^2 x = ex \text{ for all } x \in R$$

that is $\varphi(e) = \mathrm{Id}_{eR}$. Moreover, for all $a, a' \in eRe, x \in R$, we have

$$\ell_{a+a'}(ex) = (a + a')(ex) = aex + a'ex = \ell_a(ex) + \ell_{a'}(ex)$$

and

$$\ell_{aa'}(ex) = aa'ex = a(a'ex) = \ell_a(a'ex) = (\ell_a \circ \ell_{a'})(ex).$$

Therefore, $\ell_{a+a'} = \ell_a + \ell_{a'}$ and $\ell_{aa'} = \ell_a \circ \ell_{a'}$, so φ is a ring morphism.

We now prove that φ is bijective. Let $a \in eRe$ such that $\varphi_a = \mathrm{Id}_{eR}$, and write $a = ebe$ for some $b \in R$. We then have

$$\varphi_a(e) = \ell_a(e) = ae = ebe^2 = ebe = a = \mathrm{Id}_{eR}(e) = e,$$

hence φ is injective. To prove that φ is surjective, let $f \in \mathrm{End}_R(eR)$. Since f is R-linear, we get

$$f(ex) = f(e^2 x) = f(e{\cdot}ex) = f(e){\cdot}ex = f(e)ex \text{ for all } x \in R.$$

Since $f(e) \in eR$, we may write $f(e) = eb$ for some $b \in R$. Hence we get

$$f(ex) = ebex = ebe^2 x = (ebe)ex = \ell_{ebe}(ex) \text{ for all } x \in R.$$

Hence $f = \varphi(ebe)$ and φ is surjective. $\square$

The next result generalizes the ring isomorphism

$$M_n(M_m(k)) \cong M_{nm}(k).$$

LEMMA III.2.6. *Let R be a ring, and let M be a right R-module. For all $n \geq 1$, we have a ring isomorphism*

$$\mathrm{End}_R(M^n) \cong M_n(\mathrm{End}_R(M)).$$

Proof. If M is a finitely generated free R-module, that is

$$M \cong_R R^m$$

for some $m \geq 1$, then $M^n \cong_R R^{mn}$, and the lemma just says that $M_{mn}(R) \cong M_n(M_m(R))$, which is quite obvious. Unfortunately, modules over a general ring are not always free, so the proof is slightly more technical.

For $1 \leq i \leq n$, we have canonical projections

$$\pi_i : \quad \begin{aligned} M^n &\longrightarrow M \\ \mathbf{m} = (m_j)_{1 \leq j \leq n} &\longmapsto m_i \end{aligned}$$

and canonical injections

$$\iota_i : \quad \begin{aligned} M &\longrightarrow M^n \\ m &\longmapsto (0, \ldots, 0, m, 0, \ldots, 0). \end{aligned}$$

Let $f : M^n \longrightarrow M^n$ be an endomorphism. For $1 \leq i \leq n$, let $f_i = \pi_i \circ f$. Then $f_i : M^n \longrightarrow M$ is R-linear and for all $\mathbf{m} \in M^n$, we have

$$f(\mathbf{m}) = (f_1(\mathbf{m}), \ldots, f_n(\mathbf{m})).$$

Now observe that

$$f_i(\mathbf{m}) = f_i(\iota_1(\mathbf{m}) + \cdots + \iota_n(\mathbf{m})) = f_i(\iota_1(\mathbf{m})) + \cdots + f_i(\iota_n(\mathbf{m})).$$

Putting things together, we finally get

$$f = \Big(\sum_{i=1}^n \pi_i \circ f \circ \iota_1, \ldots, \sum_{i=1}^n \pi_i \circ f \circ \iota_n\Big).$$

The idea of the proof is that, according to the formula above, f should be completely determined by the maps $\pi_i \circ f \circ \iota_j$.

Notice that for $1 \leq i, j \leq n$, $\pi_i \circ f \circ \iota_j \in \mathrm{End}_R(M)$, and consider the map

$$\varphi : \quad \begin{aligned} \mathrm{End}_R(M^n) &\longrightarrow M_n(\mathrm{End}_R(M)) \\ f &\longmapsto (\pi_i \circ f \circ \iota_j)_{1 \leq i,j \leq n}. \end{aligned}$$

We are going to prove that φ is the desired ring isomorphism. We first show that it is a ring morphism. From the definitions, it is easy to see that $\pi_i \circ \iota_j = 0$ if $i \neq j$ and $\pi_i \circ \iota_i = \mathrm{Id}_M$. Thus we have

$$\varphi(\mathrm{Id}_{M^n}) = I_n.$$

Moreover, it is clear that we have

$$\varphi(f+g) = \varphi(f) + \varphi(g) \text{ for all } f, g \in \mathrm{End}_R(M^n).$$

It remains to show that $\varphi(f \circ g) = \varphi(f)\varphi(g)$. The (i,j)-coefficient of $\varphi(f)\varphi(g)$ is given by

$$\sum_{s=1}^{n}(\pi_i \circ f \circ \iota_s) \circ (\pi_s \circ g \circ \iota_j) = \pi_i \circ f \circ \Big(\sum_{s=1}^{n} \iota_s \circ \pi_s\Big) \circ g \circ \iota_j.$$

From the definitions, we have $\displaystyle\sum_{s=1}\iota_s \circ \pi_s = \mathrm{Id}_{M^n}$. Therefore, we finally get

$$\varphi(f)\varphi(g) = (\pi_i \circ f \circ g \circ \iota_j)_{1 \leq i,j \leq n} = \varphi(f \circ g).$$

We now prove that φ is bijective. First, φ is surjective. Indeed, let $(f_{ij})_{1 \leq i,j \leq n} \in \mathrm{M}_n(\mathrm{End}_R(M))$, and let $f : M^n \longrightarrow M^n$ be the map defined by

$$f = \sum_{1 \leq r,s \leq n} \iota_r \circ f_{rs} \circ \pi_s.$$

Let us compute $\varphi(f)$. We have

$$\pi_i \circ f \circ \iota_j = \sum_{1 \leq r,s \leq n} \pi_i \circ \iota_r \circ f_{rs} \circ \pi_s \circ \iota_j.$$

Since $\pi_i \circ \iota_r$ and $\pi_s \circ \iota_j$ equal zero if $i \neq r$ and $s \neq j$ respectively, we get

$$\pi_i \circ f \circ \iota_j = \pi_i \circ \iota_i \circ f_{ij} \circ \pi_j \circ \iota_j = f_{ij}.$$

Hence $\varphi(f) = (f_{ij})_{1 \leq i,j \leq n}$, so φ is surjective.

Now let $f \in \mathrm{End}_R(M^n)$ such that $\varphi(f) = 0$. By a previous computation, we know that

$$f = \Big(\sum_{i=1}^{n} \pi_i \circ f \circ \iota_1, \dots, \sum_{i=1}^{n} \pi_i \circ f \circ \iota_n\Big).$$

If $\varphi(f) = 0$, we have $\pi_i \circ f \circ \iota_j = 0$ for all $1 \leq i,j \leq n$, and consequently $f = 0$. Hence φ is also injective; this concludes the proof. $\square$

The next notion will be crucial for the study of simple algebras.

DEFINITION III.2.7. Let R be a ring. An R-module M is called **simple** if $M \neq 0$ and if it has no submodules other than 0 and M.

EXAMPLES III.2.8.

(1) A left (resp. right) ideal $I \neq 0$ of a ring R is simple (as an R-module) if and only if it is minimal for the inclusion, that is

$$\text{for every left (resp. right) ideal } I' \neq 0, I' \subset I \Rightarrow I' = I.$$

(2) Let $A = \mathrm{M}_r(D)$, where $r \geq 1$ and D is a division ring. For all $m = 1, \dots, r$, we denote by $\mathcal{L}_m$ the set of matrices whose i^{th}-row is zero whenever $i \neq m$.
   For example, if $r = 4$, $\mathcal{L}_3$ consists of matrices of the form

$$\begin{pmatrix} 0 & 0 & 0 & 0 \\ 0 & 0 & 0 & 0 \\ * & * & * & * \\ 0 & 0 & 0 & 0 \end{pmatrix}.$$

Then $\mathcal{L}_m$ is a minimal right ideal of A, hence a simple A-module, by the previous example.

The fact that $\mathcal{L}_m$ is an ideal follows from direct computations. We now prove its minimality. For, let I be a non-zero right ideal of $M_r(D)$ such that $I \subset \mathcal{L}_m$, and let $M \in I, M \neq 0$. In particular, $M \in \mathcal{L}_m$, so we may write

$$M = d_{m1}E_{m1} + \cdots + d_{mr}E_{mr}.$$

Let $j = 1, \ldots, r$ such that $d_{mj} \neq 0$. Then for all $s = 1, \ldots, r$, we have $E_{ms} = ME_{js}d_{mj}^{-1}$. Since I is a right ideal, we deduce that $E_{m1}, \ldots, E_{mr} \in I$. It easily follows that $\mathcal{L}_m \subset I$, proving that $I = \mathcal{L}_m$. Hence $\mathcal{L}_m$ is minimal.

(3) If $I \neq 0$ is a left (resp. right) ideal of R, then R/I is simple as an R-module if and only if I is maximal for the inclusion. Indeed, one can easily see that submodules of R/I have the form J/I, where J is a left (resp. right) ideal of R containing I.

$\square$

The last example shows that we are not working with an empty notion, since every ring R has a maximal left (resp. right) ideal, and so every ring R has a simple left (resp. right) R-module.

REMARK III.2.9. If $f : M \longrightarrow N$ is an R-module morphism, then $\ker(f)$ is a submodule of M and $\mathrm{Im}(f)$ is a submodule of N. $\square$

The next very simple lemma is incredibly useful, and gives us an easy way to construct division rings.

LEMMA III.2.10 (Schur's Lemma). *Let R a ring, and let M be a simple R-module M. Then $\mathrm{End}_R(M)$ is a division ring.*

Proof. We need to prove that every endomorphism $f : M \longrightarrow M$ which is not identically zero is invertible. Since $f \neq 0$, $\ker(f)$ is a submodule of M different from M and $\mathrm{Im}(f)$ is a non-zero submodule of M. Since M is simple, we then get $\ker(f) = 0$ and $\mathrm{Im}(f) = M$. Thus f is injective and surjective, hence bijective. $\square$

We would like to elucidate the structure of finitely generated modules over a (not necessarily central) simple k-algebra. We start with some easy general observations.

LEMMA III.2.11. *Let A be an arbitrary (finite dimensional) k-algebra. Then the following properties hold:*

(1) *every left (resp. right) A-module M has a natural structure of a k-vector space, induced by scalar multiplication. Moreover, M is finite dimensional over k whenever it is finitely generated as an A-module;*

(2) *A has a minimal left (resp. right) ideal;*

(3) *for all $n, m \geq 1$, we have $A^n \cong_A A^m \iff n = m$.*

Proof. The first point is obvious. To prove (2), observe that any left (resp. right) ideal of A is in particular a linear subspace of A, hence finite dimensional over k since A is. Therefore, any left (resp. right) ideal of A with minimal dimension as a k-vector space is a minimal ideal. Finally assume that A^n and A^m are isomorphic as A-modules. Then they are isomorphic as k-vector spaces, and therefore have the same dimension over k. This easily implies that $n = m$. $\qquad\square$

REMARK III.2.12. Property (3) of the lemma above may become false if we replace A by any ring or even by an infinite dimensional k-algebra. Indeed, one may show that if $A = \mathrm{End}_k(V)$, where V is any k-vector space of infinite countable dimension, then $A \cong_A A^n$ for all $n \geq 1$. $\qquad\square$

DEFINITION III.2.13. Let A be an arbitrary k-algebra, and let M be a finitely generated left (resp. right) free A-module. The last part of Lemma III.2.11 shows that $M \cong_A A^n$ for a uniquely determined integer $n \geq 1$. This integer is called the **A-rank** (or simply the rank) of M and is denoted by $\mathrm{rk}_A(M)$.

We are now ready to determine the structure of finitely generated modules over a simple k-algebra. Notice that statements below make sense in view of Lemma III.2.11.

PROPOSITION III.2.14. *Let A be a simple k-algebra, and let I be a minimal right ideal. Then the following properties hold:*

(1) *every non-zero finitely generated right A-module M is isomorphic to I^n for some $n \geq 1$;*

(2) *all finitely generated simple right A-modules, and in particular all minimal right ideals of A, are isomorphic;*

(3) *a non-zero finitely generated A-module M is free if and and only if $\dim_k(A) \mid \dim_k(M)$. In this case, we have*

$$\mathrm{rk}_A(M) = \frac{\dim_k(M)}{\dim_k(A)};$$

(4) *two non-zero finitely generated right A-modules are isomorphic if and only if they have the same dimension over k.*

Proof. Let I be a minimal right ideal of A. Let us show first how (1) implies properties $(2) - (4)$.

Let M be a finitely generated simple right A-module. In particular, M is non-trivial and therefore $M \cong_A I^n$ for some $n \geq 1$ by (1). Since M is simple, we necessarily have $n = 1$. Otherwise I^n, and thus M, would have a non-trivial submodule. Hence $M \cong_A I$; this proves (2).

Assume now that M is a non-zero finitely generated A-module. If M is free, then $M \cong_A A^n$, where $n = \mathrm{rk}_A(M)$. Since M and A^n are isomorphic as k-vector spaces, comparing dimensions then shows that $\dim_k(M) = \mathrm{rk}_A(M)\dim_k(A)$. In particular, $\dim_k(A) \mid \dim_k(M)$ and $\mathrm{rk}_A(M) = \dfrac{\dim_k(M)}{\dim_k(A)}$.

Conversely, assume that $\dim_k(A) \mid \dim_k(M)$. Since M and A are both non-zero finitely generated A-modules, we have $M \cong_A I^n$ and $A \cong_A I^m$ for some $n, m \geq 1$ by (1). The assumption then easily implies that $m \mid n$ by comparing dimensions over k. Writing $n = mr$, we then get

$$M \cong_A I^{mr} \cong_A (I^m)^r \cong_A A^r,$$

hence M is free. The previous point shows that $\operatorname{rk}_A(M) = \dfrac{\dim_k(M)}{\dim_k(A)}$.

We finally prove (4). Let M and N be two non-zero finitely generated right A-modules. Then $M \cong_A I^n$ and $N \cong_A I^m$ for some integers $n, m \geq 1$ by (1). In particular, if M and N have the same dimension as k-vector spaces, then $n \dim_k(I) = m \dim_k(I)$, and therefore $n = m$. In this case, $M \cong_A I^n \cong_A N$. Conversely, if M and N are isomorphic as R-modules, they are isomorphic as k-vector spaces and thus have the same dimension over k. This proves (4).

It remains now to prove (1). Let M be a non-zero finitely generated A-module. We claim that there exist some elements $m_1, \ldots, m_n \in M$ such that $M = m_1 \cdot I + \cdots + m_n \cdot I$.

The left ideal generated by the elements of I is a non-zero two-sided ideal of A, hence equals A by assumption. In particular, one may write

$$1 = a_1 \alpha_1 + \cdots + a_m \alpha_m \text{ for some } a_i \in A \text{ and } \alpha_i \in I.$$

Thus for all $a \in A$, we have

$$a = (a_1 \alpha_1 + \cdots + a_m \alpha_m)a = a_1(\alpha_1 a) + \cdots + a_m(\alpha_m a).$$

Since I is a right ideal, we have $\alpha_i a \in I$ for all i, and therefore we have $A = a_1 I + \cdots + a_m I$. Now by assumption, we have

$$M = x_1 \cdot A + \cdots + x_r \cdot A,$$

for some $x_1, \cdots, x_r \in A$. Putting things together, we get

$$M = \sum_{i,j} x_i \cdot (a_j I) = \sum_{i,j} (x_i \cdot a_j) \cdot I,$$

which proves the claim.

We may then write $M = m_1 \cdot I + \cdots + m_n \cdot I$, with n minimal for this property. Necessarily, $n \geq 1$ since $M \neq 0$. We now prove that we have

$$M = m_1 \cdot I \oplus \cdots \oplus m_n \cdot I.$$

Assume that $m_1 \cdot \beta_1 + \cdots + m_n \cdot \beta_n = 0$ for some $\beta_i \in I$. If one of the β_i's is non-zero, say β_n, then $\beta_n A$ is a non-zero right ideal of A contained in I (since $\beta_n \in I$). Since I is minimal, we get $\beta_n A = I$. Hence we get

$$m_n \cdot I = (m_n \cdot \beta_n) \cdot A = (-m_1 \cdot \beta_1 - \cdots - m_{n-1} \cdot \beta_{n-1}) \cdot A.$$

Now for every i, we have $(m_i \cdot \beta_i) \cdot A = m_i \cdot (\beta_i A) \subset m_i \cdot I$, so we finally obtain

$$M = m_1 \cdot I + \cdots + m_{n-1} \cdot I.$$

contradicting the minimality of n. Hence $\beta_1 = \cdots = \beta_n = 0$, proving that the sum above is direct.

It follows that the A-linear map

$$f \colon \begin{array}{ccc} I^n & \longrightarrow & M \\ (\beta_1, \ldots, \beta_n) & \longmapsto & \displaystyle\sum_{i=1}^n m_i \cdot \beta_i \end{array}$$

is an isomorphism of right A-modules, since $M = m_1 \cdot I \oplus \cdots \oplus m_n \cdot I$. This concludes the proof. $\qquad\square$

REMARK III.2.15. The previous proposition remains true if we consider left A-modules rather than right A-modules. □

COROLLARY III.2.16. *Let D be a division k-algebra. Then every non-zero finitely generated right D-module is isomorphic to D^n for some $n \geq 1$.*

Proof. If D is a division ring, then D is a minimal right ideal. Now apply the previous proposition. □

REMARK III.2.17. This corollary is well-known and boils down to the existence of a basis of a finitely generated D-vector space, which may be alternatively proved using the same arguments as in the commutative case. □

To end this section, we would like to give a nice application of Proposition III.2.14.

PROPOSITION III.2.18. *For all integers $r, s \geq 1$ and for all division k-algebras D and D', we have*

$$\mathrm{M}_r(D) \cong_k \mathrm{M}_s(D') \Rightarrow D \cong_k D' \text{ and } r = s.$$

Proof. Let $A = \mathrm{M}_r(D)$, $A' = \mathrm{M}_s(D')$ and $e = E_{11}$. Straightforward computations show that we have $e^2 = e, eAe = De = eD$ and that the map

$$D \longrightarrow eAe$$
$$d \longmapsto de$$

is a ring isomorphism. By Lemma III.2.5, we then have

$$D \cong eAe \cong \mathrm{End}_A(eA).$$

Now $I = eA$ is easily seen to be the set of matrices whose only possibly non-zero row is the first one. By Example III.2.8, this is a minimal right ideal of A. Similarly, $D' \cong \mathrm{End}_{A'}(I')$, where I' is a minimal right ideal of A'. Now if $\varphi : A \xrightarrow{\sim} A'$ is an isomorphism of k-algebras, then $\varphi(I)$ is a minimal right ideal of A'. Since all the minimal right ideals of A' are isomorphic by Proposition III.2.14, we have $I' \cong_{A'} \varphi(I)$. Therefore, we have a ring isomorphism

$$D \cong \mathrm{End}_A(I) \cong \mathrm{End}_{A'}(\varphi(I)) \cong \mathrm{End}_{A'}(I') \cong D'.$$

Since all these isomorphisms are k-linear, D and D' are isomorphic as k-algebras. Comparing dimensions over k then yields $r = s$. □

III.3. Skolem-Noether's theorem

We now study the automorphisms of a central simple k-algebra. First, we need a definition.

DEFINITION III.3.1. Let A be a central simple k-algebra. If $a \in A^\times$ is an invertible element of A, we denote by $\mathrm{Int}(a)$ the automorphism defined by

$$\mathrm{Int}(a) : \begin{array}{c} A \longrightarrow A \\ x \longmapsto axa^{-1}. \end{array}$$

An automorphism of A of the form $\mathrm{Int}(a)$ is called an **inner automorphism**.

THEOREM III.3.2 (Skolem-Noether's Theorem). *Let A and B be two simple k-algebras, and assume that A is central.*

Let $f_1, f_2 : B \longrightarrow A$ be two k-algebra morphisms. Then there exists an inner automorphism ρ of A such that $f_2 = \rho \circ f_1$. In particular, every automorphism of A is inner.

Proof. Let $i = 1, 2$. Since k is the center of A, one may check that the map $\ell_a \circ r_{f_i(b)} : A \longrightarrow A$ is k-linear for all $a \in A$ and all $b \in B$, and that the map

$$A \times B^{op} \longrightarrow \operatorname{End}_k(A)$$
$$(a, b^{op}) \longmapsto \ell_a \circ r_{f_i(b)}$$

is k-bilinear. Therefore, it induces a k-linear map

$$\varphi_i : A \otimes_k B^{op} \longrightarrow \operatorname{End}_k(A)$$

satisfying

$$\varphi_i(a \otimes b^{op}) = \ell_a \circ r_{f_i(b)} \text{ for all } a \in A, \text{ and all } b \in B.$$

We claim that φ_i is a k-algebra morphism. Since φ_i is k-linear, a distributivity argument shows that it is enough to check that

$$\varphi_i((a_1 \otimes b_1^{op})(a_2 \otimes b_2^{op})) = \varphi_i(a_1 \otimes b_1^{op}) \circ \varphi_i(a_2 \otimes b_2^{op})$$

for all $a_1, a_2 \in A$, and all $b_1, b_2 \in B$. Now for all $x \in A$, we have

$$
\begin{aligned}
\varphi_i((a_1 \otimes b_1^{op})(a_2 \otimes b_2^{op}))(x) &= \varphi_i(a_1 a_2 \otimes (b_2 b_1)^{op})(x) \\
&= (\ell_{a_1 a_2} \circ r_{f_i(b_2 b_1)})(x) \\
&= (a_1 a_2)x(f_i(b_2 b_1)) \\
&= (a_1 a_2)x(f_i(b_2)f_i(b_1)) \\
&= a_1(a_2 x f_i(b_2))f_i(b_1) \\
&= (\varphi_i(a_1 \otimes b_1^{op}) \circ \varphi_i(a_2 \otimes b_2^{op}))(x),
\end{aligned}
$$

and the claim follows. Consequently, the external law

$$(A \otimes_k B^{op}) \times A \longrightarrow A$$
$$(z, x) \longmapsto \varphi_i(z)(x)$$

endows A with the structure of a left $A \otimes_k B^{op}$-module such that

$$(a \otimes b^{op}) \bullet x = ax f_i(b) \text{ for all } a, x \in A \text{ and all } b \in B,$$

that we will denote by A_i. Since B is simple, so is B^{op}, and since A is central simple, it follows from Proposition III.1.6 that $A \otimes_k B^{op}$ is a simple k-algebra. Proposition III.2.14 (4) then shows that A_1 and A_2 are isomorphic as $A \otimes_k B^{op}$-modules. Let $\psi : A_1 \longrightarrow A_2$ be such a $A \otimes_k B$-module isomorphism. For all $a \in A$, we have

$$\psi(a) = \psi((a \otimes 1^{op}) \bullet 1) = (a \otimes 1^{op}) \bullet \psi(1) = a\psi(1).$$

Since ψ is bijective, this implies that $u = \psi(1) \in A^\times$. Notice that in particular, we get

$$\psi(f_1(b)) = f_1(b)u \text{ for all } b \in B.$$

Now for all $b \in B$, we also have

$$\psi(f_1(b)) = \psi((1 \otimes b^{op}) \bullet 1) = (1 \otimes b^{op}) \bullet \psi(1) = \psi(1)f_2(b) = uf_2(b).$$

Thus we get $\operatorname{Int}(u^{-1}) \circ f_1 = f_2$, and this concludes the proof. $\qquad\square$

III.4. Wedderburn's theorem

As pointed out in Example I.2.8 (3), $\mathrm{M}_r(D)$ is central simple for all $r \geq 1$ and every central division k-algebra D. We now prove the converse, which is known as Wedderburn's theorem.

THEOREM III.4.1 (Wedderburn's Theorem). *Any simple k-algebra A is isomorphic to $\mathrm{M}_r(D)$, for some integer $r \geq 1$ and some division k-algebra D whose center is isomorphic to the center of A.*

In particular, a central simple algebra is isomorphic to a matrix algebra over a central division k-algebra.

Moreover, the integer r and the isomorphism class of D only depend on the isomorphism class of A. More precisely, if I is a minimal right ideal of A, we have

$$D \cong_k \mathrm{End}_A(I) \ and \ A \cong_A I^r.$$

Proof. Let I be a minimal right ideal of A. Since I is a simple right A-module by Example III.2.8 (1), $D = \mathrm{End}_A(I)$ is a division ring by Schur's Lemma. Moreover, since A is a right A-module, we have $A \cong_A I^r$ for some $r \geq 1$ by Proposition III.2.14. Hence using Lemma III.2.5 with $e = 1$ and Lemma III.2.6 we get

$$A \cong \mathrm{End}_A(A) \cong \mathrm{End}_A(I^r) \cong \mathrm{M}_r(\mathrm{End}_A(I)) = \mathrm{M}_r(D).$$

One may check that all these isomorphisms are k-linear, so we get the existence part of the theorem. The uniqueness part comes directly from Proposition III.2.18 and the formula

$$\dim_k(A) = r^2 \dim_k(D).$$

Finally, we already observed in Example I.2.8 (3) that

$$Z(D) \cong_k Z(\mathrm{M}_r(D)) \cong_k Z(A).$$

This completes the proof. □

COROLLARY III.4.2. *Let A and B be two central simple k-algebras. For every integer $n \geq 1$, we have*

$$\mathrm{M}_n(A) \cong_k \mathrm{M}_n(B) \Leftrightarrow A \cong_k B.$$

Proof. Assume that $\mathrm{M}_n(A) \cong_k \mathrm{M}_n(B)$. By Wedderburn's theorem, we may write $A \cong_k \mathrm{M}_r(D)$ and $B \cong_k \mathrm{M}_s(D')$, where D, D' are central division k-algebras. We then have

$$\mathrm{M}_{nr}(D) \cong_k \mathrm{M}_{ns}(D').$$

By the uniqueness part of Wedderburn's theorem, we have $nr = ns$ and $D \cong_k D'$, which implies that $A \cong_k B$. □

COROLLARY III.4.3. *If k is algebraically closed, every central simple k-algebra is isomorphic to a matrix algebra.*

Proof. By Wedderburn's theorem, it is enough to prove that every central division k-algebra D is equal to k. Let $d \in D$. Since k is algebraically closed and $\mu_{d,k}$ is an irreducible polynomial, $\mu_{d,k}$ has degree 1. Since d is a root of $\mu_{d,k}$, we get that $d \in k$, meaning that $D = k$. □

Let A be a central simple k-algebra, and let $\overline{k}$ be an algebraic closure of k. Since $A \otimes_k \overline{k}$ is a central $\overline{k}$-simple algebra over by Corollary III.1.5 (2), the previous corollary shows that $A \otimes_k \overline{k} \cong_{\overline{k}} \mathrm{M}_n(\overline{k})$ for some $n \geq 1$. In particular, we have

$$\dim_k(A) = \dim_{\overline{k}}(A \otimes_k \overline{k}) = \dim_{\overline{k}}(\mathrm{M}_n(\overline{k})) = n^2.$$

Thus the dimension of a central simple k-algebra is the square of an integer. Therefore the following definition makes sense:

DEFINITION III.4.4. Let A be a central simple k-algebra. The **degree** of A is the integer $\deg(A) = \sqrt{\dim_k(A)}$. The **index** of A is the integer $\mathrm{ind}(A) = \deg(D)$, where D is the unique central division k-algebra associated to A by Wedderburn's theorem.

By definition, $\mathrm{ind}(A) | \deg(A)$, and $\deg(A) = \mathrm{ind}(A)$ if and only if A is a central division k-algebra.

We would like now to introduce the notion of Brauer equivalence, which is central in the theory of central simple algebras.

DEFINITION III.4.5. We say that two simple (not necessarily central) k-algebras A and B are **Brauer equivalent** if we have $A \cong_k \mathrm{M}_r(D)$ and $B \cong_k \mathrm{M}_s(D)$ for some integers $r, s \geq 1$ and some division k-algebra D. We denote it by $A \sim_k B$.

In other words, two simple k-algebras are Brauer equivalent if they correspond to isomorphic division k-algebras via Wedderburn's theorem. In particular, two Brauer equivalent simple k-algebras have the same center up to k-isomorphism by Example I.2.8 (3). Moreover, if A and B are two Brauer equivalent **central** simple k-algebras, they have the same index.

The following lemma will provide an easy way to produce examples of simple k-algebras which are Brauer equivalent to A.

LEMMA III.4.6. *Let A be a simple k-algebra, and let M be a non-zero right A-module. Then $\mathrm{End}_A(M)$ is a simple k-algebra which is Brauer equivalent to A, and we have*

$$\dim_k(M)^2 = \dim_k(A) \dim_k(\mathrm{End}_A(M)).$$

Moreover, if A is central simple, so is $\mathrm{End}_A(M)$ and we have

$$\dim_k(M) = \deg(A) \deg(\mathrm{End}_A(M)).$$

Proof. Let I be a minimal right ideal of A. Recall that every finitely generated right A-module is isomorphic to some power of I by Proposition III.2.14. Therefore, we have $M \cong_A I^s$ for some $s \geq 1$ (since M is non-zero). By Wedderburn's theorem, we know that $A \cong_k \mathrm{M}_r(D)$, where $A \cong_A I^r$ and $D \cong_k \mathrm{End}_A(I)$. We then have

$$\dim_k(A) = r^2 \dim_k(D) \text{ and } \dim_k(A) = r \dim_k(I).$$

Hence, we have $\dim_k(I) = r \dim_k(D)$. Moreover, we get

$$\mathrm{End}_A(M) \cong_k \mathrm{End}_A(I^s) \cong_k \mathrm{M}_s(\mathrm{End}_A(I)) \cong_k \mathrm{M}_s(D),$$

the second isomorphism coming from Lemma III.2.6. Hence $\mathrm{End}_A(M)$ is a simple k-algebra which is Brauer equivalent to A. Moreover, we have $\dim_k(M) = s \dim_k(I)$, and therefore

$$\dim_k(\mathrm{End}_A(M)) = s^2 \dim_k(D) = \frac{\dim_k(M)^2}{\dim_k(I)^2} \dim_k(D).$$

Thus

$$\dim_k(\mathrm{End}_A(M)) = \frac{\dim_k(M)^2}{r^2 \dim_k(D)} = \frac{\dim_k(M)^2}{\dim_k(A)}.$$

The last part comes from the fact that Brauer equivalent simple algebras have same center up to k-isomorphism, and from the definition of the degree of a central simple algebra. $\qquad\square$

REMARK III.4.7. Once again, the result is still true if we consider left A-modules rather than right A-modules. $\qquad\square$

We will see later on that every central simple k-algebra which is Brauer equivalent to A is actually isomorphic to the endomorphism ring of some A-module. Brauer equivalence will be studied in more details in Chapter V.

III.5. The centralizer theorem

In this section, we study in more detail the structure of the centralizer of simple subalgebras of a given central simple algebra. The main theorem is the following one.

THEOREM III.5.1 (Centralizer Theorem). *Let k be a field. Let A be a central simple k-algebra, and let B be a simple subalgebra of A. Then the following properties hold:*

(1) *the centralizer $C_A(B)$ of B in A is a simple subalgebra of A having same center as B. Moreover, we have*

$$\dim_k(A) = \dim_k(B)\dim_k(C_A(B)).$$

In particular, $\dim_k(B) \mid \dim_k(A)$;

(2) *we have $C_A(C_A(B)) = B$;*

(3) *if $L = Z(B)$ and $r = [L : k]$, then $A \otimes_k L \cong_k \mathrm{M}_r(B \otimes_L C_A(B))$.*
In particular, if $Z(B) = k$ we have $A \cong_k B \otimes_k C_A(B)$.

Proof.

(1) To show that $C_A(B)$ is a simple k-algebra, it is enough to show that $C_A(B) \cong_k \mathrm{End}_T(M)$, for some simple k-algebra T and some non-zero finitely generated right T-module N by Lemma III.4.6.

Let $T = B \otimes_k A^{op}$. The external product law

$$T \times A \longrightarrow A$$
$$(z, a) \longmapsto z{\cdot}a = \mathrm{Sand}(z)(a)$$

endows A with the structure of a left T-module on A (where Sand is the Sandwich morphism; see Lemma III.1.13). We claim that $C_A(B)$ is isomorphic to $\mathrm{End}_T(A)$. To prove it, notice first that by definition of the T-module structure on A an element $f \in \mathrm{End}_k(A)$ will be T-linear if and only if we have

$$f(\mathrm{Sand}(z)(a)) = \mathrm{Sand}(z)(f(a)) \text{ for all } z \in T, a \in A.$$

In other words, $\text{End}_T(A) = C_{\text{End}_k(A)}(\text{Sand}(T))$. Since Sand is an isomorphism, we get

$$\text{End}_T(A) = C_{\text{Sand}(A \otimes_k A^{op})}(\text{Sand}(T)) \cong_k C_{A \otimes_k A^{op}}(T) = C_{A \otimes_k A^{op}}(B \otimes_k A^{op}).$$

By Proposition III.1.4, we get

$$\text{End}_T(A) \cong_k C_A(B) \otimes_k C_{A^{op}}(A^{op}) \cong_k C_A(B) \otimes_k k \cong_k C_A(B).$$

This proves that $C_A(B)$ is a simple k-algebra, which is Brauer equivalent to T. In particular, $C_A(B)$ and T have the same center. Since

$$Z(T) = Z(B \otimes_k A^{op}) = Z(B) \otimes_k k \cong_k Z(B), C_A(B)$$

and B have the same center.

It remains to prove the equality between dimensions. By Lemma III.4.6, we have

$$\dim_k(A)^2 = \dim_k(T)\dim_k(\text{End}_T(A)) = \dim_k(B \otimes_k A^{op})\dim_k(C_A(B)).$$

Since $\dim_k(B \otimes_k A^{op}) = \dim_k(B)\dim_k(A)$, dividing by $\dim_k(A)$ gives the desired result.

(2) Since $C_A(B)$ is simple, applying (1) gives

$$\dim_k(C_A(B))\dim_k(C_A(C_A(B)) = \dim_k(A).$$

Since $\dim_k(B)\dim_k(C_A(B)) = \dim_k(A)$, we deduce that

$$\dim_k(B) = \dim_k(C_A(C_A(B)).$$

Now the definition easily imply that $B \subset C_A(C_A(B))$. The equality between dimensions then implies that $B = C_A(C_A(B))$.

(3) We are going to break the proof into several steps. We will prove successively the following facts.

(a) If B is a simple subalgebra of A with center k, then we have an isomorphism

$$A \cong_k B \otimes_k C_A(B).$$

(b) For any simple subalgebra B of A with center L, we have

$$C_A(L) \cong_L B \otimes_L C_A(B).$$

(c) If L is a subfield of A and $[L : k] = r$, then A is a free right $C_A(L)$-module of rank r and we have an isomorphism of L-algebras

$$A \otimes_k L \cong_L \text{End}_{C_A(L)}(A).$$

Let us show how to derive (3) from (b) and (c) ((a) being just an intermediate result which is needed to prove (b)). Let B be a simple subalgebra of A with center L. By (c), we have

$$A \otimes_k L \cong_L \text{End}_{C_A(L)}(A) = \text{End}_{C_A(L)}(C_A(L)^r) \cong_L \text{M}_r(C_A(L)),$$

the last isomorphism coming from Lemma III.2.6. Now apply (b) to get the desired isomorphism of L-algebras.

It remains to prove (a),(b) and (c).

Assume that B is simple with center k. By definition of $C_A(B)$, elements of B and $C_A(B)$ commute. Therefore, the two inclusions $B \subset A$ and $C_A(B) \subset A$ give rise to a k-algebra morphism

$$B \otimes_k C_A(B) \longrightarrow A.$$

By assumption, B is central simple, and $C_A(B)$ is simple by (1) (it is even central). Hence $B \otimes_k C_A(B)$ is a simple k-algebra by Proposition III.1.6. Moreover, (1) shows that $B \otimes_k C_A(B)$ and A have the same dimension over k. Hence the previous morphism is an isomorphism by Lemma I.2.2. This proves (a).

To establish (b), notice first that B and $C_A(L)$ have center L, and therefore may be both viewed as central simple L-algebras (the simplicity of $C_A(L)$ coming from (1)). Applying (a) then yields

$$C_A(L) \cong_L B \otimes_L C_{C_A(L)}(B).$$

Now $C_{C_A(L)}(B)$ is precisely the set of elements of A commuting with L and B, that is the set of elements of A commuting with B, since $L \subset B$. Hence $C_{C_A(L)}(B) = C_A(B)$ and this proves (b).

We finally prove (c). First, multiplication in A endows A with the structure of a right $C_A(L)$-module. Since A is finite dimensional over k, it is finitely generated as a $C_A(L)$-module. Since L is a field, L is a simple subalgebra of A (with center L). By (1), $C_A(L)$ is a central simple L-algebra and we have $\dim_k(A) = r \dim_k(C_A(L))$. By Proposition III.2.14 (3), A is a free $C_A(L)$-module of rank r. We now construct an isomorphism of L-algebras between $A \otimes_k L$ and $\mathrm{End}_{C_A(L)}(A)$.

Recall that for $a \in A$ and $\lambda \in L$, we denote by ℓ_a and r_λ the left multiplication by a and the right multiplication by λ in A respectively. These maps are both endomorphisms of the k-vector space A. Let us check that ℓ_a and r_λ are actually $C_A(L)$-linear. For all $z \in A$ and all $a' \in C_A(L)$, we have

$$\ell_a(z \cdot a') = \ell_a(za') = a(za') = (az)a' = \ell_a(z) \cdot a'.$$

Moreover, since elements of L and $C_A(L)$ commute, we have

$$r_\lambda(z \cdot a') = r_\lambda(za') = (za')\lambda = za'\lambda = z\lambda a' = r_\lambda(z) \cdot a'.$$

Hence ℓ_a and r_λ are elements of $\mathrm{End}_{C_A(L)}(A)$ which clearly commute. Thus we get a k-algebra morphism

$$A \otimes_k L \longrightarrow \mathrm{End}_{C_A(L)}(A),$$

which is easily seen to be L-linear. Using Lemma III.4.6 and (1), we have

$$\dim_k(\mathrm{End}_{C_A(L)}(A)) = \frac{\dim_k(A)^2}{\dim_k(C_A(L))} = \dim_k(A)\dim_k(L).$$

It follows that we have

$$\dim_L(\mathrm{End}_{C_A(L)}(A)) = \dim_k(A) = \dim_k(A \otimes_k L).$$

Since $A \otimes_k L$ is simple, Lemma I.2.2 shows that the map above is an isomorphism. This concludes the proof of the theorem. □

REMARK III.5.2. Property (1) does not hold if B is not simple. For example, if $A = \mathrm{M}_3(k)$, let $B = k[E_{11}]$. Since the minimal polynomial of E_{11} is $X(X-1)$, we have $B \cong_k k \times k$. In particular, B is not simple. Moreover, we have $\dim_k(B) = 2 \nmid \dim_k(A) = 9$. □

COROLLARY III.5.3. *Let A be a central simple k-algebra, and let L be a subfield of A of degree r over k. Then we have*

$$A \otimes_k L \cong_L M_r(C_A(L)).$$

Proof. Since L is a simple k-subalgebra of A and $Z_A(L) = L$ (since L is commutative), the last part of the Centralizer Theorem gives

$$A \otimes_k L \cong_L M_r(L \otimes_L C_A(L)).$$

Since $L \otimes_L C_A(L) \cong_L C_A(L)$, we are done.   □

EXERCISES

1. Let A be a central simple k-algebra. Show that the group of all k-algebra automorphisms of A is canonically isomorphic to $A^\times/k^\times$.

2. Let A be a simple k-algebra, not necessarily central. Is every automorphism of A inner?

3. The goal of this exercise is to provide an elementary and totally explicit proof of Skolem-Noether's theorem for matrix algebras.

   Let $n \geq 1$ be an integer. For $1 \leq i, j \leq n$, we denote by $E_{ij} \in M_n(k)$ the matrix whose entries are all zero, except for the entry at row i and column j, which equals 1.

   Let $\rho : M_n(k) \longrightarrow M_n(k)$ be an automorphism of k-algebras.

   (a) Show that for all $M \in M_n(k)$, M and $\rho(M)$ have the same characteristic polynomial. Deduce that the image of a projector of rank 1 is a projector of rank 1.

   (b) Deduce that there exist two non-zero vectors $C, C' \in k^n$ such that $\rho(E_{11}) = CC'^t$ and $C'^t C = 1$.

   For $j = 1, \ldots, n$, we set $C_j = \rho(E_{j1})C$.

   (c) For $1 \leq i, j, m \leq n$, check that $\rho(E_{ij})C_m = \delta_{jm}C_i$.

   (d) Compute $\rho(E_{1j})C_j$; deduce that $C_1 = C$, and that C_j is non-zero for $j = 1, \ldots, n$.

   (e) Deduce from (b) that $(C_1, \ldots, C_n)$ is a k-basis of k^n.

   (f) Let $P \in M_n(k)$ be the matrix whose columns are $C_1, \ldots, C_n$. Use the previous questions to show that P is invertible and that $\rho(E_{ij})P = PE_{ij}$ for $1 \leq i, j \leq n$.

   (g) Deduce that $\rho = \text{Int}(P)$.

4. Let $r \geq 1$ be an integer, let D be a central division k-algebra, and let $A = M_r(D)$. For every subset S of $\{1, \ldots, r\}$, let $\mathcal{L}_S$ be the set of matrices $M \in M_r(D)$ whose i^{th} row is zero whenever $i \notin S$.

   (a) Show that $\mathcal{L}_S$ is a right ideal of A.

(b) Show that every right ideal is isomorphic to some $\mathcal{L}_S$.

(c) Deduce that every right ideal of A is principal.

(d) Describe the left ideals of A.

5. Let $A = \mathrm{M}_n(k)$, and let B the subalgebra of upper triangular matrices.

(a) Show that $C_A(B)$ is the subalgebra of diagonal matrices, and that $C_A(C_A(B)) = A$.

(b) Deduce that B is not simple.

(c) Recover the result of the previous question by exhibiting a non-trivial two-sided ideal of B.

6. Let $n \geq 1$ be an integer, let k be a field satisfying $\mathrm{char}(k) \nmid n$, and containing ζ_n, a primitive n^{th}-root of 1. For $a, b \in k^\times$, we denote by $\{a, b\}_{n,\zeta_n}$ the k-algebra generated by two elements e and f subject to the relations

$$e^n = a, f^n = b, ef = \zeta_n fe.$$

This is a central simple k-algebra of degree n.

Let n, m be two coprime integers. Assume that $\mathrm{char}(k) \nmid nm$, and that $\zeta_{nm} \in k^\times$. For all $a, b \in k^\times$, show that we have

$$\{a, b\}_{nm,\zeta_{nm}} \cong_k \{a, b\}_{n,\zeta_{nm}^m} \otimes_k \{a, b\}_{m,\zeta_{nm}^n}.$$

Hint: Identify the k-subalgebra B generated by e^m and f^m, and compute its centralizer.

CHAPTER IV

Splitting fields of central simple algebras

The goal of this chapter is to prove the existence of splitting fields of central simple algebras with good properties. We will then use the results obtained to extend the notion of determinant to central simple algebras. This will lead to the notion of reduced norm.

Recall from Definition I.2.10 that a **subfield** of a given central simple k-algebra A is a subalgebra of A which is also a field. In particular, L contains k.

IV.1. Splitting fields

We start by introducing the notion which will be studied in this chapter.

DEFINITION IV.1.1. Let A be a central simple k-algebra of degree n. A field L is called a **splitting field** of A if it contains k and if we have an L-algebra isomorphism $A \otimes_k L \cong_L \mathrm{M}_n(L)$. In this situation, we also say that A splits over L, or that L splits A.

EXAMPLE IV.1.2. By Corollary III.4.3, any algebraic closure $\bar{k}$ of k is a splitting field of every central simple k-algebra. □

REMARK IV.1.3. Let A be a central simple k-algebra. By Wedderburn's Theorem, we may write $A \cong_k \mathrm{M}_r(D)$, for some central division k-algebra D and some integer $r \geq 1$. Then for any field extension L/k, L splits A if and only if L splits D.

Indeed, let $n = \deg_k(A)$ and $d = \deg_k(D)$, so that $n = rd$. We have

$$A \otimes_k L \cong_L \mathrm{M}_r(D \otimes_k L)$$

by Lemma I.1.11 (2). If L splits A, we have

$$A \otimes_k L \cong_L \mathrm{M}_n(L) = \mathrm{M}_{rd}(L),$$

and therefore

$$\mathrm{M}_{rd}(L) \cong_L \mathrm{M}_r(D \otimes_k L).$$

By Corollary III.4.2, we get $D \otimes_k L \cong_L \mathrm{M}_d(L)$, so L splits D.

Conversely, if L splits D, then $D \otimes_k L \cong_L \mathrm{M}_d(L)$, and we have

$$A \otimes_k L \cong_L \mathrm{M}_r(\mathrm{M}_d(L)) = \mathrm{M}_n(L),$$

so L splits A. □

In this section, we will establish the existence of splitting fields of central simple algebras with various properties. We first investigate under which conditions a subfield L of a central simple k-algebra A of degree n is a splitting field.

The first lemma shows that there is some restriction on the degree of a subfield.

LEMMA IV.1.4. *Let A be a central simple k-algebra, and let L be a subfield of A. Then $[L : k] \mid \deg(A)$.*

Proof. Since L is a field containing k (by definition of a subfield), L is a simple k-subalgebra of A. By the Centralizer Theorem, we then have

$$[L : k] \dim_k(C_A(L)) = \dim_k(A) = \deg(A)^2.$$

Moreover, $C_A(L)$ is an L-vector space (since $L \subset C_A(L)$), and we have

$$\dim_k(C_A(L)) = \dim_L(C_A(L))[L : k].$$

Therefore, $[L : k]^2 \mid \deg(A)^2$. This implies immediately the desired result. $\square$

This lemma tells us in particular that the maximal possible degree of a subfield of a central simple k-algebra of degree n is n. Such subfields deserve a special name.

DEFINITION IV.1.5. Let A be a central simple k-algebra of degree n. A **maximal subfield** of A is a subfield of degree n over k.

The main interest of these particular subfields is given by the following proposition.

PROPOSITION IV.1.6. *Let A be a k-algebra, and let L be a subfield of A. Then there exists a unique L-algebra morphism*

$$f : A \otimes_k L \longrightarrow \operatorname{End}_L(A)$$

defined on elementary tensors by

$$f(a \otimes \lambda)(z) = az\lambda \text{ for all } a, z \in A, \lambda \in L.$$

Moreover, if A is central simple and $[L : k] = \deg(A)$, then f is an isomorphism; in other words, any maximal subfield is a splitting field of A.

Proof. Let us consider the k-algebra morphism

$$\phi : \begin{array}{l} A \longrightarrow \operatorname{End}_L(A) \\ a \longmapsto \ell_a \end{array}$$

of Lemma I.2.11. We now define

$$\iota : \begin{array}{l} L \longrightarrow \operatorname{End}_L(A) \\ \lambda \longmapsto \operatorname{Id}_A \lambda. \end{array}$$

We claim that the map ι is a k-algebra morphism and that the images of ι and ϕ commute.

Indeed, for all $c \in k, z \in A, \lambda, \lambda' \in L$, we have

$$\iota(\lambda + \lambda' c)(z) = z(\lambda + \lambda' c) = z\lambda + z\lambda' c = (\iota(\lambda) + \iota(\lambda')c)(z),$$

$$\iota(\lambda\lambda')(z) = z(\lambda\lambda') = z(\lambda'\lambda) = (\iota(\lambda) \circ \iota(\lambda'))(z).$$

Hence ι is a k-algebra morphism. Moreover, we have

$$(\phi(a) \circ \iota(\lambda))(z) = \phi(a)(z\lambda) = az\lambda,$$

and

$$(\iota(\lambda) \circ \phi(a))(z) = \iota(\lambda)(az) = az\lambda.$$

If follows that we have a unique morphism of k-algebras

$$f : A \otimes_k L \longrightarrow \operatorname{End}_L(A)$$

satisfying

$$f(a \otimes \lambda) = \phi(a) \circ \iota(\lambda) \text{ for all } a \in A, \lambda \in L.$$

In other words, we have

$$f(a \otimes \lambda)(z) = az\lambda \text{ for all } a, z \in A, \lambda \in L.$$

We now prove that the map $f : A \otimes_k L \longrightarrow \text{End}_L(A)$ is L-linear. For all $a, z \in A, \lambda, \lambda' \in L$, we have

$$f((a \otimes \lambda)\lambda')(z) = f(a \otimes \lambda\lambda')(z) = az\lambda\lambda' = (f(a \otimes \lambda)\lambda')(z).$$

Hence $f((a \otimes \lambda)\lambda') = f(a \otimes \lambda)\lambda'$. Thus f is L-linear, and hence a morphism of L-algebras.

It remains to check that f is bijective whenever A is central simple and $[L : k] = \deg(A)$. Since A is central simple, so is $A \otimes_k L$ by Corollary III.1.8 (2). Notice in this case that $\dim_k(A) = n^2$ and that by assumption $[L : k] = n$. Hence A is a right L-vector space of dimension n. Therefore, $\dim_L(\text{End}_L(A)) = n^2$. Since we also have

$$\dim_L(A \otimes_k L) = \dim_k(A) = n^2,$$

it is enough to apply Lemma I.2.2 to conclude.   □

REMARK IV.1.7. A maximal subfield does not necessarily exist. For example, if k is algebraically closed, there is no proper field extension of k, so L does not exist if $A \neq k$. However, we are going to prove that such an L always exists if A is a division algebra.   □

PROPOSITION IV.1.8. *Let D be a central division k-algebra, and let L be a subfield of D. Then the following properties are equivalent:*

(1) *L is a splitting field of D;*
(2) *$C_D(L) = L$;*
(3) *L is a maximal subfield of D.*

Proof.

(1) $\Rightarrow$ (2). If L splits D, then $D \otimes_k L \cong_L \text{M}_n(L)$, where $n = \deg(D)$. By Corollary III.5.3, we then have

$$\text{M}_r(C_D(L)) \cong_L \text{M}_n(L),$$

where $r = [L : k]$. If we set $s = \deg(C_D(L))$ (this makes sense since $C_D(L)$ is a central simple L-algebra), we get by comparing the degrees that $n = rs$. We then have

$$\text{M}_r(C_D(L)) \cong_L \text{M}_r(\text{M}_s(L)),$$

and therefore $C_D(L) \cong_L \text{M}_s(L)$ by Corollary III.4.2. But since D is a division ring, so is $C_D(L)$. Hence we must have $s = 1$, so $C_D(L)$ has dimension 1 over L. Since $L \subset C_D(L)$, we get $L = C_D(L)$.

(2) $\Rightarrow$ (3). If $C_D(L) = L$, we have $[L : k]^2 = \dim_k(D) = \deg(D)^2$ by Theorem III.5.1 (1). Therefore, $[L : k] = \deg(D)$, and we are done.

(3) $\Rightarrow$ (1). This is a particular case of Proposition IV.1.6.   □

COROLLARY IV.1.9. *Every central division algebra has a maximal subfield. In particular, every central simple k-algebra A has a subfield of degree* ind(A) *over k which splits A.*

Proof. Let D be a central division k-algebra of degree d, and let L be a subfield of D of largest possible degree. We claim that $L = C_D(L)$. Otherwise, it would exist an element $x \in C_D(L)$, $x \notin L$. But $L(x)$ would be a subfield of D of strictly larger degree (since D is a division ring), contradicting the maximality of $[L : k]$. Hence $C_D(L) = L$ and L is then a maximal subfield of D by Proposition IV.1.8.

Now if A is a central simple k-algebra, we may write $A \cong_k \mathrm{M}_r(D)$ by Wedderburn's Theorem. Let L be a maximal subfield of D. Then L splits D by Proposition IV.1.6 and therefore A by Remark IV.1.3. Since $[L : k] = \deg(D) = \mathrm{ind}(A)$, this concludes the proof. $\square$

REMARKS IV.1.10.

(1) If A is a central simple k-algebra and L is a maximal subfield, then $C_A(L) = L$. Indeed, $L \subset C_A(L)$ and an application of the Centralizer Theorem shows that $[C_A(L) : k] = n = [L : k]$. Therefore, $C_A(L) = L$.

(2) Let D be a central division k-algebra, and let L be a subfield of D. Then L is a maximal subfield of D if and only if it is maximal for the inclusion.

Indeed, if L is a maximal subfield, it has maximal degree by Lemma IV.1.4 and therefore is maximal for the inclusion. Conversely, if L is maximal for the inclusion, it has the largest possible degree among the subfields of D, and the arguments in the proof of Proposition IV.1.8 show that L is a maximal subfield.

(3) This result is not true anymore for arbitrary central simple k-algebras. For example, $\overline{k}$ is a subfield of $\mathrm{M}_n(\overline{k})$ which is maximal for the inclusion, but is not a maximal subfield in the sense of Definition IV.1.5.

(4) The second remark also implies that every subfield K of a central division k-algebra D is contained in a maximal subfield, since it is contained in a subfield which is maximal for the inclusion. In particular, every element $d \in D$ is contained in a maximal subfield of D (since $d \in k(d)$).

$\square$

We now use the previous corollary to give a short proof of one of the most famous results of Wedderburn.

THEOREM IV.1.11 (Wedderburn). *Every finite division ring is commutative.*

Proof. Let D be a finite division ring, and let k be the center of D; D is then a central division k-algebra. Notice that k is a finite field by assumption. Let $n = \deg(D)$, and fix a maximal subfield L of D. By Remark IV.1.10 (3), every element $d \in D$ is contained in a maximal subfield L_d. Since $[L_d : k] = [L : k] = n$, the theory of finite fields shows that L_d and L are isomorphic as k-algebras. Let

$$f_d : L_d \xrightarrow{\sim} L$$

be such an isomorphism. If ι is the inclusion $L \subset D$ and ι_d is the inclusion $L_d \subset D$, Skolem-Noether's Theorem shows that there exists $x_d \in D^\times$ such that

$$\iota_d = \mathrm{Int}(x_d) \circ (\iota \circ f_d).$$

In particular, we have

$$d = x_d f_d(d) x_d^{-1} \in x_d L x_d^{-1} \text{ for all } d \in D.$$

This implies that $D^\times$ is the union of some conjugates of $L^\times$. We are now going to show that $D^\times = L^\times$, that is $D = L$, which will conclude the proof since L is commutative. For, we use the following:

Claim: Let G be a finite group, and let H be a proper subgroup of G. Then G is not the union of conjugates of H.

To prove the claim, let $r = [G : H]$, and let $g_1 H, \ldots, g_r H$ be a set of representative cosets. Let $g \in G$, and write $g = g_i h$ for some i and $h \in H$. Then we have

$$gHg^{-1} = g_i(hHh^{-1})g_i^{-1} = g_i H g_i^{-1},$$

so the number of distinct subgroups of G conjugate to H is at most r. Now $g_i H g_i^{-1}$ contains the neutral element of G for all i, and therefore, since $r \geq 2$ by assumption on H, we have

$$\left| \bigcup_{i=1}^{r} g_i H g_i^{-1} \right| < r|H| = |G|.$$

Hence any union of subgroups conjugate to H is a proper subset of G. □

Let us go back to the study of splitting fields of central simple algebras.

A central simple k-algebra A may have splitting fields of degree larger than $\deg(A)$, and therefore not contained in A by Lemma IV.1.4. However, we have the following result.

PROPOSITION IV.1.12. *Let A be a central simple k-algebra of degree n, and let L be a splitting field of A of finite degree m over k. Then there exists a central simple k-algebra A' of degree m such that:*

(1) $M_m(A) \cong_k M_n(A')$;

(2) *L is isomorphic to a maximal subfield of A'.*

Proof. It follows from Lemma III.1.11 (4) and the fact that L splits A that we have an isomorphism of L-algebras

$$A^{op} \otimes_k L \cong_L M_n(L).$$

Let $m = [L : k]$. We are now going to embed L and A^{op} into $M_{mn}(k)$ as follows. First of all, notice that we have

$$A^{op} \otimes_k L \cong_L M_n(L) \cong_L \mathrm{End}_L(L^n).$$

Since $\mathrm{End}_L(L^n) \subset \mathrm{End}_k(L^n) \cong_k M_{mn}(k)$, we get an injective k-algebra morphism

$$\varphi : A^{op} \otimes_k L \hookrightarrow M_{mn}(k).$$

Now let $\iota_1 : A^{op} \hookrightarrow A^{op} \otimes_k L$ and $\iota_2 : L \hookrightarrow A^{op} \otimes_k L$ be the two canonical injections, and set

$$f_j = \varphi \circ \iota_j, j = 1, 2.$$

By definition, f_1, f_2 are injective k-algebra morphisms, with commuting images.

Set $A' = C_{\mathrm{M}_{mn}(k)}(f_1(A^{op}))$ and $L' = f_2(L)$. Since A is a central simple k-algebra, so is $f_1(A^{op})$. The first part of the Centralizer Theorem then shows that A' is a central simple k-algebra as well. Moreover, we have by definition $L' \subset \mathrm{M}_{mn}(k)$, and since the images of f_1 and f_2 commute, we get $L' \subset A'$. Notice that L' is isomorphic to L since f_2 is injective. Since f_1 is also injective, we have $f_1(A^{op}) \cong_k A^{op}$, and therefore the Centralizer Theorem gives

$$\dim_k(A) \dim_k(A') = m^2 n^2,$$

that is $\dim_k(A') = m^2$, or equivalently $\deg(A') = [L : k]$.

It remains to prove (1). The last part of the Centralizer Theorem shows that we have

$$\mathrm{M}_{nm}(k) \cong_k f_1(A^{op}) \otimes_k A' \cong_k A^{op} \otimes_k A'.$$

Tensoring by A and using Lemmas III.1.13 and I.1.11, as well as the commutativity of the tensor product, we get

$$\mathrm{M}_{nm}(A) \cong_k (A \otimes_k A^{op}) \otimes_k A' \cong_k \mathrm{M}_{n^2}(A').$$

Now use Corollary III.4.2 to conclude. □

REMARK IV.1.13. Using Wedderburn's Theorem, we may easily see that condition (1) is equivalent to saying that A and A' are Brauer equivalent. We leave the details for now since we will come back to this in the next chapter. □

COROLLARY IV.1.14. *Let A be a central simple k-algebra, and let L be a field extension satisfying $[L : k] = \deg(A)$. Then L is a splitting field of A if and only if it is isomorphic to a maximal subfield of A.*

Proof. The assumption and the previous proposition imply that L is isomorphic to a subfield of a central simple k-algebra A' satisfying

$$\mathrm{M}_n(A') \cong_k \mathrm{M}_n(A).$$

Now Corollary III.4.2 implies that $A' \cong_k A$, and we are done. Conversely, if L is isomorphic to a maximal subfield of A, then it is a splitting field by Proposition IV.1.6. □

COROLLARY IV.1.15. *Let A be central simple k-algebra. For any splitting field L of finite degree over k, we have*

$$\mathrm{ind}(A) \mid [L : k].$$

Proof. Write $A \cong_k \mathrm{M}_r(D)$ for some central division k-algebra D. Let $d = \deg(D) = \mathrm{ind}(A)$. Let L be a splitting field of A of finite degree m over k. By Remark IV.1.3, L is a splitting field of D. By Proposition IV.1.12, there exists a central simple k-algebra A' of degree m such that

$$\mathrm{M}_d(A') \cong_k \mathrm{M}_m(D),$$

and L is isomorphic to a maximal subfield of A'. Write $A' \cong_k \mathrm{M}_s(D')$ for some central division k-algebra D'. We then get

$$\mathrm{M}_{ds}(D') \cong_k \mathrm{M}_m(D).$$

By Wedderburn's Theorem, we obtain $D' \cong_k D$, that is

$$A' \cong_k \mathrm{M}_s(D).$$

In particular, we have $\deg(A') = m = sd$, and therefore $d \mid m$. In other words, $\mathrm{ind}(A) \mid [L : k]$. $\qquad\qquad\qquad\qquad\qquad\qquad\qquad\qquad\qquad\qquad\qquad\qquad\square$

We are now going to investigate the existence of separable and Galois splitting fields. We start with a lemma.

LEMMA IV.1.16. *Let D be a central division k-algebra, and assume that every subfield of D is purely inseparable over k. Then $D = k$.*

Proof. If $\mathrm{char}(k) = 0$, then for every $d \in D$, $k(d)/k$ is separable. The hypothesis implies that this extension has degree 1 over k, that is $d \in k$. We then have $D = k$ in this case. Now assume that $\mathrm{char}(k) = p > 0$. The assumption implies that every maximal subfield L of D (which exists, by Corollary IV.1.9) is purely inseparable over k. In particular, we have

$$\deg(D) = [L : k] = p^r,$$

for some $r \geq 0$.

Assume that $r \geq 1$. By Corollary III.4.3, there exists an isomorphism of $\overline{k}$-algebras

$$f : D \otimes_k \overline{k} \xrightarrow{\sim} \mathrm{M}_{p^r}(\overline{k}).$$

Let $d \in D$. Since $k(d)/k$ is purely inseparable by assumption, there exists $s \geq 0$ such that $d^{p^s} = a$ for some $a \in k$. We then have

$$f(d \otimes 1)^{p^s} = f((d \otimes 1)^{p^s}) = f(a \otimes 1) = af(1 \otimes 1) = aI_{p^r} = (\sqrt[p^s]{a}I_{p^r})^{p^s},$$

that is

$$(f(d \otimes 1) - \sqrt[p^s]{a}I_{p^r})^{p^s} = 0 \in \mathrm{M}_{p^r}(\overline{k}),$$

since $\mathrm{char}(k) = p$. Since the trace of a nilpotent matrix is zero, we get

$$\mathrm{tr}(f(d \otimes 1) - \sqrt[p^s]{a}I_{p^r}) = \mathrm{tr}(f(d \otimes 1)) - p^r\sqrt[p^s]{a} = 0.$$

Since $r \geq 1$, we get

$$\mathrm{tr}(f(d \otimes 1)) = 0 \text{ for all } d \in D.$$

Since the elements $d \otimes 1, d \in D$ span $D \otimes_k \overline{k}$ as a $\overline{k}$-vector space and f is surjective, we deduce from the previous relation that

$$\mathrm{tr}(M) = 0 \text{ for all } M \in \mathrm{M}_{p^r}(\overline{k}),$$

which is a contradiction. Therefore $r = 0$, meaning that $D = k$ also in this case. $\qquad\qquad\qquad\qquad\qquad\qquad\qquad\qquad\qquad\qquad\qquad\qquad\qquad\quad\square$

We are now able to prove the following refinement of Corollary IV.1.9.

THEOREM IV.1.17. *Every central division k-algebra has a separable maximal subfield. In particular, every central simple k-algebra A has a separable splitting field of finite degree over k.*

Proof. Let D be a central division k-algebra, and let L be the largest subfield of D which is separable over k. We already observed earlier that the centralizer $C_D(L)$ is a central division L-algebra. If L' is a subfield of $C_D(L)$, then L'/L is purely inseparable. Otherwise, any separable element $x \in L', x \notin L$ would generate a nontrivial separable extension $L'(x)/L$. But then $L'(x)$ would be a subfield separable over k strictly larger than L, contradicting the choice of L. By the previous lemma, we get $C_D(L) = L$, and therefore L is a maximal subfield by Proposition IV.1.8. $\qquad\qquad\qquad\qquad\qquad\qquad\qquad\qquad\qquad\qquad\qquad\qquad\qquad\qquad\qquad\quad\square$

COROLLARY IV.1.18. *Every central simple k-algebra has a Galois splitting field of finite degree over k.*

Proof. Let A be a central simple k-algebra of degree n, and write $A \cong_k M_r(D)$, where D is a division central k-algebra. By the previous theorem, D has a maximal separable subfield L, which splits D (by Proposition IV.1.6) and therefore A. Let L' be the Galois closure of L in $\overline{k}$. We have

$$A \otimes_k L' \cong_{L'} (A \otimes_k L) \otimes_L L' \cong_{L'} M_n(L) \otimes_L L' \cong_{L'} M_n(L'),$$

so L' splits A. □

REMARK IV.1.19. One may wonder if every central division k-algebra D has a Galois maximal subfield. The answer is negative, but producing a counterexample would be beyond the scope of this book. The reader will refer for example to [**3**], where the first construction of division algebras having no Galois maximal subfield is exposed, or to [**18**] for a totally explicit example.

Central simple algebras having a Galois maximal subfield are called crossed products, and will be studied in a forthcoming chapter. □

At this stage, it is worth summarizing the various characterizations of central simple algebras we have established.

THEOREM IV.1.20. *Let k be a field, let $\overline{k}$ be a fixed algebraic closure of k. For any finite dimensional k-algebra A, the following properties are equivalent:*

(1) *A is a central simple k-algebra;*

(2) *$A \otimes_k \overline{k} \cong_{\overline{k}} M_n(\overline{k})$;*

(3) *there exists a field extension L/k such that $A \otimes_k L \cong_L M_n(L)$ for some $n \geq 1$.*
In this case, L/k may be chosen of finite degree over k, finite separable or Galois.

Proof. (1) $\Rightarrow$ (2) is Corollary III.1.5 (2), and (1) $\Rightarrow$ (3) is just a summary of the results of this section. Now (2)$\Rightarrow$(1) and (3)$\Rightarrow$(1) follows from Corollary III.1.8 (2). □

IV.2. The reduced characteristic polynomial

If k is a field and $M \in M_n(k)$, recall that the characteristic polynomial χ_M of M is defined by

$$\chi_M = \det(XI_n - M) \in k[X].$$

We would like in this section to extend this notion to elements of arbitrary central simple algebras. A first natural idea would be to reduce to the split case as follows: if A is a central simple k-algebra of degree n and $a \in A$, let $P_a \in k[X]$ be the characteristic polynomial of the endomorphism $\ell_a \in \mathrm{End}_k(A)$.

Let us compute P_a in the split case: if $A = M_n(k)$ and $a = M \in M_n(k)$, it is not difficult to see that the matrix of ℓ_M in the canonical basis of $M_n(k)$ is given by

$$\begin{pmatrix} M & & \\ & \ddots & \\ & & M \end{pmatrix}.$$

Hence we get $P_M = \chi_M^n$.

This definition is then not completely satisfactory, since we do not recover the characteristic polynomial of a matrix in the split case. Instead of using this approach, we are going to exploit the existence of splitting fields.

Let A be a central simple k-algebra, let L be a splitting field of A, and let

$$f : A \otimes_k L \xrightarrow{\sim} M_n(L)$$

be an isomorphism of L-algebras.

LEMMA IV.2.1. *Keeping the notation above, for every $a \in A$, the characteristic polynomial of $f(a \otimes 1)$ does not depend on the choice of L or f, and has coefficients in k.*

Proof. We proceed in several steps.

(1) Given a splitting field L, let us consider two L-algebras isomorphisms

$$f, g : A \otimes_k L \xrightarrow{\sim} M_n(L).$$

Let $a \in A$. Since $g \circ f^{-1}$ is an automorphism of $M_n(L)$, by Skolem-Noether's Theorem, we have

$$g \circ f^{-1} = \mathrm{Int}(M) \text{ for some } M \in \mathrm{GL}_n(L).$$

We then have $g(a \otimes 1) = Mf(a \otimes 1)M^{-1}$, and therefore $f(a \otimes 1)$ and $g(a \otimes 1)$ have the same characteristic polynomial. Thus, the characteristic polynomial of $f(a \otimes 1)$ does not depend on the choice of f.

If L is a splitting field of A, for any $a \in A$, we set

$$\chi_{a,L} = \chi_{f(a \otimes 1)} \in L[X],$$

where $f : A \otimes_k L \xrightarrow{\sim} M_n(L)$ is an isomorphism of L-algebras. The considerations above show that this definition makes sense.

(2) Let L/k and E/k be two field extensions of k, and let $\varphi : L \longrightarrow E$ be an morphism of k-algebras. If $P \in L[X]$, we denote by $\varphi \cdot P$ the polynomial of $E[X]$ obtained by applying φ to each coefficient of P. If $M \in M_n(L)$, we denote by $\tilde{\varphi}(M)$ the matrix of $M_n(E)$ obtained by applying φ to each coefficient of M. It is not difficult to check that

$$\tilde{\varphi} : M_n(L) \longrightarrow M_n(E)$$

is a k-algebra morphism, and that we have

$$\chi_{\tilde{\varphi}(M)} = \varphi \cdot \chi_M.$$

We are going to prove the following: if L is a splitting field of A, so is E and we have

$$\chi_{a,E} = \varphi \cdot \chi_{a,L}.$$

Let us fix an isomorphism of L-algebras

$$f : A \otimes_k L \xrightarrow{\sim} M_n(L).$$

We first consider two special cases.

First case: φ is an isomorphism of k-algebras.

In this case, set $g = \tilde{\varphi} \circ f \circ (\mathrm{Id}_A \otimes \varphi^{-1})$. By definition, the map

$$g : A \otimes_k E \longrightarrow \mathrm{M}_n(E)$$

is an isomorphism of k-algebras, and the reader may check easily that it is also E-linear (this comes from the fact that f is an isomorphism of L-algebras, and from the definition of $\tilde{\varphi}$).

Therefore E is a splitting field for A, and Step (1) implies that we have

$$\chi_{a,E} = \chi_{g(a \otimes 1)}.$$

Now we have $g(a \otimes 1) = \tilde{\varphi}(f(a \otimes 1))$, and therefore we get

$$\chi_{a,E} = \varphi \cdot \chi_{a,L},$$

as claimed.

Second case: $L \subset E$ and φ is the inclusion. In this case, the equality we need to prove reads

$$\chi_{a,E} = \chi_{a,L}.$$

Consider the successive E-algebra isomorphisms

$$A \otimes_k E \cong_E (A \otimes_k L) \otimes_L E \cong_E \mathrm{M}_n(L) \otimes_L E \cong_E \mathrm{M}_n(E),$$

the second one being induced by $f \otimes \mathrm{Id}_E$. The composite isomorphism

$$g : A \otimes_k E \xrightarrow{\ \sim\ } \mathrm{M}_n(E)$$

satisfies

$$g(a \otimes 1) = f(a \otimes 1) \text{ for all } a \in A.$$

Therefore, E is a splitting field of A and we get

$$\chi_{a,E} = \det(X I_n - g(a \otimes 1)) = \det(X I_n - f(a \otimes 1)) = \chi_{a,L}.$$

Let us go back to the general case. By the two previous cases, we have

$$\chi_{a,E} = \chi_{a,\varphi(L)} = \varphi \cdot \chi_{a,L},$$

since φ is injective and induces an isomorphism between L and $\varphi(L)$.

(3) We are now able to finish the proof. Fix a finite Galois extension L/k which splits A (such an extension exists by Corollary IV.1.18). By the previous point, we have

$$\sigma \cdot \chi_{a,L} = \chi_{a,L} \text{ for all } \sigma \in \mathrm{Gal}(L/k).$$

In other words, every coefficient of $\chi_{a,L}$ is fixed under the action of $\mathrm{Gal}(L/k)$. Hence $\chi_{a,L} \in k[X]$.

Now let L' be a splitting field of A. Then there exists a field E and two k-algebra morphisms

$$\varphi : L \longrightarrow E \text{ and } \varphi' : L' \longrightarrow E.$$

Indeed, we can set $E = L \otimes_k L'/\mathfrak{m}$, where $\mathfrak{m}$ is any maximal ideal of the commutative k-algebra $L \otimes_k L'$. Then we define

$$\varphi(\lambda) = \overline{\lambda \otimes 1} \text{ and } \varphi'(\lambda') = \overline{1 \otimes \lambda'} \text{ for all } \lambda \in L, \lambda' \in L'.$$

The maps φ and φ' are clearly k-algebra morphisms.

By the previous point, we get

$$\chi_{a,E} = \varphi \cdot \chi_{a,L} = \varphi' \cdot \chi_{a,L'}.$$

Since $\chi_{a,L} \in k[X]$ and φ is k-linear we get

$$\chi_{a,L} = \varphi' \cdot \chi_{a,L'}.$$

Since φ' is k-linear as well, we get

$$\varphi' \cdot \chi_{a,L} = \chi_{a,L} = \varphi' \cdot \chi_{a,L'},$$

and by injectivity of φ', we obtain easily that $\chi_{a,L'} = \chi_{a,L} \in k[X]$. This concludes the proof since L' was an arbitrary splitting field of A. □

DEFINITION IV.2.2. Let A be a central simple k-algebra of degree n, and let $a \in A$. The **reduced characteristic polynomial** of $a \in A$ is the polynomial $\mathrm{Prd}_A(a)$ defined by

$$\mathrm{Prd}_A(a) = \chi(f(a \otimes 1)) = \det(X I_n - f(a \otimes 1)),$$

where

$$f : A \otimes_k L \xrightarrow{\sim} \mathrm{M}_n(L)$$

is an isomorphism of L-algebras. By the previous lemma, the polynomial $\mathrm{Prd}_A(a)$ lies in $k[X]$ and does not depend on the choice of L or the isomorphism f.

We now study briefly the properties of the reduced characteristic polynomial.

LEMMA IV.2.3. *Let k be a field and let A be a central simple k-algebra. Then we have the following properties:*

(1) *if $A = \mathrm{M}_n(k)$, then $\mathrm{Prd}_A(M) = \chi_M$ for all $M \in A$;*

(2) *let L/k be an arbitrary field extension. Then we have*

$$\mathrm{Prd}_{A \otimes_k L}(a \otimes 1) = \mathrm{Prd}_A(a) \text{ for all } a \in A;$$

(3) *if $\rho : A \xrightarrow{\sim} A'$ is an isomorphism of central simple k-algebras, then we have*

$$\mathrm{Prd}_{A'}(\rho(a)) = \mathrm{Prd}_A(a) \text{ for all } a \in A;$$

(4) *if $\deg(A) = n$, we have the equality*

$$\chi_{\ell_a} = \mathrm{Prd}_A(a)^n;$$

(5) *for all $a \in A$, we have $\mathrm{Prd}_A(a)(a) = 0$.*

Proof.

(1) Since we have a canonical isomorphism of k-algebras

$$\mathrm{M}_n(k) \otimes_k k \cong_k \mathrm{M}_n(k)$$

sending $M \otimes 1$ to M, we have immediately the desired result.

(2) Let E be a splitting field of A containing L (for example, one can take for E the algebraic closure of L). Let

$$f : A \otimes_k E \xrightarrow{\sim} \mathrm{M}_n(E)$$

be an isomorphism of E-algebras, and consider the isomorphism of E-algebras defined by

$$g : (A \otimes_k L) \otimes_L E \xrightarrow{\sim} A \otimes_k E \xrightarrow{\sim} \mathrm{M}_n(E),$$

where the last map is f. We then have by definition

$$\mathrm{Prd}_{A \otimes_k L}(a \otimes 1) = \chi_{g((a \otimes 1) \otimes 1)} = \chi_{f(a \otimes 1)} = \mathrm{Prd}_A(a).$$

(3) If $f : A' \otimes_k L \xrightarrow{\sim} M_n(L)$ is an isomorphism of L-algebras, then

$$f \circ (\rho \otimes \mathrm{Id}_L) : A \otimes_k L \xrightarrow{\sim} M_n(L)$$

is also an isomorphism of L-algebras. Now apply the definition.

(4) If $A = M_n(k)$, by (1), the desired equality rewrites

$$\chi_{\ell_M} = \chi_M^n \text{ for all } M \in M_n(k),$$

which we already observed at the beginning of this section. If now A is an arbitrary central simple k-algebra, let L be a splitting field and let

$$f : A \otimes_k L \xrightarrow{\sim} M_n(L)$$

be an isomorphism of L-algebras. For $a \in A$, we have by the split case

$$\chi_{\ell_{f(a \otimes 1)}} = \chi_{f(a \otimes 1)}^n = \mathrm{Prd}_A(a)^n.$$

But the isomorphism f induces an isomorphism

$$\tilde{f} : \begin{array}{c} \mathrm{End}_L(A \otimes_k L) \xrightarrow{\sim} \mathrm{End}_L(M_n(L)) \\ u \longmapsto f \circ u \circ f^{-1}. \end{array}$$

It is easy to check that we have

$$\tilde{f}(\ell_z) = \ell_{f(z)} \text{ for all } z \in A \otimes_k L,$$

so we get

$$\chi_{\ell_{f(a \otimes 1)}} = \chi_{\tilde{f}(\ell_{a \otimes 1})} = \chi_{\ell_{a \otimes 1}}$$

using the previous point. Now the canonical isomorphism

$$\mathrm{End}_L(A \otimes_k L) \cong_k \mathrm{End}_k(A) \otimes_k L$$

maps $\ell_{a \otimes 1}$ to $\ell_a \otimes 1$, so we finally get

$$\chi_{\ell_{f(a \otimes 1)}} = \chi_{\ell_a \otimes 1} = \chi_{\ell_a}.$$

Putting things together yields the result.

(5) Let $f : A \otimes_k L \xrightarrow{\sim} M_n(L)$ be an isomorphism of L-algebras. Let $a \in A$, and write

$$\chi_{f(a \otimes 1)} = X^n + \lambda_{n-1} X^{n-1} + \cdots + \lambda_0, \lambda_i \in k.$$

By Cayley-Hamilton's theorem, we have $\chi_{f(a \otimes 1)}(f(a \otimes 1)) = 0$. Since f is a morphism of L-algebras, this can be rewritten as

$$f((a^n + \lambda_{n-1} a^{n-1} + \cdots + \lambda_0) \otimes 1) = 0.$$

By injectivity of f, we get

$$(a^n + \lambda_{n-1} a^{n-1} + \cdots + \lambda_0) \otimes 1 = 0,$$

and since the map

$$A \longrightarrow A \otimes_k L$$
$$a \longmapsto a \otimes 1$$

is injective, we get

$$a^n + \lambda_{n-1} a^{n-1} + \cdots + \lambda_0 = 0.$$

But $\chi_{f(a \otimes 1)} = \mathrm{Prd}_A(a)$ by definition; this concludes the proof. $\square$

We now study more closely the reduced characteristic polynomial of a division algebra. Recall that if D is a central division k-algebra and $d \in D$, the k-subalgebra generated by d is a subfield $k(d)$ of D and $[k(d) : k] \mid \deg(D)$ by Lemma IV.1.4. Hence the following statement makes sense.

LEMMA IV.2.4. *Let D be a central division k-algebra. For all $d \in D$, we have*

$$\mathrm{Prd}_D(d) = \mu_{d,k}^s,$$

where $\mu_{d,k}$ is the minimal polynomial of d over k and $s = \dfrac{\deg(D)}{[k(d) : k]}$.

Proof. Let $n = \deg(D), r = [k(d) : k]$ and $m = \dfrac{n^2}{r}$, and let $(e_1, \ldots, e_m)$ be a basis of D viewed as a left $k(d)$-vector space. Then

$$(e_1, de_1, \ldots, d^{r-1}e_1, \ldots, e_m, de_m, \ldots, d^{r-1}e_m)$$

is a k-basis of D. The matrix of ℓ_d in this basis is easily seen to be

$$\begin{pmatrix} C_{\mu_{d,k}} & & \\ & \ddots & \\ & & C_{\mu_{d,k}} \end{pmatrix},$$

where $C_{\mu_{d,k}}$ is the companion matrix associated to μ_d. Hence we get

$$\chi_{\ell_d} = \mu_{d,k}^m.$$

By the previous lemma, we also have $\chi_{\ell_d} = \mathrm{Prd}_D(d)^n$. Hence μ_d is the unique monic irreducible factor of $\mathrm{Prd}_D(d)$, and since $\mathrm{Prd}_D(d)$ is monic, we get

$$\mathrm{Prd}_D(d) = \mu_{d,k}^s$$

for some $s \geq 1$. The desired result now follows by comparing degrees. $\square$

COROLLARY IV.2.5. *Let D be a central division k-algebra. Then two elements of D are conjugate if and only if they have the same reduced characteristic polynomial.*

Proof. Let $d, d' \in D$, and assume that $\mathrm{Prd}_D(d) = \mathrm{Prd}_D(d')$. The previous result implies that we have $\mu_d = \mu_{d'}$. Hence there is a unique homomorphism of k-algebras $\rho : k(d) \longrightarrow D$ satisfying

$$\rho(d) = d'.$$

By Skolem-Noether's theorem there exists $d_0 \in D^\times$ such that

$$\rho = \mathrm{Int}(d_0) \circ \iota,$$

where ι is the inclusion $k(d) \subset D$. In particular, we have

$$d' = \rho(d) = d_0 d d_0^{-1}.$$

Conversely, assume that $d' = d_0 d d_0^{-1}$ for some $d_0 \in D^\times$. Let

$$f : D \otimes_k L \overset{\sim}{\longrightarrow} \mathrm{M}_n(L)$$

be an isomorphism of L-algebras. Then $f(d \otimes 1)$ and $f(d' \otimes 1)$ are conjugate by $f(d_0 \otimes 1)$, so they have the same characteristic polynomial. By definition, this means that d and d' have the same reduced characteristic polynomial; this concludes the proof. $\square$

REMARK IV.2.6. The previous lemma and its corollary are obviously false for non-division central simple k-algebras, as the case of a matrix algebra already shows.   □

Thanks to the reduced characteristic polynomial, we are now able to generalize the notions of trace and determinant to central simple algebras.

DEFINITION IV.2.7. Let A be a central simple k-algebra, and let $a \in A$. Write

$$\mathrm{Prd}_A(a) = X^n - s_1 X^{n-1} + s_2 X^{n-2} + \cdots + (-1)^n s_n.$$

The coefficients s_1 and s_n are called respectively **the reduced trace** and **the reduced norm** of a. They are denoted respectively by

$$\mathrm{Trd}_A(a) \text{ and } \mathrm{Nrd}_A(a).$$

REMARK IV.2.8. Let A be a central simple k-algebra, let L/k be a splitting field of A, and let

$$f : A \otimes_k L \xrightarrow{\sim} \mathrm{M}_n(L)$$

be an isomorphism of L-algebras. Then by definition, we have

$$\mathrm{Trd}_A(a) = \mathrm{tr}(f(a \otimes 1)) \text{ and } \mathrm{Nrd}_A(a) = \det(f(a \otimes 1))$$

for all $a \in A$.

Assume now that A has a maximal subfield L. In this case, there exists an L-algebra isomorphism

$$f : A \otimes_k L \to \mathrm{End}_L(A)$$

satisfying

$$f(a \otimes 1) = \ell_a \text{ for all } a \in A,$$

where $\ell_a \in \mathrm{End}_L(A)$ is the endomorphism of left multiplication by a in the right L-vector space A. After the choice of an L-basis of A, we get an L-algebra isomorphism

$$f' : A \otimes_k L \to \mathrm{M}_n(L)$$

satisfying

$$f'(a \otimes 1) = M_a \text{ for all } a \in A,$$

where M_a is the representative matrix of ℓ_a in the fixed L-basis of A. In particular, we get

$$\mathrm{Nrd}_A(a) = \det(M_a) \text{ and } \mathrm{Trd}_A(a) = \mathrm{tr}(M_a) \text{ for all } a \in A.$$

□

EXAMPLE IV.2.9. Let $Q = (a, b)_k$ be a quaternion algebra, and we let $q = x + yi + zj + tij \in Q$. Then $L = k(i) \cong_k k(\sqrt{a})$ is a maximal subfield of Q. Using the previous remark and Lemma II.1.10, we get easily

$$\mathrm{Prd}_Q(q) = X^2 - 2xX + (x^2 - ay^2 - bz^2 + abt^2).$$

Hence we get

$$\mathrm{Trd}_A(q) = 2x = q + \gamma(q) \text{ and } \mathrm{Nrd}_A(q) = q\gamma(q),$$

where $\gamma(q) = x - yi - zj - tij$. In particular, we recover the reduced norm of a quaternion algebra defined earlier (see Definition II.1.2).   □

We end this section by proving the main properties of the reduced norm and the reduced trace.

LEMMA IV.2.10. *Let A be a central simple k-algebra of degree n. Then the following properties hold:*

(1) *for all $n \geq 1$, $\mathrm{Trd}_{\mathrm{M}_n(k)} = \mathrm{tr}$ and $\mathrm{Nrd}_{\mathrm{M}_n(k)} = \det$;*

(2) *the map*

$$\mathrm{Trd}_A : A \longrightarrow k$$

is a non-zero linear form;

(3) *for all $a, a' \in A$ and all $\lambda \in k$, we have*

$$\mathrm{Nrd}_A(aa') = \mathrm{Nrd}_A(a)\mathrm{Nrd}_A(a') \text{ and } \mathrm{Nrd}_A(\lambda) = \lambda^n.$$

Moreover,

$$\mathrm{Nrd}_A(a) \neq 0 \iff a \in A^{\times};$$

(4) *let L/k be a field extension. For all $a \in A$, we have*

$$\mathrm{Trd}_{A \otimes_k L}(a \otimes 1) = \mathrm{Trd}_A(a) \text{ and } \mathrm{Nrd}_{A \otimes_k L}(a \otimes 1) = \mathrm{Nrd}_A(a);$$

(5) *let $\rho : A \overset{\sim}{\longrightarrow} A'$ be an isomorphism of central simple k-algebras. For all $a \in A$, we have*

$$\mathrm{Trd}_{A'}(\rho(a)) = \mathrm{Trd}_A(a) \text{ and } \mathrm{Nrd}_{A'}(\rho(a)) = \mathrm{Nrd}_A(a).$$

Proof. Points $(1), (4)$ and (5) come directly from the properties of the reduced characteristic polynomial described in Lemma IV.2.3. Let us prove (2). Assume that Trd_A is identically zero, and let

$$f : A \otimes_k L \overset{\sim}{\longrightarrow} \mathrm{M}_n(L)$$

be an isomorphism of L-algebras. Then, by Remark IV.2.8, we would have

$$\mathrm{tr}(f(a \otimes 1)) = 0 \text{ for all } a \in A.$$

Since the elements $a \otimes 1$ span $A \otimes_k L$ as an L-vector space, their images span $\mathrm{M}_n(L)$, and we get that tr is identically zero, which is absurd. Hence Trd_A is not identically zero. The linearity of the reduced trace comes from the linearity of the trace of matrices.

It remains to prove (3). Let $f : A \otimes_k L \overset{\sim}{\longrightarrow} \mathrm{M}_n(L)$ be an L-algebra isomorphism. Then by Remark IV.2.8, we have

$$\mathrm{Nrd}_A(aa') = \det(f(aa' \otimes 1)) = \det(f((a \otimes 1)(a' \otimes 1))).$$

Hence we get

$$\mathrm{Nrd}_A(aa') = \det(f(a \otimes 1)) \det(f(a' \otimes 1)) = \mathrm{Nrd}_A(a)\mathrm{Nrd}_A(a').$$

If $\lambda \in k$, we have

$$f(\lambda \otimes 1) = \lambda f(1_{A \otimes L}) = \lambda I_n,$$

so $\mathrm{Nrd}_A(\lambda) = \lambda^n$. Finally, let $a \in A^{\times}$. Then we have

$$\mathrm{Nrd}_A(a)\mathrm{Nrd}_A(a^{-1}) = \mathrm{Nrd}_A(aa^{-1}) = 1,$$

so $\mathrm{Nrd}_A(a) \neq 0$. Conversely, if $\mathrm{Nrd}_A(a) \neq 0$, then $f(a \otimes 1)$ is an invertible matrix, and so $a \otimes 1 \in A \otimes_k L$ is invertible. Arguing as at the end of the proof of Skolem-Noether's theorem, we get that $a \in A^{\times}$. $\square$

IV.3. The minimum determinant of a code

When designing codes, one has in mind a particular class of channels, on which the codes will be used. The properties that a code has to satisfy depend on what channel is considered. Classical error-correcting codes, intended for discrete channels, usually consider the minimum Hamming distance as a design criterion. When dealing with space-time codes, the "equivalent" of the minimum Hamming distance is given by the **minimum determinant** of the code.

As explained in Section I.3, in the case of a fully diverse space-time codebook $\mathcal{C}$, we have the following estimation for the probability error

$$\mathbb{P}(\mathbf{X}' \to \mathbf{X}'') \leq \frac{c(\rho)^{-n^2}}{|\det(\mathbf{X}' - \mathbf{X}'')|^{2n}},$$

where M is the number of antennas at the transmitter and at the receiver. It is therefore natural to introduce the following definition.

DEFINITION IV.3.1. Let $\mathcal{C} \subset \mathrm{M}_m(\mathbb{C})$ be a space-time codebook. The minimum determinant of the code $\mathcal{C}$ is given by

$$\delta_{min}(\mathcal{C}) = \inf_{\mathbf{X}' \neq \mathbf{X}'' \in \mathcal{C}} |\det(\mathbf{X}'' - \mathbf{X}')|^2.$$

Notice that if $\mathcal{C}$ is finite, we have

$$\delta_{min}(\mathcal{C}) = \min_{\mathbf{X}' \neq \mathbf{X}'' \in \mathcal{C}} |\det(\mathbf{X}'' - \mathbf{X}')|^2,$$

which is the classical definition [52] of the minimum determinant of a code.

With this definition, one may obtain the following uniform upper bound for the probability error :

$$\mathbb{P}(\mathbf{X} \longrightarrow \hat{\mathbf{X}}) \leq \frac{c(\rho)^{-2n}}{\delta_{min}(\mathcal{C})^n},$$

for all $\mathbf{X} \neq \hat{\mathbf{X}} \in \mathcal{C}$.

The code design then consists in finding codes that maximize the minimum determinant. Recall that making the minimum determinant arbitrarily big simply by scaling every codeword by a constant is not a valid option.

Let A be a simple k-algebra, and let L be a subfield of A. Recall that we assume $L \subset \mathbb{C}$. Let $\mathcal{C}$ be a codebook based on A, that is

$$\mathcal{C} = \{\mathbf{X} = \varphi_{A,L}(a) \mid a \in \mathcal{A}\},$$

where $\mathcal{A}$ is a subset of A and

$$\varphi_{A,L} : \begin{array}{c} A \longrightarrow \mathrm{M}_m(L) \\ a \longmapsto M_a \end{array}$$

with $m = \dim_L(A)$ and M_a denotes the matrix of multiplication by a, as in Definition I.3.2. In this case, the minimum determinant can be rewritten as

$$\delta_{min}(\mathcal{C}) = \inf_{a' \neq a'' \in \mathcal{A}} |\det(M_{a'} - M_{a''})|^2.$$

Note that the minimum determinant is well-defined since we have assumed that $k \subset \mathbb{C}$.

Recall from Remark I.3.3 that if $\mathbf{X}' \neq \mathbf{X}'' \in \mathcal{C}$, then there exist $a' \neq a'' \in \mathcal{A}$ such that $\mathbf{X}' = \varphi_{A,L}(a)$ and $\mathbf{X}'' = \varphi_{A,L}(a'')$. Thus

$$\mathbf{X}' - \mathbf{X}'' = \varphi_{A,L}(a' - a'') = M_{a'-a''}$$

for $a = a' - a'' \in A$. Therefore if $\mathcal{C}$ is an algebra based code, the definition of the minimum determinant can be expressed as

$$\delta_{min}(\mathcal{C}) = \inf_{a' \neq a'' \in \mathcal{A}} |\det(M_{a'-a''})|^2.$$

If moreover $\mathcal{A}$ is a subgroup of A, this simplifies to give

$$\delta_{min}(\mathcal{C}) = \inf_{a \in \mathcal{A}\backslash\{0\}} |\det(M_a)|^2.$$

In particular, for any code $\mathcal{C}$ based on A, we obtain

$$\delta_{min}(\mathcal{C}) \geq \inf_{a \in A\backslash\{0\}} |\det(M_a)|^2,$$

with equality if $\mathcal{C} = \varphi_{A,L}(A)$.

EXAMPLE IV.3.2. Let F/k be a field extension of degree m generated by an element θ with minimal polynomial $X^m - \lambda$, and let $\mathcal{C}_{F,k}$ be the code introduced in Example I.3.4. We then have:

$$\mathcal{C}_{F,k} = \left\{ \begin{pmatrix} a_0 & \lambda a_{m-1} & \lambda a_{m-2} & \cdots & \lambda a_1 \\ a_1 & a_0 & \lambda a_{m-1} & \cdots & \lambda a_2 \\ a_2 & a_1 & a_0 & \cdots & \lambda a_3 \\ \vdots & \vdots & \vdots & \cdots & \vdots \\ a_{m-1} & a_{m-2} & a_{m-3} & \cdots & a_0 \end{pmatrix}, \ a_0, \ldots, a_{m-1} \in k \right\}.$$

If $\mathcal{C}$ is a subset of $\mathcal{C}_{F,k}$ built on a subset of $\mathcal{A}$ of F, we have

$$\begin{aligned} \delta_{min}(\mathcal{C}) &= \inf_{\mathbf{X}' \neq \mathbf{X}'' \in \mathcal{C}} |\det(\mathbf{X}' - \mathbf{X}'')|^2 \\ &= \inf_{a' \neq a'' \in \mathcal{A}} |N_{F/k}(a' - a'')|^2 \end{aligned}$$

where $N_{F/k}$ denotes the norm map.

If $\mathcal{C} = \mathcal{C}_{F/k}$ (that is, if $\mathcal{A} = F$), we clearly have $\delta_{min}(\mathcal{C}_{F,k}) = 0$, since $N_{F/k}(\frac{1}{n}) = \frac{1}{n^m}$ for all $n \geq 1$.

If the subset $\mathcal{A}$ used to build $\mathcal{C}$ is finite, then $\delta_{min}(\mathcal{C}) \neq 0$ since the minimum is over non-zero elements of F. However, nothing prevents $\delta_{min}(\mathcal{C})$ to be arbitrarily close to zero when the size of the codebook increases. For example, if $a_0, \ldots, a_{m-1} \in \{\frac{k}{n} \mid k = 0, \ldots, n-1\}$, we obtain a codebook with m^n elements with a non-zero minimum determinant. However, since $N_{F/k}(\frac{1}{n}) = \frac{1}{n^m}$, we have

$$\delta_{min}(\mathcal{C}_{F,k}) \leq \frac{1}{n^{2m}}.$$

$\square$

We now consider the case where A is not only simple, but is furthermore a central k-algebra of degree n with a maximal subfield L. By Remark IV.2.8, we have

$$\det(M_a) = \operatorname{Nrd}_A(a) \text{ for all } a \in A.$$

Recall that $\operatorname{Nrd}_A(a)$ denotes the reduced norm of $a \in A$ as defined in Definition IV.2.7.

Thus the minimum determinant of $\mathcal{C}$ can be rewritten as

$$\delta_{min}(\mathcal{C}) = \inf_{a' \neq a'' \in \mathcal{A}} |\operatorname{Nrd}_A(a' - a'')|^2.$$

If furthermore $\mathcal{A}$ is a subgroup of A, the formula simplifies similarly as before, to give

$$\delta_{min}(\mathcal{C}) = \inf_{a \in \mathcal{A} \setminus \{0\}} |\operatorname{Nrd}_A(a)|^2.$$

Note that by Lemma IV.2.1, the reduced norm of an element of A lies in k. We thus immediately get the following result, under the above assumptions:

LEMMA IV.3.3. *Let A be a central division k-algebra, let k be a subfield of $\mathbb{C}$ which is closed under complex conjugation, and let $\mathcal{C}$ be a code based on A. Then its minimum determinant $\delta_{min}(\mathcal{C})$ belongs to $k \cap \mathbb{R}^+$.*

EXAMPLE IV.3.4. Consider a division quaternion algebra $Q = (a, b)_k$, where k is a subfield of $\mathbb{C}$ which is closed under complex conjugation. Set $L = k(\sqrt{a})$, and let $\mathcal{C}_{Q,L}$ be the following codebook built on Q (see Section II.3), defined by

$$\mathcal{C}_{Q,L} = \left\{ \begin{pmatrix} x + y\sqrt{a} & b(z + t\sqrt{a}) \\ z - t\sqrt{a} & x - y\sqrt{a} \end{pmatrix}, \ x, y, z, t \in k \right\}.$$

We have that

$$\delta_{min}(\mathcal{C}_{Q,L}) = \inf_{q \in Q^\times} |\operatorname{Nrd}_Q(q)|^2.$$

In this case, there are several ways to see that $\delta_{min}(\mathcal{C})$ belongs to $k \cap \mathbb{R}^+$.

(1) From Example IV.2.9, we know that

$$\operatorname{Nrd}_Q(q) = x^2 - ay^2 - bz^2 + abt^2,$$

for $q = x + yi + zj + tij \in Q$. It is now immediate that $\operatorname{Nrd}_Q(q) \in k$ since $x, y, z, t, a, b \in k$.

(2) Starting again from the definition, the reduced norm of $q \in Q$ is given by

$$\begin{aligned} \operatorname{Nrd}_Q(q) &= \det \begin{pmatrix} x + y\sqrt{a} & b(z + t\sqrt{a}) \\ z - t\sqrt{a} & x - y\sqrt{a} \end{pmatrix} \\ &= N_{k(\sqrt{a})/k}(x + y\sqrt{a}) - bN_{k(\sqrt{a})/k}(z + t\sqrt{a}), \end{aligned}$$

where $N_{k(\sqrt{a})/k}$ denotes the norm map. Now $N_{k(\sqrt{a})/k}(k(\sqrt{a})) \subset k$, and since $b \in k$, this implies that $\operatorname{Nrd}_Q(q)$ is in k.

If $\mathcal{C} \subset \mathcal{C}_{Q,L}$ is built on a subset $\mathcal{A}$ of Q, the minimum determinant of $\mathcal{C}$ is given by

$$\delta_{min}(\mathcal{C}) = \inf_{x+yi+zj+tij \in \mathcal{A}} |N_{k(\sqrt{a})/k}(x + y\sqrt{a}) - bN_{k(\sqrt{a})/k}(z + t\sqrt{a})|^2,$$

or equivalently

$$\delta_{min}(\mathcal{C}) = \inf |x^2 - ay^2 - bz^2 + abt^2|^2,$$

where $x + yi + zj + tij$ runs through the non-zero elements of $\mathcal{A}$.

If $\mathcal{C} = \mathcal{C}_{Q,L}$, then we have $\delta_{min}(\mathcal{C}_{Q,L}) = 0$ since $\text{Nrd}_Q(\frac{1}{n}) = \frac{1}{n^2}$ for all $n \geq 1$.

If the information symbols x, y, z, t belong to a finite subset of k, then $\mathcal{A}$ is finite and $\delta_{min}(\mathcal{C}_{Q,L}) \neq 0$ if Q is a division algebra. However nothing prevents $\delta_{min}(\mathcal{C}_{Q,L})$ to be arbitrary close to zero if the size of the constellation increases, as before.

$\square$

We can see from the above examples that without any restriction in choosing $\mathcal{C} \subset \mathcal{C}_{A,L}$, it is unlikely to actually bound the minimum determinant away from zero. Note that practical codes contain a finite number of codewords, so that if this number is small enough, one can get a bound on the minimum determinant by choosing carefully a small number of matrices. However good codebooks usually require a large number of matrices. As already mentioned before, scaling the codebook by a constant is not a valid solution. We will thus discuss now a more general solution, which will work also for infinite codebooks.

In the case where L/k is a number field extension, a natural restriction consists of considering $\mathcal{O}_L \subset L$, the ring of integers of L.

EXAMPLE IV.3.5. Let k be a number field which is closed under complex conjugation, let F/k be a field extension of degree m and consider the code $\mathcal{C}$ described in Example IV.3.2 above. Let us now explain how we can obtain a subcode $\mathcal{C}$ of $\mathcal{C}_{F,k}$, by considering the ring of integers $\mathcal{O}_F$ of F. Recall that $\mathcal{C}_{F,k}$ is obtained via $\varphi_{F,k}$ as follows:

$$\varphi_{F,k}: \begin{array}{c} F \longrightarrow \mathrm{M}_m(k) \\ a \longmapsto M_a. \end{array}$$

We define $\mathcal{C}_{\mathcal{O}_F,k}$ similarly, by restricting $\varphi_{F,k}$ to $\mathcal{O}_F$, so that

$$\mathcal{C}_{\mathcal{O}_F,k} = \{M_a \mid a \in \mathcal{O}_F\} \subset \mathcal{C}_{F,k}.$$

Let us now recall how encoding was done for $\mathcal{C}_{F,k}$. Let $a_0, \ldots, a_{m-1} \in k$ be the m information symbols to be sent. Here the set of possible information symbols is k. They are encoding into an element $a \in F$ using a k-basis of F as follows: $a = a_0 + a_1\theta + \cdots + a_{m-1}\theta^{m-1}$. The corresponding codeword is given by M_a. We want to obtain a similar encoding procedure for $\mathcal{C}_{\mathcal{O}_F,k}$. For that, we need a mapping that sends the information symbols to an element of $\mathcal{O}_F$, and that for any choice of information symbols in the set of possible information symbols. For it to be possible, we need to restrict the set of information symbols to $\mathcal{O}_k \subset k$. If $\mathcal{O}_F$ is a free $\mathcal{O}_k$-module, we may define the encoding for $\mathcal{C}_{\mathcal{O}_F,k}$ as follows: the m information symbols $a_0, \ldots, a_{m-1} \in \mathcal{O}_k$ are encoded into an element of $a \in \mathcal{O}_F$ using an $\mathcal{O}_k$-basis of $\mathcal{O}_F$.

Let us now assume that $\mathcal{O}_k$ is a principal ideal domain, so that there exists an $\mathcal{O}_k$-basis of $\mathcal{O}_F$. We define a new code $\mathcal{C}_{\mathcal{O}_F,k}$ similarly to $\mathcal{C}$, by restricting ourselves to elements of $\mathcal{O}_F$. In order to compute the code explicitly, we consider the case where $\mathcal{O}_F$ is monogeneous, that is of the form $\mathcal{O}_k[\alpha]$ with α satisfying

$$\alpha^m = -\mu_0 - \mu_1\alpha - \cdots - \mu_{m-1}\alpha^{m-1}$$

for some $\mu_i \in \mathcal{O}_k$, $i = 0, \ldots, m-1$. Thus we have

$$
\mathcal{C}_{\mathcal{O}_F,k} = \left\{ \begin{pmatrix} a_0 & -\mu_0 a_{m-1} & \cdots \\ a_1 & a_0 - \mu_1 a_{m-1} & \cdots \\ a_2 & a_1 - \mu_2 a_{m-1} & \cdots \\ \vdots & \vdots & \cdots \\ a_{m-1} & a_{m-2} - \mu_{m-1} a_{m-1} & \cdots \end{pmatrix}, \; a_0, \ldots, a_{m-1} \in \mathcal{O}_k \right\}.
$$

The minimum determinant is now given by

$$
\delta_{min}(\mathcal{C}_{\mathcal{O}_F,k}) = \inf_{x \in \mathcal{O}_F \setminus \{0\}} |N_{F/k}(x)|^2.
$$

Since $N_{F/k}(\mathcal{O}_F) \subset \mathcal{O}_k$, we have

$$
\delta_{min}(\mathcal{C}_{\mathcal{O}_F,k}) \in \mathcal{O}_k \cap \mathbb{R}^+.
$$

Note that if $k = \mathbb{Q}$ or $k = \mathbb{Q}(\sqrt{-d})$, where d is a positive square free integer, then $\mathcal{O}_k \cap \mathbb{R}^+ = \mathbb{N}$ and

$$
\delta_{min}(\mathcal{C}_{\mathcal{O}_F,k}) \geq 1,
$$

since the infimum is taken over the non-zero elements of $\mathcal{O}_F$. Note that $N_{F/k}(1) = 1$ and therefore

$$
\delta_{min}(\mathcal{C}_{\mathcal{O}_F,k}) = 1.
$$

This contrasts with the case where the norm was in k, where no lower bound could be given. Note that this bound holds for an infinite family of matrices. $\qquad\square$

EXAMPLE IV.3.6. Let us now give a concrete particular case [**37**] of the above construction. Let $\zeta = e^{i\pi/2^{r-1}} \in \mathbb{C}$, with $r \geq 3$, and let $\mathbb{Q}(\zeta)$ denote the corresponding cyclotomic field. Its ring of integers is given by $\mathbb{Z}[\zeta]$, with $\mathbb{Z}$-basis given by $(1, \zeta, \ldots, \zeta^{2^{r-1}-1})$. We now consider the field extension $\mathbb{Q}(\zeta)/\mathbb{Q}(i)$, which has degree $m = 2^{r-2}$ over $\mathbb{Q}(i)$. Moreover, the family $(1, \zeta, \ldots, \zeta^{2^{r-2}-1})$ is a $\mathbb{Z}[i]$-basis of its ring of integers.

If $x = \sum_{l=0}^{m-1} x_l \zeta^l$, we have

$$
x\zeta = \sum_{l=0}^{m-1} x_l \zeta^{l+1} = \sum_{l=0}^{m-2} x_l \zeta^{l+1} + i x_{m-1} = i x_{m-1} + \sum_{k=1}^{m-1} x_{k-1} \zeta^k
$$

since $\zeta^m = \zeta^{2^{r-2}} = e^{i\pi/2} = i$. Similarly,

$$
x\zeta^2 = \sum_{l=0}^{m-3} x_l \zeta^{l+2} + i x_{m-2} + i x_{m-1}\zeta = i x_{m-2} + i x_{m-1}\zeta + \sum_{k=2}^{m-1} x_{k-2} \zeta^k.
$$

More generally,

$$
\begin{aligned}
x\zeta^t &= \sum_{l=0}^{m-1} x_l \zeta^{l+t} \\
&= \sum_{l=0}^{m-t-1} x_l \zeta^{l+t} + i \sum_{l'=1}^{t} x_{m-l'} \zeta^{t-l'} \\
&= i \sum_{l'=1}^{t} x_{m-l'} \zeta^{t-l'} + \sum_{k=t}^{m-1} x_{k-t} \zeta^k.
\end{aligned}
$$

We thus have that M_x is given by

$$
\begin{pmatrix}
x_0 & ix_{m-1} & ix_{m-2} & \cdots & ix_1 \\
x_1 & x_0 & ix_{m-1} & & ix_2 \\
x_2 & x_1 & x_0 & & ix_3 \\
\vdots & \vdots & \vdots & & \vdots \\
x_{m-2} & x_{m-3} & x_{m-4} & & ix_{m-1} \\
x_{m-1} & x_{m-2} & x_{m-3} & \cdots & x_0
\end{pmatrix},
$$

and

$$
\mathcal{C}_{\mathbb{Z}[\varsigma],\mathbb{Q}(i)} = \left\{
\begin{pmatrix}
x_0 & ix_{m-1} & ix_{m-2} & \cdots & ix_1 \\
x_1 & x_0 & ix_{m-1} & & ix_2 \\
x_2 & x_1 & x_0 & & ix_3 \\
\vdots & \vdots & \vdots & & \vdots \\
x_{m-2} & x_{m-3} & x_{m-4} & & ix_{m-1} \\
x_{m-1} & x_{m-2} & x_{m-3} & \cdots & x_0
\end{pmatrix}, \; x_0, \ldots, x_{m-1} \in \mathbb{Z}[i]
\right\}.
$$

Now, the minimum determinant is given by

$$
\delta_{min}(\mathcal{C}_{\mathbb{Z}[\theta],\mathbb{Q}(i)}) = \inf_{x \in \mathbb{Z}[\varsigma] \setminus \{0\}} |N_{\mathbb{Q}(\varsigma)/\mathbb{Q}(i)}(x)|^2 \in \mathbb{N},
$$

since $N_{\mathbb{Q}(\varsigma)/\mathbb{Q}(i)}(x) \in \mathbb{Z}[i]$. Thus

$$
\delta_{min}(\mathcal{C}_{\mathbb{Z}[\varsigma],\mathbb{Q}(i)}) = 1.
$$

$\square$

Let us consider the case of a central division k-algebra A, where k is a number field, and let let L be a subfield of A. For any ideal I of $\mathcal{O}_L$, we consider a new codebook $\mathcal{C}_{A,I}$, defined as follows. Let $(e_1, \ldots, e_n)$ be an L-basis of A, where A is considered as a right L-vector space, that is $A = e_1 L \oplus e_2 L \oplus \cdots \oplus e_n L$. The code $\mathcal{C}_{A,I}$ is defined by restricting ourselves to elements in $\Lambda_{A,I} = e_1 I \oplus e_2 I \oplus \cdots \oplus e_n I$:

$$
\mathcal{C}_{A,I} = \{\mathbf{X} = M_a \mid a \in \Lambda_{A,I}\}.
$$

PROPOSITION IV.3.7. *Let k be a number field which is closed under complex conjugation. Let $k^+ = k \cap \mathbb{R}$ be the subfield of k fixed by the complex conjugation, let A be a central division k-algebra A of degree n, and let I be an ideal of $\mathcal{O}_L$, where L/k is a maximal subfield of A. Then there exists an integer $c > 0$ such that we have*

$$
|\det(M_a)|^2 \in \frac{1}{c}\mathcal{O}_{k^+} \text{ for all } a \in \Lambda_{A,I}.
$$

Proof. Since every element $x \in L$ may be written as $x = \dfrac{y}{s}$ for some $y \in \mathcal{O}_L$ and $s \in \mathbb{Z}$, the set

$$
\{m \in \mathbb{Z} \mid mM_{e_i} \in \mathrm{M}_n(\mathcal{O}_L) \text{ for } i = 1, \ldots, n\}
$$

is a non-zero ideal of $\mathbb{Z}$, hence generated by a unique positive integer $r \geq 1$. We deduce that, for every $a = e_1 a_1 + \cdots + e_n a_n \in \Lambda_{A,I}$, we have

$$
M_a = M_{e_1} a_1 + \cdots + M_{e_n} a_n \in \frac{1}{r}\mathrm{M}_n(\mathcal{O}_L).
$$

It easily follows that $r^n \det(M_a) = r^n \mathrm{Nrd}_A(a) \in \mathcal{O}_L \cap k = \mathcal{O}_k$, and thus we obtain that

$$\det(M_a) \in \frac{1}{r^n} \mathcal{O}_k.$$

Hence for all $a \in \Lambda_{A,I}$, there exists $x \in \mathcal{O}_k$ such that

$$\det(M_a) = \frac{x}{r^n}.$$

By assumption on k, complex conjugation induces an automorphism of k, hence an automorphism of $\mathcal{O}_k$. Therefore, $|x|^2 = \overline{x}x \in \mathcal{O}_k$, and thus $|x|^2 \in \mathcal{O}_k \cap \mathbb{R} = \mathcal{O}_{k^+}$. Consequently,

$$|\det(M_a)|^2 = \frac{|x|^2}{r^{2n}} \in \frac{1}{r^{2n}} \mathcal{O}_{k^+} \text{ for all } a \in \Lambda_{A,I}.$$

Setting $c = r^{2n}$ then yields the conclusion.   □

Contrary to what happens in Example IV.3.5, this proposition still does not ensure that $\delta_{min}(\mathcal{C})$ is different from 0. For example, if $k = \mathbb{Q}(\sqrt{2})$, then $\mathcal{O}_k$ contains elements whose modulus can arbitrary close to zero (e.g. $x = (-1+\sqrt{2})^n$). However, if k is either $\mathbb{Q}$ or an imaginary quadratic field, we have the following result.

COROLLARY IV.3.8. *Assume $k = \mathbb{Q}$ or $\mathbb{Q}(\sqrt{-d})$ for some squarefree positive integer d, and let $\mathcal{C} \subset \mathcal{C}_{A,I}$ be a codebook coming from a central division k-algebra A of degree n, where L/k is a maximal subfield of A and I is an ideal of $\mathcal{O}_L$. Then there exists an integer $c > 0$ such that we have*

$$\delta_{min}(\mathcal{C}) \geq \frac{1}{c}.$$

Proof. The proof of the previous proposition shows that there exists an integer $c > 0$ such that $c|\det(\mathbf{X})|^2 \in \mathcal{O}_{k^+}$ for all $\mathbf{X} \in \mathcal{C}$. The assumptions imply that $\mathcal{O}_{k^+} = \mathbb{Z}$. Hence $c|\det(\mathbf{X})|^2$ is a non-negative integer, and since A is division, $c|\det(\mathbf{X})|^2 \neq 0$, and therefore greater than 1. Thus we have

$$|\det(\mathbf{X})|^2 \geq \frac{1}{c} \text{ for all } \mathbf{X} \in \mathcal{C} \setminus \{0\}.$$

This leads immediately to the conclusion.   □

REMARK IV.3.9. The assumption on k is not very restrictive since in practice, a signal is represented by an element of $\mathbb{Q}(i)$ or $\mathbb{Q}(j)$, where $j = e^{2i\pi/3}$ (see [**14**] for more details).   □

EXAMPLE IV.3.10. Let k be an number field which is closed under complex conjugation. As in Example IV.3.4, let us consider again the quaternion algebra $Q = (a, b)_k$, with its associated code $\mathcal{C}_{Q,k(\sqrt{a})}$, whose minimum determinant is given by

$$\delta_{min}(\mathcal{C}) = \inf |N_{k(\sqrt{a})/k}(x + y\sqrt{a}) - b N_{k(\sqrt{a})/k}(z + t\sqrt{a})|^2,$$

where (x, y, z, t) runs through the non-zero elements of k^4. Set $L = k(\sqrt{a})$. The code $\mathcal{C}_{Q,\mathcal{O}_L}$ is defined by restricting the elements of L to elements of $\mathcal{O}_L$, thus we have that

$$\delta_{min}(\mathcal{C}_{Q,\mathcal{O}_L}) = \inf |N_{k(\sqrt{a})/k}(u) - b N_{k(\sqrt{a})/k}(v)|^2$$

where (u, v) runs through the non-zero elements of $\mathcal{O}_L^2$. Since $b \in k$, one may write $b = \frac{b'}{c}$, for some elements $b', c \in \mathcal{O}_k$ (which are not necessarily unique).

Now $N_{L/k}(\mathcal{O}_L) \subset \mathcal{O}_k$. It follows that $cN_{k(\sqrt{a})/k}(u) - b'N_{k(\sqrt{a})/k}(v) \in \mathcal{O}_k$, and therefore

$$|cN_{k(\sqrt{a})/k}(u) - b'N_{k(\sqrt{a})/k}(v)|^2 \in \mathcal{O}_k \cap \mathbb{R} = \mathcal{O}_{k+}.$$

Now if $k = \mathbb{Q}$ or k is a quadratic imaginary field, we have $\mathcal{O}_{k+} = \mathbb{Z}$. Therefore

$$\delta_{min}(\mathcal{C}_{Q,\mathcal{O}_L}) = \inf \frac{1}{|c|^2}|cN_{k(\sqrt{a})/k}(u) - b'N_{k(\sqrt{a})/k}(v)|^2 \in \frac{1}{|c|^2}\mathbb{Z}$$

and

$$\delta_{min}(\mathcal{C}_{Q,\mathcal{O}_L}) \geq \frac{1}{|c|^2},$$

as soon as Q is a division k-algebra. □

EXAMPLE IV.3.11. We now give a concrete particular case [5] of the above example. Let ζ_8 denote a primitive 8-th root of unity, and consider the quaternion algebra $Q = (i, 1 + 2i)_{\mathbb{Q}(i)}$, and $L = \mathbb{Q}(\zeta_8)$ with associated codebook $\mathcal{C}_{Q,L}$ given by

$$\mathcal{C}_{Q,L} = \left\{ \begin{pmatrix} x + y\zeta_8 & (1 + 2i)(z + t\zeta_8) \\ z - t\zeta_8 & x - y\zeta_8 \end{pmatrix}, \ x, y, z, t \in \mathbb{Q}(i) \right\}.$$

By Example II.1.8, Q is a division algebra. Now, since $1, \zeta_8$ form a $\mathbb{Z}[i]$-basis of $L = \mathbb{Q}(i)(\zeta_8)$, we obtain a new codebook as follows:

$$\mathcal{C}_{Q,\mathcal{O}_L} = \left\{ \begin{pmatrix} x + y\zeta_8 & (1 + 2i)(z + t\zeta_8) \\ z - t\zeta_8 & x - y\zeta_8 \end{pmatrix}, \ x, y, z, t \in \mathbb{Z}[i] \right\}.$$

Since $\zeta_8 \in \mathcal{O}_L$, one may take $c = 1$ in the previous example. Therefore we have

$$\delta_{min}(\mathcal{C}_{Q,\mathcal{O}_L}) \geq 1,$$

and taking $(x, y, z, t) = (1, 0, 0, 0)$ shows that

$$\delta_{min}(\mathcal{C}_{Q,\mathcal{O}_L}) = 1.$$

□

Note that it has been thought for a long time that in order to obtain a lower bound on the minimum determinant, it was necessary to take a codebook of finite cardinality. The fact that one could construct an infinite codebook with fixed minimum determinant has been found independently in [5], where codes over quaternion algebras are proposed, and in [58], where codes are built by parametrization of rotations. In [5], the properties of the reduced norm of quaternion algebras was already exploited, and the authors introduced the terminology **non-vanishing determinant** to describe the fact that the minimum determinant is lower bounded even with a codebook of infinite size. In [58] the same property was shown "by hand", by computing explicitly determinants of 2×2 matrices. However, the result that we have derived in this section is true for central division simple algebras in general, and thus hold for any dimension. A similar theory is also available on maximal orders of some central simple algebras [56].

Exercises

1. Let D be a central division k-algebra of degree n, and let $x \in D$. For any subfield M/k of D containing x, show that we have

$$\mathrm{Trd}_D(x) = \frac{n}{[M:k]}\mathrm{Tr}_{M/k}(x) \quad \text{and} \quad \mathrm{Nrd}_D(x) = N_{M/k}(x)^{\frac{n}{[M:k]}}.$$

Deduce that for any maximal subfield L/k containing x, we have

$$\mathrm{Trd}_D(x) = \mathrm{Tr}_{L/k}(x) \quad \text{and} \quad \mathrm{Nrd}_D(x) = N_{L/k}(x).$$

2. Let D be a central division k-algebra, and let $L = k(\sqrt{a})$ be a quadratic field extension (not necessarily separable). Show that L is isomorphic to a subfield of D if and only if $D \otimes_k L$ is not a division k-algebra.

   Hint: If $D \otimes_k L$ is not a division algebra, let $x_1, x_2 \in D$ such that $x_1 \otimes 1 + x_2 \otimes \sqrt{a}$ is non-zero and non-invertible. Check that $x_2 \neq 0$, and set

$$z = (x_2^{-1}x_1 \otimes 1 + 1 \otimes \sqrt{a})(x_2^{-1}x_1 \otimes 1 - 1 \otimes \sqrt{a}).$$

   Show that z is a non-invertible element of $D \otimes_k 1$. Deduce that $z = 0$ and conclude.

3. Let A be a central simple k-algebra, and let $a \in A$. By reducing to the split case (i.e. the case where A is a split central simple k-algebra), show that $C_A(C_A(k[a])) = k[a]$.

4. Let A be a central simple k-algebra. The **trace form** of A is the symmetric bilinear form

$$T_A: \begin{array}{c} A \times A \longrightarrow k \\ (a, a') \longmapsto \mathrm{Trd}_A(aa') \end{array}$$

   (a) By reducing to the split case, show that T_A is non-singular, that is

$$\mathrm{Trd}_A(aa') = 0 \text{ for all } a' \in A \Rightarrow a = 0.$$

   (b) Let A, B two central simple k-algebras. Show that if $A \cong_k B$, then $T_A \cong_k T_B$.

   (c) Let A, B two central simple k-algebras. Show that

$$\mathrm{Trd}_{A \otimes_k B}(a \otimes b) = \mathrm{Trd}_A(a)\mathrm{Trd}_B(b) \text{ for all } a \in A, b \in B.$$

   (d) Assume that for all $a \in A, \mathrm{Trd}_A(a^2) = 0 \Rightarrow a = 0$. Show that A is a division algebra.

   Hint: Assume that A is not a division algebra. Use (b) to reduce to the case where $A = \mathrm{M}_r(k) \otimes_k D$, for some $r \geq 2$ and D is a division algebra.

5. Let p be a prime number, $p \equiv 5[8]$. Let $Q = (i, p)_{\mathbb{Q}(i)}$. Since $p \equiv 1[4]$, one may write $p = u^2 + v^2, u, v \in \mathbb{Z}$.

   (a) Show that Q is a division algebra.

   Hint: If Q is split, show that there exist $x, y, z \in \mathbb{Z}[i]$, where x, z are not divisible by $u + iv$, such that $iz^2 = x^2 - py^2$. Using the fact that $\mathbb{Z}[i]/(u + iv) \simeq \mathbb{F}_p$, show that $\mathbb{F}_p^\times$ contains an element of order 8, and deduce a contradiction.

   Let $L = \mathbb{Q}(i)(\sqrt{p})$.

(b) Use Proposition B.2.28 to show that $\mathcal{O}_L = \mathbb{Z}[i, \theta]$, where $\theta = \dfrac{1 + \sqrt{p}}{2}$.

Let I be the ideal of $\mathcal{O}_L$ generated by $\dfrac{p-1}{2} + \theta$ and $u + iv$.

(c) Show that $\dfrac{p-1}{2} + \theta$ and $u + iv$ form a $\mathbb{Z}[i]$-basis of I.

(d) Deduce that for all $a \in I$, we have $N_{L/\mathbb{Q}(i)}(a) \in (u + iv)\mathbb{Z}[i]$.

(e) Use the previous results to show that $\delta_{min}(\mathcal{C}_{A,I}) = p$.

CHAPTER V

The Brauer group of a field

In this chapter, we define the so-called Brauer group of a field k, which encodes all the information on central simple k-algebras (this vague statement will be made more precise later on).

V.1. Definition of the Brauer group

Recall from Chapter III that two simple k-algebras A and B are Brauer equivalent if $A \cong_k \mathrm{M}_r(D)$ and $B \cong_k \mathrm{M}_s(D)$ for some $r, s \geq 1$ and some central division k-algebra D. We denote it by $A \sim_k B$.

Since two Brauer equivalent k-algebras have the same center, it is easy to see that it induces an equivalence relation on the set of central simple k-algebras. The equivalence class of A will be denoted by $[A]$.

The next lemma gives the basic properties of Brauer equivalence.

LEMMA V.1.1. *Let A, A', B and B' be central simple k-algebras, and let L/k be a field extension. Then the following properties hold:*

(1) *the central simple k-algebras A and A' are Brauer equivalent if and only if there exist two integers $r, s \geq 1$ such that $\mathrm{M}_s(A) \cong_k \mathrm{M}_r(A')$;*

(2) *$A \cong_k B$ if and only if $A \sim_k B$ and $\deg(A) = \deg(B)$;*

(3) *if $A \sim_k A'$ and $B \sim_k B'$, then $A \otimes_k B \sim_k A' \otimes_k B'$;*

(4) *if $A \sim_k A'$, then $A \otimes_k L \sim_L A' \otimes_k L$;*

(5) *there exists a central division k-algebra D such that $A \sim_k D$, which is unique up to isomorphism.*

Proof.

(1) Assume that A and A' are Brauer equivalent. Then there exist two integers $r, s \geq 1$ and a central division k-algebra D such that $A \cong_k \mathrm{M}_r(D)$ and $A' \cong_k \mathrm{M}_s(D)$. In this case, we have

$$\mathrm{M}_s(A) \cong_k \mathrm{M}_{rs}(D) \cong_k \mathrm{M}_r(A').$$

Conversely, assume that there exist two integers $r, s \geq 1$ such that $\mathrm{M}_s(A) \cong_k \mathrm{M}_r(A')$. By Wedderburn's Theorem, we may write

$$A \cong_k \mathrm{M}_n(D), A' \cong_k \mathrm{M}_m(D'),$$

for some integers $n, m \geq 1$ and some central division k-algebras D and D'. Thus we have

$$\mathrm{M}_{sn}(D) \cong_k \mathrm{M}_{rm}(D'),$$

and by the uniqueness part of Wedderburn's Theorem, we get that $D \cong_k D'$. Therefore, we have $A \cong_k M_n(D)$ and $A' \cong_k M_m(D)$, meaning that A and A' are Brauer equivalent.

(2) If A and B are isomorphic, they have the same degree. Moreover, they are also Brauer equivalent by the uniqueness part of Wedderburn's Theorem. Conversely assume that $A \sim_k B$ and $\deg(A) = \deg(B)$. We then have $A \cong_k M_r(D)$ and $B \cong_k M_s(D)$ for some integers $r, s \geq 1$ and some central division k-algebra D. Since we have $\deg(A) = r \deg(D)$ and $\deg(B) = s \deg(D)$, we get $r = s$ and therefore $A \cong_k B$.

(3) By (1), we have $M_s(A) \cong_k M_r(A')$ and $M_m(B) \cong_k M_n(B')$ for some integers $m, n, r, s \geq 1$. Using Lemma I.1.11, Remark I.1.12 and the commutativity of the tensor product several times, we get

$$\begin{aligned} M_{ms}(A \otimes_k B) &\cong_k M_s(A) \otimes_k M_m(B) \\ M_{nr}(A' \otimes_k B') &\cong_k M_r(A') \otimes_k M_n(B'). \end{aligned}$$

On the other hand, we also have

$$M_s(A) \otimes_k M_m(B) \cong_k M_r(A') \otimes_k M_n(B'),$$

and thus

$$M_{ms}(A \otimes_k B) \cong_k M_{nr}(A' \otimes_k B'),$$

implying by (1) that $A \otimes_k B$ and $A' \otimes_k B'$ are Brauer equivalent central simple k-algebras.

(4) By (1), we have an isomorphism of k-algebras $M_r(A) \cong_k M_s(A')$ for some integers $r, s \geq 1$. We then get an isomorphism of L-algebras

$$M_r(A) \otimes_k L \cong_L M_s(A') \otimes_k L.$$

By Lemma I.1.11, we have isomorphisms of L-algebras

$$M_r(A) \otimes_k L \cong_L (M_r(k) \otimes_k A) \otimes_k L \cong_L (M_r(k) \otimes_k L) \otimes_L (A \otimes_k L),$$

as well as an isomorphism of L-algebras

$$M_r(k) \otimes_k L \cong_L M_r(L).$$

Hence we have

$$M_r(A) \otimes_k L \cong_L M_r(L) \otimes_L (A \otimes_k L) \cong_L M_r(A \otimes_k L).$$

Similarly, $M_s(A') \otimes_k L \cong_L M_s(A' \otimes_k L)$, and therefore we finally get

$$M_r(A \otimes_k L) \cong_L M_s(A' \otimes_k L).$$

By (1), this implies that $A \otimes_k L$ and $A' \otimes_k L$ are Brauer equivalent L-algebras.

(5) By Wedderburn's Theorem, we have $A \cong_k M_r(D)$ for some integer $r \geq 1$ and some central division k-algebra D. By definition, this means that $A \sim_k D$. Now if $A \sim_k D'$ for some other central division k-algebra D', then $D \sim_k D'$, so that there exist two integers $m, n \geq 1$ such that $M_n(D) \cong_k M_m(D')$. By Wedderburn's Theorem, we get that D and D' are isomorphic k-algebras. This concludes the proof. $\qquad\square$

Let $\mathrm{Br}(k)$ be the set of equivalence classes of central simple k-algebras under the Brauer equivalence relation. For $[A], [B] \in \mathrm{Br}(k)$, we set

$$[A] + [B] = [A \otimes_k B].$$

By Lemma V.1.1 (3), this operation is well-defined. The properties of the tensor product imply that it is associative and commutative. By Lemma I.1.11 (1), the class $[k](= [\mathrm{M}_n(k)])$ is a neutral element for this operation. Finally, $[A^{op}]$ is an inverse for $[A]$ by Lemma III.1.13.

Therefore, $(\mathrm{Br}(k), +)$ is an abelian group.

DEFINITION V.1.2. The group $\mathrm{Br}(k)$ is called the **Brauer group** of k.

Notice that by Lemma V.1.1 (5), any class $[A] \in \mathrm{Br}(k)$ may be represented by a central division k-algebra D, which is unique up to isomorphism. Hence the map $D \longmapsto [D]$ yields a one-to-one correspondence between the set of isomorphism classes of central division k-algebras and the Brauer group $\mathrm{Br}(k)$. Therefore, knowing $\mathrm{Br}(k)$ is equivalent to knowing all central division k-algebras up to isomorphism, or equivalently to knowing all central simple k-algebras up to isomorphism by Wedderburn's Theorem. However, the main interest of $\mathrm{Br}(k)$ is to carry naturally a group structure.

We now give some computations of Brauer groups.

EXAMPLE V.1.3. If k is algebraically closed, then $\mathrm{Br}(k) = 0$ since every central simple k-algebra is split in this case by Corollary III.4.3. □

EXAMPLE V.1.4. We know from Theorem II.2.1 that $\mathbb{R}$ and $\mathbb{H}$ are the only central division $\mathbb{R}$-algebras up to isomorphism. Since every Brauer class $[A] \in \mathrm{Br}(\mathbb{R})$ is represented by a division $\mathbb{R}$-algebra, unique up to isomorphism, $\mathrm{Br}(\mathbb{R})$ consists of two classes, namely 0 and $[\mathbb{H}]$. In particular, $\mathrm{Br}(\mathbb{R})$ is an abelian group of order 2, and therefore we have a group isomorphism

$$\mathrm{Br}(\mathbb{R}) \cong \mathbb{Z}/2\mathbb{Z},$$

where the non-trivial class is the Brauer class of the Hamilton quaternion algebra $\mathbb{H}$.

Notice that we also may recover the fact that $[\mathbb{H}]$ is an element of order 2 as follows. First of all, $\mathbb{H}$ is a division $\mathbb{R}$-algebra, which is not isomorphic to $\mathbb{R}$, and thus $[\mathbb{H}] \neq 0$ in $\mathrm{Br}(\mathbb{R})$. Moreover, using Proposition II.1.9 and Example II.1.6 immediately yields the following isomorphisms of $\mathbb{R}$-algebras

$$\mathbb{H} \otimes_{\mathbb{R}} \mathbb{H} \cong_k \mathrm{M}_2((1, -1)_{\mathbb{R}}) \cong_k \mathrm{M}_2(\mathrm{M}_2(\mathbb{R})) \cong_k \mathrm{M}_4(\mathbb{R}).$$

Therefore, we get

$$2[\mathbb{H}] = [\mathbb{H} \otimes_{\mathbb{R}} \mathbb{H}] = [\mathrm{M}_4(\mathbb{R})] = [\mathbb{R}] = 0 \in \mathrm{Br}(\mathbb{R}).$$

Consequently, $[\mathbb{H}]$ has order 2 in $\mathrm{Br}(\mathbb{R})$. □

EXAMPLE V.1.5. Let k be a finite field. Then $\mathrm{Br}(k) = 0$.

Indeed, let $[A] \in \mathrm{Br}(k)$, and let D be a central division k-algebra such that $[A] = [D]$ (which exists by Lemma V.1.1 (5)). Then D is a finite division ring, hence is commutative by Theorem IV.1.11. Since D is central, it implies that $D = k$. Thus $[A] = 0$. □

Lemma V.1.1 (4) and the properties of tensor product with respect to scalar extension imply immediately the following lemma:

LEMMA V.1.6. *For any field extension L/k, the map*

$$\mathrm{Res}_{L/k} \colon \begin{array}{c} \mathrm{Br}(k) \longrightarrow \mathrm{Br}(L) \\ [A] \longmapsto [A \otimes_k L] \end{array}$$

is a well-defined group morphism. Moreover, we have $\mathrm{Res}_{k/k} = \mathrm{Id}_k$, and if $k \subset L \subset E$, we have

$$\mathrm{Res}_{E/k} = \mathrm{Res}_{E/L} \circ \mathrm{Res}_{L/k}.$$

DEFINITION V.1.7. The map $\mathrm{Res}_{L/k}$ is called the **restriction map** from k to L.

DEFINITION V.1.8. The kernel of the restriction map $\mathrm{Res}_{L/k}$ is called **the relative Brauer group** of L/k, and is denoted by $\mathrm{Br}(L/k)$. By definition, $[A] \in \mathrm{Br}(L/k)$ if and only if A is split by L.

V.2. Brauer equivalence and bimodules

In this section, we give an interpretation of Brauer equivalence in terms of bimodules. As an application, given a central simple k-algebra A of degree n, we will construct a central simple k-algebra of degree $\binom{n}{r}$ which is Brauer equivalent to $A^{\otimes r}$, for all $r = 1, \ldots, n$. These algebras will be proven useful when studying the index of powers of A. We first introduce the notion of a bimodule.

DEFINITION V.2.1. If R, S are two rings, an $R - S$-**bimodule** is an abelian group $(M, +)$ which has the structure of a left R-module

$$R \times M \longrightarrow M$$
$$(a, x) \longmapsto a \bullet x$$

and the structure of a right S-module

$$M \times S \longrightarrow M$$
$$(x, b) \longmapsto x * b$$

such that

$$(a \bullet x) * b = a \bullet (x * b) \text{ for all } a \in R, b \in S, x \in M.$$

EXAMPLE V.2.2. If M is a right S-module, then M carries a natural structure of an $\mathrm{End}_S(M) - S$-bimodule. Indeed, the external law

$$\mathrm{End}_S(M) \times M \longrightarrow M$$
$$(f, x) \longmapsto f(x)$$

endows M with a structure of a left $\mathrm{End}_S(M)$-module. Moreover, for all $f \in \mathrm{End}_S(M), x \in M$ and $b \in S$, we have

$$(f \bullet x) * b = f(x) * b = f(x * b) = f \bullet (x * b),$$

since f is S-linear.   □

We may now state and prove the key result of this section, which shows in particular that the converse of Lemma III.4.6 is true.

PROPOSITION V.2.3. *Let A, B be two central simple k-algebras. Then the following properties are equivalent:*

(1) *A and B are Brauer equivalent;*

(2) *there exists a $B - A$-bimodule M of dimension $\deg(A)\deg(B)$ over k. In this case, $B \cong_k \operatorname{End}_A(M)$;*

(3) *there exists a finitely generated non-zero right A-module M, with $B \cong_k \operatorname{End}_A(M)$.*

If we assume furthermore that $\deg(B) \leq \deg(A)$, then the conditions above are also equivalent to:

(4) *there exists a non-zero idempotent $e \in A$ such that $B \cong_k eAe$. In this case, $B \cong_k \operatorname{End}_A(eA)$, where eA is considered as an A-module.*

Proof.

Let us prove first the equivalence of $(1), (2)$ and (3).

$(1) \Rightarrow (2)$ Assume that A and B are Brauer equivalent. Therefore, we have two isomorphisms of k-algebras

$$f : A \xrightarrow{\sim} \mathrm{M}_r(D) \text{ and } g : B \xrightarrow{\sim} \mathrm{M}_s(D),$$

where D is a central division k-algebra. Let $M = \mathrm{M}_{s,r}(D)$ be the abelian group of $s \times r$ matrices with entries in D. One may check that the external products

$$\begin{array}{cc} M \times A \longrightarrow M & B \times M \longrightarrow M \\ (T,a) \longmapsto T * a = Tf(a) & (b,T) \longmapsto b \bullet T = g(b)T \end{array} \quad \text{and}$$

endow M with a structure of a $B - A$-bimodule. Now for all $b \in B$, the map

$$\ell_b : \begin{array}{c} M \longrightarrow M \\ x \longmapsto b \bullet x \end{array}$$

is an endomorphism of the right A-module M.

Indeed, for all $a \in A, T \in M$ we have

$$b \bullet (T * a) = (b \bullet T) * a$$

since M is a $B - A$-bimodule, that is

$$\ell_b(T * a) = \ell_b(T) * a.$$

Hence, we get a map

$$\varphi : \begin{array}{c} B \longrightarrow \operatorname{End}_A(M) \\ b \longmapsto \ell_b \end{array}$$

which is easily seen to be a k-algebra morphism. By Lemma III.4.6, $\operatorname{End}_A(M)$ is a central simple k-algebra and we have

$$\deg(\operatorname{End}_A(M)) = \frac{\dim_k(M)}{\deg(A)} = \frac{sr\deg(D)^2}{r\deg(D)} = s\deg(D) = \deg(B).$$

Thus B and $\operatorname{End}_A(M)$ have the same dimension over k, and since B is simple, it follows from Lemma I.2.2 that φ is an isomorphism.

$(2) \Rightarrow (3)$ Any $B - A$-bimodule is a right A-module by definition.

$(3) \Rightarrow (1)$ This comes from Lemma III.4.6.

Assume now that $\deg(B) \leq \deg(A)$. We now prove that (1) and (4) are equivalent. If $B \cong_k eAe$ for some non-zero idempotent $e \in A$, then $B \cong_k \operatorname{End}_A(eA)$ by Lemma III.2.5 (the k-linearity being left to the reader). Therefore it implies that A and B are Brauer equivalent by the previous points. Conversely, assume that A and B are Brauer equivalent. By definition, we have two isomorphisms of k-algebras

$$f : A \xrightarrow{\sim} \mathrm{M}_r(D) \text{ and } g : B \xrightarrow{\sim} \mathrm{M}_s(D),$$

where D is a central division k-algebra. Since $\deg(B) \leq \deg(A)$, we have $s \leq r$. Then the matrix

$$E = \begin{pmatrix} I_s & 0 \\ 0 & 0 \end{pmatrix} \in \mathrm{M}_r(D)$$

is a non-zero idempotent of $\mathrm{M}_r(D)$ satisfying

$$E\mathrm{M}_r(D)E = \begin{pmatrix} \mathrm{M}_s(D) & 0 \\ 0 & 0 \end{pmatrix} \cong_k \mathrm{M}_s(D).$$

Let $e = f^{-1}(E) \in A$. Then e is a non-zero idempotent, and we have

$$eAe \cong_k f(eAe) = E\mathrm{M}_r(D)E \cong_k \mathrm{M}_s(D) \cong_k B.$$

This concludes the proof.   □

REMARK V.2.4. The correspondence between Brauer equivalence and bimodules can be made a bit more precise. The reader may refer to [**29**], section 1.B for more details. Notice however that in [**29**] the authors are dealing with left A-modules rather than right A-modules.   □

COROLLARY V.2.5. *Let A be a central simple k-algebra. Then A is split if and only if A has a right ideal of dimension $\deg(A)$ over k.*

Proof.   Let $n = \deg(A)$. Assume first that A is a split k-algebra, and let $f : A \xrightarrow{\sim} \mathrm{M}_n(k)$ be an isomorphism of k-algebras. The set $\mathcal{L}_n$ of matrices whose $n-1$ first columns are zero is a right ideal of $\mathrm{M}_n(k)$ of dimension n by Example III.2.8 (2), and thus $f^{-1}(\mathcal{L}_n)$ is a right ideal of A of dimension n. Conversely, assume that I is a right ideal of A of dimension n. One may verify that the external product

$$k \times I \longrightarrow I$$
$$(\lambda, x) \longmapsto \lambda x$$

endows I with the structure of a left k-vector space which is compatible with its natural right A-module structure, or in other words that I is a $k - A$-bimodule of dimension n (this follows from the fact that k commutes with the elements of A). By the previous proposition, A is Brauer equivalent to k. In other words, $A \cong_k \mathrm{M}_n(k)$. This concludes the proof.   □

Let A be a central simple k-algebra of degree n, and let $1 \leq r \leq n$ be an integer. In the sequel, we will denote by $A^{\otimes r}$ the tensor product of r copies of A. To end this section, we would like to construct a central simple k-algebra $\lambda^r A$ of degree $\begin{pmatrix} n \\ r \end{pmatrix}$, which is Brauer equivalent to $A^{\otimes r}$. The existence of such an algebra will be crucial for the study of the structure of the Brauer group. The previous proposition shows that we necessarily have

$$\lambda^r A = \operatorname{End}_{A^{\otimes r}}(M),$$

where M is a right $A^{\otimes r}$-module. Since the simplest examples of right $A^{\otimes r}$-modules are the right ideals of $A^{\otimes r}$, we are going to look for $\lambda^r A$ under the form

$$\operatorname{End}_{A^{\otimes r}}(s_{r,A} A^{\otimes r}) \text{ for some } s_{r,A} \in A^{\otimes r},$$

since any right ideal of a central simple k-algebra is principal (see Chapter III, Exercise 3).

The construction of $s_{r,A}$ will be quite involved, and will take the rest of this section. Let us summarize the main steps of the construction. First, we will introduce the so-called Goldman element of A and study its main properties. This element will allow us to construct a canonical group morphism

$$\gamma_{r,A} : \mathfrak{S}_r \longrightarrow (A^{\otimes r})^\times,$$

where $\mathfrak{S}_r$ is the symmetric group on r letters.

The element $s_{r,A} \in A^{\otimes r}$ will be then given by the formula

$$s_{r,A} = \sum_{\sigma \in \mathfrak{S}_r} \varepsilon(\sigma) \gamma_{r,A}(\sigma) \in A^{\otimes r}.$$

The last step will be to check that $\operatorname{End}_{A^{\otimes r}}(s_{r,A} A^{\otimes r})$ has degree $\begin{pmatrix} n \\ r \end{pmatrix}$, that we will do by reducing to the split case (that is, the case where A is a split central simple algebra). Before starting the construction by itself, we need a lemma.

LEMMA V.2.6. *Let V be a finite dimensional k-vector space, and let $r \geq 1$. Then we have a unique isomorphism of k-algebras*

$$\eta_r : \operatorname{End}_k(V)^{\otimes r} \xrightarrow{\sim} \operatorname{End}_k(V^{\otimes r})$$

satisfying

$$\eta_r(f_1 \otimes \cdots \otimes f_r) = f_1 \otimes \cdots \otimes f_r \text{ for all } f_i \in \operatorname{End}_k(V),$$

where the map $f_1 \cdots \otimes \cdots f_r \in \operatorname{End}_k(V^{\otimes r})$ is defined as in Lemma A.2.1. In other words, for all $f_1, \ldots, f_r \in End_k(V)$, and all $v_1, \ldots, v_r \in V$, we have

$$\eta_r(f_1 \otimes \cdots \otimes f_r)(v_1 \otimes \cdots \otimes v_r) = f_1(v_1) \otimes \cdots \otimes f_r(v_r).$$

Proof. For $i = 1, \ldots, r$, the maps

$$\varphi_i : \begin{array}{l} \operatorname{End}_k(V) \longrightarrow \operatorname{End}_k(V^{\otimes r}) \\ f_i \longmapsto \operatorname{Id}_V \cdots \otimes \cdots \operatorname{Id}_V \otimes f_i \otimes \cdots \otimes \operatorname{Id}_V \end{array}$$

are clearly k-algebra morphisms with commuting images (here, once again, $\operatorname{Id}_V \cdots \otimes \cdots \operatorname{Id}_V \otimes f_i \otimes \cdots \otimes \operatorname{Id}_V$ is the endomorphism of $V^{\otimes r}$ defined as in Lemma A.2.1), so they induce a map

$$\eta_r : \operatorname{End}_k(V)^{\otimes r} \longrightarrow \operatorname{End}_k(V^{\otimes r})$$

satisfying

$$\eta_r(\operatorname{Id}_V \otimes \cdots \operatorname{Id}_V \otimes f_i \otimes \cdots \otimes \operatorname{Id}_V) = \operatorname{Id}_V \otimes \cdots \operatorname{Id}_V \otimes f_i \otimes \cdots \otimes \operatorname{Id}_V,$$

for all $f_i \in \operatorname{End}_k(V)$. Since η is a k-algebra morphism and since we have

$$f_1 \otimes \cdots \otimes f_r = (f_1 \otimes \operatorname{Id}_V \otimes \cdots \otimes \operatorname{Id}_V) \cdots (\operatorname{Id}_V \otimes \cdots \otimes \operatorname{Id}_V \otimes f_r),$$

it follows that η_r has the desired property. The uniqueness of η_r follows from the fact that elementary tensors span the tensor product. By Lemma I.2.2, η is an isomorphism, since $\mathrm{End}_k(V)^{\otimes r}$ is simple, and since $\mathrm{End}_k(V^{\otimes r})$ and $\mathrm{End}_k(V)^{\otimes r}$ have the same dimension over k.   □

We now define the Goldman element.

Let $u : A \longrightarrow A^{op}$ be the isomorphism of k-vector spaces which maps a onto a^{op}. The k-linear map

$$\mathrm{Sand}_A \circ (\mathrm{Id}_A \otimes u) : A^{\otimes 2} \longrightarrow \mathrm{End}_k(A)$$

is an isomorphism of k-vector spaces, which maps an elementary tensor $a \otimes a'$ onto the endomorphism

$$A \longrightarrow A$$

$$z \longmapsto aza'.$$

Since we can consider the k-linear map $\mathrm{Trd}_A : A \longrightarrow k$ as an endomorphism of A (since $k \subset A$), the following definition makes sense.

DEFINITION V.2.7. The **Goldman element** of A is the unique element $g_A \in A^{\otimes 2}$ which is mapped onto Trd_A via the previous isomorphism. In other words, g_A is the unique element $g_A = \sum_{i=1}^{m} a_i \otimes a_i' \in A^{\otimes 2}$ satisfying

$$\sum_{i=1}^{m} a_i a a_i' = \mathrm{Trd}_A(a) \text{ for all } a \in A.$$

The Goldman element will play a crucial role in our construction. We now study its basic properties.

LEMMA V.2.8. Let A be a central simple k-algebra and let $g_A \in A^{\otimes 2}$ be the Goldman element of A. Then the following properties hold:

(1) if $A = \mathrm{End}_k(V)$, let $(e_1, \ldots, e_n)$ be a k-basis of V and let e_{ij} be the endomorphism of V defined by

$$e_{ij}(e_s) = \delta_{sj} e_i \text{ for } s = 1, \ldots, n.$$

Then $g_A = \sum_{i,j} e_{ij} \otimes e_{ji}$. Moreover, we have

$$\eta_2(g_A)(v \otimes v') = v' \otimes v \text{ for all } v, v' \in V;$$

(2) let L/k be an arbitrary field extension. Then $g_{A \otimes_k L} = \rho_2(g_A \otimes 1)$, where

$$\rho_2 : (A^{\otimes 2}) \otimes_k L \xrightarrow{\sim} (A \otimes_k L)^{\otimes 2}$$

is the canonical isomorphism of L-algebras;

(3) if $\varphi : A \xrightarrow{\sim} A'$ is an isomorphism of k-algebras, then we have $g_{A'} = (\varphi \otimes \varphi)(g_A)$;

(4) we have $g_A^2 = 1 \in A^{\otimes 2}$. In particular, $g_A \in A^{\times}$.

Proof.

(1) This is an easy computation. Details are left to the reader.

(2) Write $g_A = \sum_{i=1}^{m} a_i \otimes a_i'$. By definition of ρ_2, we have

$$\rho_2(g_A \otimes 1) = \sum_{i=1}^{m} (a_i \otimes 1) \otimes (a_i' \otimes 1).$$

By definition of the Goldman element, we have for all $a \in A$

$$\sum_{i=1}^{m} (a_i \otimes 1)(a \otimes 1)(a_i' \otimes 1) = \left(\sum_{i=1}^{m} a_i a a_i'\right) \otimes 1 = \mathrm{Trd}_A(a).$$

By Lemma IV.2.10 (4), we get

$$\sum_{i=1}^{m} (a_i \otimes 1)(a \otimes 1)(a_i' \otimes 1) = \mathrm{Trd}_{A \otimes_k L}(a \otimes 1) \text{ for all } a \in A.$$

Since $A \otimes_k L$ is spanned as an L-vector space by the elements of the form $a \otimes 1, a \in A$, we deduce that

$$\sum_{i=1}^{m} (a_i \otimes 1) z (a_i' \otimes 1) = \mathrm{Trd}_{A_L}(z) \text{ for all } z \in A \otimes_k L,$$

meaning that $\rho_2(g_A \otimes 1)$ is the Goldman element of $A \otimes_k L$.

(3) If $g_A = \sum_{i=1}^{m} a_i \otimes a_i'$, we have

$$(\varphi \otimes \varphi)(g_A) = \sum_{i=1}^{m} \varphi(a_i) \otimes \varphi(a_i').$$

For all $a \in A$, we have

$$\sum_{i=1}^{m} \varphi(a_i) \varphi(a) \varphi(a_i') = \varphi\left(\sum_{i=1}^{m} a_i a a_i'\right) = \varphi(\mathrm{Trd}_A(a)) = \mathrm{Trd}_A(a),$$

since φ is k-linear. By Lemma IV.2.10 (5), we get

$$\sum_{i=1}^{m} \varphi(a_i) \varphi(a) \varphi(a_i') = \mathrm{Trd}_{A'}(\varphi(a)) \text{ for all } a \in A.$$

Since φ is surjective, this proves that $(\varphi \otimes \varphi)(g_A)$ is the Goldman element of A'.

(4) Let $f : A \otimes_k L \xrightarrow{\sim} \mathrm{End}_L(V)$ be an isomorphism of L-algebras, where V is an L-vector space. Since the map

$$A^{\otimes 2} \longrightarrow A^{\otimes 2} \otimes_k L$$
$$a \longmapsto a \otimes 1$$

is injective, it is enough to prove that $g_A^2 \otimes 1 = 1$. Since f is an isomorphism of L-algebras, so is $(f \otimes f) \circ \rho_2$ (where ρ_2 is the map described in (2)). Hence this is equivalent to proving the equality

$$((f \otimes f) \circ \rho_2)(g_A^2 \otimes 1) = ((f \otimes f) \circ \rho_2)(g_A \otimes 1)^2 = 1.$$

By (2) and (3), $((f \otimes f) \circ \rho_2)(g_A \otimes 1)$ is the Goldman element of $\mathrm{End}_L(V)$, it remains to prove that the equality holds in the split case, which is easy to check using (1). $\qquad \square$

We are now ready to state and prove the next proposition.

PROPOSITION V.2.9. *Let A be a central simple k-algebra. For $r \geq 1$, there is a canonical group morphism*

$$\gamma_{r,A} : \mathfrak{S}_r \longrightarrow (A^{\otimes r})^\times$$

such that in the special case $A = \mathrm{End}_k(V)$, we have

$$\eta_r(\gamma_{r,\mathrm{End}_k(V)}(\sigma))(v_1 \otimes \cdots \otimes v_r) = v_{\sigma^{-1}(1)} \otimes \cdots \otimes v_{\sigma^{-1}(r)},$$

for all $\sigma \in \mathfrak{S}_r$ and all $v_i \in V$. Moreover, the following properties hold:

(1) *for every field extension L/k and every $\sigma \in \mathfrak{S}_r$, we have*

$$\gamma_{r,A\otimes_k L}(\sigma) = \rho_r(\gamma_{r,A}(\sigma) \otimes 1),$$

where

$$\rho_r : (A^{\otimes r}) \otimes_k L \xrightarrow{\sim} (A \otimes_k L)^{\otimes r}$$

is the canonical isomorphism;

(2) *for every k-algebra isomorphism $\varphi : A \xrightarrow{\sim} A'$, and every $\sigma \in \mathfrak{S}_r$, we have*

$$\gamma_{r,A'}(\sigma) = \varphi^{\otimes r}(\gamma_{r,A}(\sigma)).$$

Proof. Let g_A be the Goldman element of A. For $i = 1, \ldots, r-1$, let τ_i be the transposition $(i\, i+1)$, and set

$$\gamma_{r,A}(\tau_i) = 1 \otimes \cdots \otimes 1 \otimes g_A \otimes 1 \otimes \cdots \otimes 1 \text{ for all } i = 1, \ldots, r-1.$$

Since the transpositions $\tau_i, i = 1, \ldots, r-1$ generate the group $\mathfrak{S}_r$, we may choose a decomposition $\sigma = \tau_{i_1} \cdots \tau_{i_s}$ and set

$$\gamma_{r,A}(\sigma) = \gamma_{r,A}(\tau_{i_1}) \cdots \gamma_{r,A}(\tau_{i_s}).$$

We have to prove that $\gamma_{r,A}(\sigma)$ does not depends on the choice of this decomposition. Once this is done, it will follows easily that $\gamma_{r,A}$ is a group morphism. The fact that $\gamma_{r,A}(\sigma)$ is invertible will then be immediate from the definition since the Goldman element is invertible.

We start with the split case. Assume that $A = \mathrm{End}_k(V)$, for some k-vector space V of dimension n over k. By Lemma A.2.2, for every $\sigma \in \mathfrak{S}_r$, there exists a unique k-linear map

$$f_\sigma : V^{\otimes r} \longrightarrow V^{\otimes r}$$

satisfying

$$f_\sigma(v_1 \otimes \cdots \otimes v_r) = v_{\sigma^{-1}(1)} \otimes \cdots \otimes v_{\sigma^{-1}(r)} \text{ for all } v_i \in V.$$

It is not difficult to see that we have

$$f_{\sigma\tau} = f_\sigma \circ f_\tau \text{ for all } \sigma, \tau \in \mathfrak{S}_r.$$

The properties of the Goldman element then show that the equality

$$\eta_r(\gamma_{r,A}(\tau_i))(v_1 \otimes \cdots \otimes v_r) = v_{\tau_i^{-1}(1)} \otimes \cdots \otimes v_{\tau_i^{-1}(r)}$$

holds for all $i = 1, \ldots, r-1$ and all $v_1, \ldots, v_r \in V$. Since elementary tensors span the tensor product, we get

$$\eta_r(\gamma_{r,A}(\tau_i)) = f_{\tau_i} \text{ for all } i = 1, \ldots, r-1.$$

Now if we choose a decomposition $\sigma = \tau_{i_1} \cdots \tau_{i_s} \in \mathfrak{S}_r$, we have

$$
\begin{aligned}
\eta_r(\gamma_{r,A}(\sigma)) &= \eta_r(\gamma_{r,A}(\tau_{i_1}) \cdots \gamma_{r,A}(\tau_{i_s})) \\
&= \eta_r(\gamma_{r,A}(\tau_{i_1})) \circ \cdots \circ \eta_r(\gamma_{r,A}(\tau_{i_s})) \\
&= f_{\tau_{i_1}} \circ \cdots \circ f_{\tau_{i_s}} \\
&= f_{\tau_{i_1} \cdots \tau_{i_s}}.
\end{aligned}
$$

Therefore, we have $\eta_r(\gamma_{r,A}(\sigma)) = f_\sigma$ for all $\sigma \in \mathfrak{S}_r$. Now since η_r is an isomorphism, we get

$$
\gamma_{r,A}(\sigma) = \eta_r^{-1} \circ f_\sigma \text{ for all } \sigma \in \mathfrak{S}_r.
$$

The map

$$
\mathfrak{S}_r \longrightarrow \operatorname{End}_k(V^{\otimes r})
$$
$$
\sigma \longmapsto f_\sigma
$$

is a well-defined group morphism, and the same is true for the map η_r^{-1}. It follows that $\gamma_{r,A}$ is well-defined in the split case. Applying this equality above to elementary tensors then gives the first part of the proposition.

Assume now that A is an arbitrary central simple k-algebra. We are going to show that $\gamma_{r,A}$ is well-defined by reducing to the split case. For, let L be a splitting field of A, and let

$$
f : A \otimes_k L \xrightarrow{\sim} \operatorname{End}_L(V)
$$

be an isomorphism of L-algebras, where V is an L-vector space. The image of $\gamma_{r,A}(\tau_i) \otimes 1 \in A^{\otimes r} \otimes_k L$ under the isomorphism

$$
f^{\otimes r} \circ \rho_r : A^{\otimes r} \otimes_k L \xrightarrow{\sim} \operatorname{End}_L(V)^{\otimes r}
$$

is $\gamma_{r,\operatorname{End}_L(V)}(\tau_i)$, as we can see by applying Lemma V.2.8. Since $\gamma_{r,\operatorname{End}_L(V)}$ is well-defined, it follows from the previous observation that the map $\gamma_{r,A} \otimes \operatorname{Id}_L$ is a well-defined group morphism. Therefore, the same is true for $\gamma_{r,A}$.

To prove properties (1) and (2), observe that they are true for the transpositions τ_i in view of the properties of the Goldman element described in Lemma V.2.8. Since the transpositions $\tau_i, i = 1, \ldots, r-1$ generate $\mathfrak{S}_r$, these two properties then hold for every permutation. This concludes the proof. $\qquad\square$

We are now ready to define the element $s_{r,A}$. For $2 \leq r \leq \deg(A)$, we set

$$
s_{r,A} = \sum_{\sigma \in \mathfrak{S}_r} \varepsilon(\sigma) \gamma_{r,A}(\sigma) \in A^{\otimes r},
$$

where $\varepsilon(\sigma)$ denotes the sign of the permutation σ. We also set $s_{1,A} = 1$.

We then have the result we are looking for.

PROPOSITION V.2.10. *Let A be a central simple k-algebra of degree n. For all $r = 1, \ldots, n$, the k-algebra $\lambda^r A = \operatorname{End}_{A^{\otimes r}}(s_{r,A} A^{\otimes r})$ is a central simple k-algebra of degree $\binom{n}{r}$ which is Brauer equivalent to $A^{\otimes r}$.*

Proof. If $r = 1$, we have $\operatorname{End}_A(A) \cong_k A$ by Lemma III.2.5, so there is nothing to do. Assume now that $r \geq 2$. By Proposition V.2.3 (1), $\operatorname{End}_{A^{\otimes r}}(s_{r,A} A^{\otimes r})$ is a central simple k-algebra Brauer equivalent to $A^{\otimes r}$ of degree $\dfrac{\dim_k(s_{r,A} A^{\otimes r})}{n^r}$.

Let $f : A \otimes_k L \xrightarrow{\sim} \mathrm{End}_k(V)$ be an isomorphism of L-algebras, where V is an L-vector space. From the properties of $\gamma_{r,A}$ and the definition of $s_{r,A}$, we get an isomorphism of L-vector spaces

$$s_{r,A} A^{\otimes r} \otimes_k L \cong_k s_{r,\mathrm{End}_L(V)} \mathrm{End}_L(V)^{\otimes r}.$$

Since $\dim_L(s_{r,A} A^{\otimes r} \otimes_k L) = \dim_k(s_{r,A} A^{\otimes r})$, it is enough to compute this dimension in the split case.

Assume that $A = \mathrm{End}_k(V)$, for some k-vector space V. Applying the canonical isomorphism

$$\eta_r : \mathrm{End}_k(V)^{\otimes r} \xrightarrow{\sim} \mathrm{End}_k(V^{\otimes r}),$$

we see that it is enough to compute $\dim_k(u_r \mathrm{End}_k(V^{\otimes r}))$, where

$$u_r = \sum_{\sigma \in \mathfrak{S}_r} \varepsilon(\sigma) f_\sigma.$$

Notice that we have an isomorphism of k-vector spaces

$$u_r \mathrm{End}_k(V^{\otimes r}) \cong_k \mathrm{Hom}_k(V^{\otimes r}, \mathrm{Im}(u_r)).$$

Indeed, we have a canonical injective k-linear map

$$\Theta : u_r \mathrm{End}_k(V^{\otimes r}) \longrightarrow \mathrm{Hom}_k(V^{\otimes r}, \mathrm{Im}(u_r)),$$

obtained by considering an element of $u_r \mathrm{End}_k(V^{\otimes r})$ as a k-linear map with values in $\mathrm{Im}(u_r)$. To prove the surjectivity, let us fix a basis $(z_1, \ldots, z_m)$ of $V^{\otimes r}$ and let $g \in \mathrm{Hom}_k(V^{\otimes r}, \mathrm{Im}(u_r))$. By assumption, $g(z_j) = u_r(y_j)$, for some $y_j \in V^{\otimes r}$. Now let $f \in \mathrm{End}_k(V^{\otimes r})$ be the unique endomorphism defined by

$$f(z_j) = y_j \text{ for } j = 1, \ldots, m.$$

Then $g = \Theta(u_r \circ f)$ and thus Θ is an isomorphism.

It follows from Lemma III.4.6 that we have $\deg(\lambda^r A) = \dim_k(\mathrm{Im}(u_r))$. We are going to show that $\dim_k(\mathrm{Im}(u_r)) = \binom{n}{r}$.

Let $(e_1, \ldots, e_n)$ be a k-basis of V. We claim that the elements

$$u_r(e_{i_1} \otimes \cdots \otimes e_{i_r}), 1 \leq i_1 < \cdots < i_r \leq n$$

form a k-basis of $\mathrm{Im}(u_r)$. They are k-linearly independent since they involve pairwise disjoint sets of basis elements of $V^{\otimes r}$. We now proceed to show that these elements span $\mathrm{Im}(u_r)$. Notice that for all $\tau \in \mathfrak{S}_r$, we have

$$u_r \circ f_\tau = \sum_{\sigma \in \mathfrak{S}_r} \varepsilon(\sigma) f_\sigma \circ f_\tau = \sum_{\sigma \in \mathfrak{S}_r} \varepsilon(\sigma) f_{\sigma\tau} = \sum_{\sigma' \in \mathfrak{S}_r} \varepsilon(\sigma'\tau^{-1}) f_{\sigma'} = \varepsilon(\tau) u_r.$$

This shows in particular that $\mathrm{Im}(u_r)$ is generated by the elements

$$u_r(e_{j_1} \otimes \cdots \otimes e_{j_r}), 1 \leq j_1 \leq \cdots \leq j_r \leq n.$$

To conclude, it is enough to show that $u_r(v_1 \otimes \cdots \otimes v_r) = 0$ as soon as $v_i = v_j$ for some $i \neq j$. Assume that $v_i = v_j$ and let $\tau = (i\, j)$. Then we have

$$\mathfrak{S}_r = \mathfrak{A}_r \cup \mathfrak{A}_r \tau,$$

where $\mathfrak{A}_r$ denotes the alternating group on r letters, and therefore

$$u_r(v_1 \otimes \cdots \otimes v_r) = \sum_{\sigma \in \mathfrak{A}_r} f_\sigma(v_1 \otimes \cdots \otimes v_r) - \sum_{\sigma \in \mathfrak{A}_r} f_\sigma(f_\tau(v_1 \otimes \cdots \otimes v_r)).$$

Since $v_i = v_j$ and $\tau = (i\,j)$, we have $f_\tau(v_1 \otimes \cdots \otimes v_r) = v_1 \otimes \cdots \otimes v_r$, and therefore we get $u_r(v_1 \otimes \cdots \otimes v_r) = 0$.

This shows that the elements

$$u_r(e_{i_1} \otimes \cdots \otimes e_{i_r}), 1 \leq i_1 < \cdots < i_r \leq n$$

span $\text{Im}(u_r)$. Hence $\dim_k(\text{Im}(u_r)) = \binom{n}{r}$, and we are done. $\qquad\square$

V.3. Index and exponent

In this section, we prove that every element $[A] \in \text{Br}(k)$ has finite order, and relate this order to the index of A. We start with the following result.

THEOREM V.3.1. *Let A, B be two central simple k-algebras, and let L/k be a field extension. Then the following properties hold:*

(1) *if $[A] = [B]$ (that is if A and B are Brauer equivalent), then $\text{ind}(A) = \text{ind}(B)$;*

(2) *$\text{ind}(A) \mid \deg(A)$ and equality holds if and only if A is a division algebra;*

(3) *$\text{ind}(A) = gcd\{[K : k] \mid K$ splitting field of A, $[K : k] < +\infty$ $\}$;*

(4) *$\text{ind}(A \otimes_k L) \mid \text{ind}(A)$;*

(5) *if L/k has finite degree, we have $\text{ind}(A) \mid [L : k]\text{ind}(A \otimes_k L)$;*

(6) *if $[L : k]$ is prime to $\text{ind}(A)$, then $\text{ind}(A \otimes_k L) = \text{ind}(A)$. In particular, if A is a division algebra, so is $A \otimes_k L$;*

(7) *$\text{ind}(A \otimes_k B) \mid \text{ind}(A)\text{ind}(B)$;*

(8) *if $gcd(\text{ind}(A), \text{ind}(B)) = 1$, then $\text{ind}(A \otimes_k B) = \text{ind}(A)\text{ind}(B)$. In particular, if A and B are division algebras of coprime degrees, then $A \otimes_k B$ is a division algebra;*

(9) *we have $\text{ind}(A)[A] = 0 \in \text{Br}(k)$, and thus $\deg(A)[A] = 0 \in \text{Br}(k)$. In particular, $\text{Br}(k)$ is a torsion group;*

(10) *for all $r \geq 1$, $\text{ind}(A^{\otimes r}) \mid \text{ind}(A)$;*

(11) *if $gcd(r, \text{ind}(A)) = 1$, $\text{ind}(A^{\otimes r}) = \text{ind}(A)$;*

(12) *for all $r \mid \text{ind}(A)$, $\text{ind}(A^{\otimes r}) \mid \dfrac{\text{ind}(A)}{r}$.*

Proof.

(1) This follows from the definition of Brauer equivalence.

(2) This follows from the definition of the index.

(3) This follows from Corollary IV.1.9 and Corollary IV.1.15.

(4) Write $A \cong_k \text{M}_r(D)$ and $D \otimes_k L \cong_k \text{M}_s(D')$, where D is a central division k-algebra and D' is a central division L-algebra. Then we have

$$A \otimes_k L \cong_L \text{M}_r(D \otimes_k L) \cong_L \text{M}_{rs}(D'),$$

so $\text{ind}(A \otimes_k L) = \deg(D')$. By definition of D', $\deg(D')$ divides the degree of the L-algebra $D \otimes_k L$, which is nothing but the degree of the k-algebra D. Hence, we have $\deg(D') \mid \deg(D)$, that is

$$\text{ind}(A \otimes_k L) \mid \text{ind}(A).$$

(5) By Corollary IV.1.9, the L-algebra $A \otimes_k L$ has a splitting field L' of degree $\mathrm{ind}(A \otimes_k L)$ over L. Since $A \otimes_k L' \cong_{L'} (A \otimes_k L) \otimes_L L'$ and $A \otimes_k L$ is split by L', it follows that L' is a splitting field of A, so $\mathrm{ind}(A) \mid [L' : k]$ by Corollary IV.1.15. Since we have

$$[L' : k] = [L : k][L' : L] = [L : k]\mathrm{ind}(A \otimes_k L),$$

the result follows.

(6) It follows from the two previous points that A and $A \otimes_k L$ have the same index if $[L : k]$ is prime to $\mathrm{ind}(A)$. Now if A is a division algebra, we have $\mathrm{ind}(A) = \deg(A)$. Now we conclude using the fact that $A \otimes_k L$ has degree $\deg(A)$ over L.

(7) By Lemma V.1.1 (5), there exist central division k-algebras D and D' such that $A \sim_k D$ and $B \sim_k D'$. By point (3) of the same lemma, we have $A \otimes_k B \sim_k D \otimes_k D'$. By (1), we then have

$$\mathrm{ind}(A \otimes_k B) = \mathrm{ind}(D \otimes_k D').$$

By (2), we then get $\mathrm{ind}(A \otimes_k B) \mid \deg(D \otimes_k D')$. Now we have

$$\deg(D \otimes_k D') = \deg(D)\deg(D') = \mathrm{ind}(A)\mathrm{ind}(B),$$

the last equality following from the definition of the index. This concludes the proof.

(8) Let L be a splitting field of A of degree $\mathrm{ind}(A)$ over k. Then we have

$$\mathrm{ind}((A \otimes_k B) \otimes_k L) \mid \mathrm{ind}(A \otimes_k B)$$

by (4). But we have

$$\mathrm{Res}_{L/k}([A \otimes_k B]) = \mathrm{Res}_{L/k}([A]) + \mathrm{Res}_{L/k}([B]) = \mathrm{Res}_{L/k}([B]),$$

and therefore $\mathrm{ind}((A \otimes_k B) \otimes_k L) = \mathrm{ind}(B \otimes_k L)$ by (1).

Since $[L : k] = \mathrm{ind}(A)$ is prime to $\mathrm{ind}(B)$, we get

$$\mathrm{ind}((A \otimes_k B) \otimes_k L) = \mathrm{ind}((A \otimes_k (B \otimes_k L)) = \mathrm{ind}(B)$$

by (6), and therefore $\mathrm{ind}(B) \mid \mathrm{ind}(A \otimes_k B)$. By permuting the roles of A and B, we see that $\mathrm{ind}(A) \mid \mathrm{ind}(A \otimes_k B)$, and since $\mathrm{ind}(A)$ and $\mathrm{ind}(B)$ are coprime, we get

$$\mathrm{ind}(A)\mathrm{ind}(B) \mid \mathrm{ind}(A \otimes_k B).$$

Using (7), we then get the desired equality. If now A and B are division algebras we get

$$\mathrm{ind}(A \otimes_k B) = \mathrm{ind}(A)\mathrm{ind}(B) = \deg(A)\deg(B) = \deg(A \otimes_k B),$$

so $A \otimes_k B$ is also a division algebra.

(9) Let A be a central simple k-algebra, and write $A \cong_k \mathrm{M}_r(D)$, where D is a central division k-algebra.

By definition, we have $\mathrm{ind}(A) = \deg(D)$ and $[A] = [D]$, so we have to prove that $d[D] = 0 \in \mathrm{Br}(k)$, where $d = \deg(D)$. By Proposition V.2.10, $D^{\otimes d}$ is Brauer equivalent to $\lambda^d D$, which is a central simple k-algebra of degree $\binom{d}{d} = 1$. In other words, $D^{\otimes d}$ is Brauer equivalent to k, that is

$$d[D] = [D^{\otimes d}] = [k] = 0 \in \mathrm{Br}(k).$$

(10) By Corollary IV.1.9, A has a splitting field L of degree $\mathrm{ind}(A)$ over k. Hence we get

$$\mathrm{Res}_{L/k}([A^{\otimes r}]) = \mathrm{Res}_{L/k}(r[A]) = r\mathrm{Res}_{L/k}([A]) = r[A \otimes_k L] = 0,$$

so L splits $A^{\otimes r}$ (another way to see this is to use the isomorphism $A^{\otimes r} \otimes_k L \cong_k (A \otimes_k L)^{\otimes r}$). By Corollary IV.1.15, $\mathrm{ind}(A^{\otimes r}) \mid \mathrm{ind}(A)$.

(11) By (9), the order e of $[A]$ in $\mathrm{Br}(k)$ divides $\mathrm{ind}(A)$. By assumption on r, r is then coprime to e. Therefore, $r[A]$ and $[A]$ generate the same subgroup in $\mathrm{Br}(k)$. Hence there exists $s \geq 1$ such that $[A] = s(r[A])$, that is

$$[A] = [(A^{\otimes r})^{\otimes s}].$$

In particular, $\mathrm{ind}((A^{\otimes r})^{\otimes s}) = \mathrm{ind}(A)$, and therefore by (10) applied to the algebra $A^{\otimes r}$, we get

$$\mathrm{ind}(A) \mid \mathrm{ind}(A^{\otimes r}).$$

By (10) again, we have $\mathrm{ind}(A^{\otimes r}) \mid \mathrm{ind}(A)$, hence $\mathrm{ind}(A^{\otimes r}) = \mathrm{ind}(A)$.

(12) If $r = 1$, there is nothing to do, so we may assume that $r \geq 2$. We are going to prove our result by induction on the number of (non necessarily distinct) prime factors of r.

Assume first that $r = p$ is a prime number. Since two Brauer equivalent central simple k-algebras have the same index by (1), we have

$$\mathrm{ind}(A) = \mathrm{ind}(D) \text{ and } \mathrm{ind}(A^{\otimes p}) = \mathrm{ind}(D^{\otimes p}).$$

Thus, one may assume that A is a division k-algebra D without any loss of generality. Let $d = \deg(D) = \mathrm{ind}(D)$. By (10), we have $\mathrm{ind}(D^{\otimes p}) \mid d$.

By Proposition V.2.10, there exists a central simple k-algebra B of degree $\dbinom{d}{p}$ which is Brauer equivalent to $D^{\otimes p}$. Therefore, B and $D^{\otimes p}$ have the same index by (1), and applying (2) to the algebra B then gives

$$\mathrm{ind}(D^{\otimes p}) \mid \binom{d}{p}.$$

Thus we get

$$\mathrm{ind}(D^{\otimes p}) \mid gcd(\binom{d}{p}, d).$$

Write $d = p^m d', p \nmid d'$. Then clearly p^{m-1} is the highest power of p dividing $\dbinom{d}{p}$. Since on the other hand d' also divides $\dbinom{d}{p}$, it follows that we have

$$gcd(\binom{d}{p}, d) = p^{m-1}d' = \frac{d}{p}.$$

This concludes the proof in this case.

Now assume that the result is true for every central simple k-algebra A, and every integer $r \geq 2$ which is the product of t prime factors such that $r \mid \mathrm{ind}(A)$.

Assume now that $r \mid \mathrm{ind}(A)$ is the product of $t+1$ prime factors, and write $r = ps$, where p is prime and s has t prime factors. Assume first that $p \nmid \mathrm{ind}(A^{\otimes s})$. By induction, $\mathrm{ind}(A^{\otimes s}) \mid \dfrac{\mathrm{ind}(A)}{s}$, hence we may write

$$\mathrm{ind}(A) = sq\,\mathrm{ind}(A^{\otimes s}) \text{ for some } q \geq 1.$$

Since $r \mid \mathrm{ind}(A)$, we get that $p \mid q\,\mathrm{ind}(A^{\otimes s})$, and thus $p \mid q$ by assumption on p. It follows that $\mathrm{ind}(A^{\otimes s}) \mid \dfrac{\mathrm{ind}(A)}{r}$. Now by (11), $\mathrm{ind}(A^{\otimes s}) = \mathrm{ind}(A^{\otimes ps}) = \mathrm{ind}(A^{\otimes r})$ since p is prime to $\mathrm{ind}(A^{\otimes s})$, and we get that $\mathrm{ind}(A^{\otimes r}) \mid \dfrac{\mathrm{ind}(A)}{r}$ in this case.

Finally, assume that $p \mid \mathrm{ind}(A^{\otimes s})$. Then by the case $r = p$, we have $\mathrm{ind}(A^{\otimes ps}) \mid \dfrac{\mathrm{ind}(A^{\otimes s})}{p}$. Using the induction hypothesis, we get $\mathrm{ind}(A^{\otimes ps}) \mid \dfrac{\mathrm{ind}(A)}{ps}$, that is $\mathrm{ind}(A^{\otimes r}) \mid \dfrac{\mathrm{ind}(A)}{r}$. This concludes the proof. $\qquad\square$

Using the properties of the index, one may show the following variant of Corollary IV.1.14.

PROPOSITION V.3.2. *Let D be a central division k-algebra, and let K/k be an extension of degree $r \geq 2, r \mid \deg(D)$. If K is isomorphic to a subfield of D, then $D \otimes_k K$ is not a division algebra. Moreover if r is prime, then the converse holds as well.*

Proof. Let $n = \deg(D)$. Notice that n is also the degree of $D \otimes_k K$ over K. Assume that K is isomorphic to a subfield K' of D. Since K and K' are isomorphic k-algebras, we have $D \otimes_k K \cong_k D \otimes_k K'$. Since $D \otimes_k K$ is a central simple K-algebra, we may write $D \otimes_k K \cong_K \mathrm{M}_r(\Delta)$, where Δ is a central division K-algebra, by Wedderburn's Theorem. Similarly, we have $D \otimes_k K' \cong_{K'} \mathrm{M}_r(\Delta')$, where Δ' is a central division K'-algebra. Since $D \otimes_k K \cong_k D \otimes_k K'$, we get that $\Delta \cong_k \Delta'$ by the uniqueness part of Wedderburn's Theorem. Comparing dimensions shows that we have

$$[K : k]\deg(\Delta)^2 = [K' : k]\deg(\Delta')^2,$$

the degree on the left-hand side being taken over K, and the degree on the right-hand side being taken over K'. Since $[K : k] = [K' : k]$, it follows easily that the index of $D \otimes_k K$ over K is equal to the index of $D \otimes_k K'$ over K'.

Therefore, we may assume without any loss of generality that K is a subfield of D. Let L be a maximal subfield of D containing K (see Remark IV.1.10 (4)). Then L splits D by Proposition IV.1.6, and therefore L splits $D \otimes_k K$. Hence we have $\mathrm{ind}(D \otimes_k K) \mid [L : K]$ by Theorem V.3.1 (3). Since $[K : k] = r \geq 2$, we have $[L : K] < [L : k]$, and therefore

$$\mathrm{ind}(D \otimes_k K) < [L : k].$$

Since $[L : k] = n = \deg(D \otimes_k K)$, we get

$$\mathrm{ind}(D \otimes_k K) < \deg(D \otimes_k K),$$

and therefore $D \otimes_k K$ is not a division K-algebra by point (2) of the previous theorem.

Assume now that r is prime and that $D \otimes_k K$ is not a division algebra. Now $\text{ind}(D \otimes_k K) \mid n$ and $n \mid \text{ind}(D \otimes_k K)r$ by points (4) and (5) of the previous theorem. Write $n = \text{ind}(D \otimes_k K)m$, and $\text{ind}(D \otimes_k K)r = ns$. We easily have $ms = r$, and since r is prime, we get $m = 1$ or r. If $m = 1$, we get $n = \text{ind}(D \otimes_k K)$, which is impossible since it would mean that $D \otimes_k K$ is a division K-algebra. Hence $m = r = [K : k]$, and therefore we get

$$n = \text{ind}(D \otimes_k K)[K : k].$$

Now by Corollary IV.1.9, there exists a subfield L of $D \otimes_k K$ of degree $\text{ind}(D \otimes_k K)$ over K which splits $D \otimes_k K$. Hence L splits D, since we have

$$D \otimes_k L \cong_L (D \otimes_k K) \otimes_K L$$

and L splits $D \otimes_k K$. Therefore, we have

$$[L : k] = \text{ind}(D \otimes_k K)[K : k] = n = \deg(D).$$

By Corollary IV.1.14, L is isomorphic to a subfield of D, and thus so is K. $\square$

Part (9) of Theorem V.3.1 proves that every element of $\text{Br}(k)$ has finite order. Therefore, the following definition makes sense:

DEFINITION V.3.3. Let A be a central simple k-algebra. The **exponent** of A is the order of $[A]$ in the Brauer group of k. It is denoted by $\exp(A)$.

REMARK V.3.4. Notice that two Brauer equivalent k-algebras have the same exponent, and that $\exp(A) = 1$ if and only if A is split. $\square$

We now prove a very useful property of the exponent.

THEOREM V.3.5. *Let A be a central simple k-algebra. Then*

$$\exp(A) \mid \text{ind}(A).$$

Moreover, $\exp(A)$ and $\text{ind}(A)$ have the same prime factors.

Proof. The first part comes from the definition of the exponent and the fact that $\text{ind}(A)[A] = 0$ by Theorem V.3.1 (9). To prove the second part, let p be a prime factor of $\text{ind}(A)$, and write

$$\text{ind}(A) = p^e s, e \geq 1, p \nmid s.$$

Let L/k be Galois field extension of finite degree which splits A (such an L exists by Corollary IV.1.18). By Corollary IV.1.15, we have

$$p \mid \text{ind}(A) \mid [L : k],$$

so we may write $[L : k] = p^m q, m \geq e, p \nmid q$. Let H be a p-Sylow subgroup of $\text{Gal}(L/k)$, and let $K = L^H$. Then we have $[K : k] = q$ and $[L : K] = p^m$. Since L/k splits A, L/K splits $A \otimes_k K$ and therefore

$$\text{ind}(A \otimes_k K) \mid [L : K].$$

Hence $\text{ind}(A \otimes_k K)$ is a power of p. Since $\text{ind}(A \otimes_k K) \mid \text{ind}(A)$ by Theorem V.3.1 (4), we get $\text{ind}(A \otimes_k K) = p^f, f \leq e$. On the other hand, using Theorem V.3.1 (5) gives

$$\text{ind}(A) = p^e s \mid qp^f.$$

Since q and s are prime to p, we get $e \leq f$, and thus $\mathrm{ind}(A \otimes_k K) = p^e$. In particular $A \otimes_k K$ is not split since $e \geq 1$. Consequently, we have $\exp(A \otimes_k K) > 1$. By the first point, we have
$$\exp(A \otimes_k K) \mid p^e,$$
and therefore $p \mid \exp(A \otimes_k K)$. Notice now that $\exp(A \otimes_k K) \mid \exp(A)$, since we have
$$0 = \mathrm{Res}_{K/k}(\exp(A)[A]) = \exp(A)\mathrm{Res}_{K/k}([A]) = \exp(A)[A \otimes_k K].$$
Putting things together, we obtain that $p \mid \exp(A)$. Hence, any prime divisor of $\mathrm{ind}(A)$ is a prime divisor of $\exp(A)$. Conversely, since $\exp(A)$ divides $\mathrm{ind}(A)$, any prime divisor of $\exp(A)$ also divides $\mathrm{ind}(A)$. This concludes the proof. $\qquad\square$

REMARK V.3.6. Given two integers d and e having same prime factors and such that $e \mid d$, Brauer constructed a central simple algebra A over a suitable field such that $\mathrm{ind}(A) = d$ and $\exp(A) = e$ (see [**25**, Chapter, Section 2.8] for example). $\qquad\square$

We now end this chapter by proving the existence of a unique decomposition of a central simple k-algebra into a product of primary components.

THEOREM V.3.7. *Let $n_1, n_2 \geq 1$ be two coprime integers, and let A be a central simple k-algebra of degree $n_1 n_2$. There exist two central simple k-algebras A_1 and A_2, uniquely determined up to isomorphism, such that:*

(1) $\deg(A_i) = n_i, i = 1, 2$;

(2) $A \cong_k A_1 \otimes_k A_2$.

Moreover, A is a division algebra if and only if A_1 and A_2 are.

Proof. Since n_1 and n_2 are coprime and $\exp(A) \mid \mathrm{ind}(A) \mid \deg(A)$, one may write
$$\exp(A) = e_1 e_2, \text{ with } e_i \mid n_i$$
in a unique way. Assume first that A_1 and A_2 exist. In particular, we have $[A] = [A_1] + [A_2]$. Since $\exp(A_i) \mid \mathrm{ind}(A_i) \mid n_i$, it follows that $\exp(A_1)$ and $\exp(A_2)$ are coprime. Hence $[A_1]$ and $[A_2]$ have coprime orders in the abelian group $\mathrm{Br}(k)$, and thus the order of $[A] = [A_1] + [A_2]$ is the product of the orders of $[A_1]$ and $[A_2]$, that is
$$\exp(A) = \exp(A_1)\exp(A_2).$$
Since $\exp(A_1)$ and $\exp(A_2)$ are coprime and $\exp(A_i) \mid n_i$, we get that $\exp(A_i) = e_i, i = 1, 2$.

Now e_i and e_2 are coprime, so there exist two integers $u_1, u_2 \in \mathbb{Z}$ such that $u_1 e_1 + u_2 e_2 = 1$.

We then get
$$u_1 e_1 [A] = u_1 e_1 [A_1] + u_1 e_1 [A_2] = u_1 e_1 [A_2] = (1 - u_2 e_2)[A_2] = [A_2]$$
and similarly
$$u_2 e_2 [A] = [A_1].$$
Hence the classes $[A_1]$ and $[A_2]$ are uniquely determined by $[A]$. Since a central simple k-algebra is uniquely determined up to isomorphism by its Brauer class and its degree (by Lemma V.1.1 (2)), it follows that A_1 and A_2 are unique up to isomorphism. More precisely, keeping the notation above, if the decomposition we

are looking for exists, then A_1 (resp. A_2) is the unique central simple k-algebra of degree n_1 (resp. n_2) which is Brauer equivalent to $A^{\otimes u_2 e_2}$ (resp. $A^{\otimes u_1 e_1}$).

We now prove that such algebras A_1 and A_2 effectively exist. Let D_1 and D_2 be the central division k-algebras which are Brauer equivalent to $A^{\otimes u_2 e_2}$ and $A^{\otimes u_1 e_1}$ respectively, and let d_1 and d_2 their respective degrees. Elementary group theory shows that we have

$$\exp(A^{\otimes u_2 e_2}) = \frac{\exp(A)}{gcd(u_2 e_2, \exp(A))} = \frac{e_1 e_2}{gcd(u_2 e_2, e_1 e_2)} = e_1,$$

since u_2 and e_1 are coprime. Similarly, $\exp(D_2) = e_2$. Since D_i is Brauer equivalent to a tensor power of A, we have

$$\mathrm{ind}(D_i) \mid \mathrm{ind}(A) \mid n_1 n_2$$

by Theorem V.3.1 (1) and (10). Since $\exp(D_i) \mid \mathrm{ind}(D_i)$ by Theorem V.3.5, we get

$$e_i \mid d_i \mid n_1 n_2 \text{ and } e_i \mid n_i \text{ for } i = 1, 2.$$

Since e_i and d_i have the same prime factors by Theorem V.3.5 again, it follows that we have $d_i \mid n_i$ for $i = 1, 2$, since n_1 and n_2 are coprime. Write $n_i = d_i r_i$ for $i = 1, 2$ and set

$$A_i = \mathrm{M}_{r_i}(D_i).$$

Then $\deg(A_i) = n_i$ for $i = 1, 2$ and we have

$$[A_1 \otimes_k A_2] = [A_1] + [A_2] = [D_1] + [D_2] = [A^{\otimes u_2 e_2}] + [A^{\otimes u_1 e_1}],$$

and therefore

$$[A_1 \otimes_k A_2] = (u_2 e_2 + u_1 e_1)[A] = [A].$$

Since A and $A_1 \otimes_k A_2$ have the same degree, it follows from Lemma V.1.1 (2) that $A \cong_k A_1 \otimes_k A_2$.

If A_1 and A_2 are division k-algebras, so is $A_1 \otimes_k A_2$ by Theorem V.3.1 (8). The converse will follow from the following observation: if A is a finite dimensional k-algebra, A is a not a division algebra if and only if it has zero divisors. Indeed, assume A is not a division algebra, and let $a \in A \setminus \{0\}$ be a non-invertible element. Then the k-linear map

$$r_a \colon \begin{array}{c} A \longrightarrow A \\ x \longmapsto xa \end{array}$$

is not surjective since 1_A does not have a preimage, and hence has non-trivial kernel since A has finite dimension over k. Therefore $a \in A$ is a zero divisor of A. Conversely, if A has zero divisors, then A is not a division algebra.

Now if one of the A_i's, say A_1, is not a division algebra, then it has a zero divisor $a_1 \in A_1 \setminus \{0\}$. Since the map

$$\begin{array}{c} A_1 \longrightarrow A_1 \otimes_k A_2 \\ a_1' \longmapsto a_1' \otimes 1_{A_2} \end{array}$$

is injective, it follows that $a_1 \otimes 1_{A_2}$ is a zero divisor of $A \cong_k A_1 \otimes_k A_2$. Hence A is not a division algebra. $\qquad\square$

The following very important corollary follows from the previous theorem by induction.

COROLLARY V.3.8 (Primary decomposition theorem). *Let A be a central simple k-algebra of degree n, and let $n = p_1^{n_1} \cdots p_r^{n_r}$ be the decomposition of n into distinct prime powers. Then there exist r central simple k-algebras $A_1, \cdots, A_r$, uniquely determined up to isomorphism, such that*

(1) $\deg(A_i) = p_i^{n_i}$, *for all* $i = 1, \ldots, r$;
(2) $A \cong_k A_1 \otimes_k \cdots \otimes_k A_r$.

Moreover, A is a division algebra if and only if $A_1, \ldots, A_r$ are division algebras.

REMARK V.3.9. Assume that D is a central division k-algebra of degree $p_1^{n_1} \cdots p_r^{n_r}$. Then the corollary above gives the existence of a unique decomposition

$$D \cong_k D_1 \otimes_k \cdots \otimes_k D_r,$$

where D_i is a central division k-algebra of degree $p_i^{n_i}$.   □

EXERCISES

1. A C^1-**field** is a field k satisfying the following property: every homogeneous polynomial in n variables of degree $1 \leq d < n$ has a non-trivial zero. Examples of C_1-fields are finite fields, or field extensions of transcendence degree 1 of an algebraically closed field (See for example [**16**, §66.2] for a proof).

   Show that if k is a C^1-field k, then $\mathrm{Br}(k) = 0$.

   Hint: Let D be a division k-algebra. Interpret the function

   $$f: \begin{array}{c} D \longrightarrow k \\ d \longmapsto \mathrm{Nrd}_D(d) \end{array}$$

   as an homogeneous polynomial in the coordinates of d in a fixed k-basis of D.

2. Let k be a field of characteristic different from 2. Show that the map

   $$k^\times/k^{\times 2} \times k^\times/k^{\times 2} \longrightarrow \mathrm{Br}(k)$$
   $$(\overline{a}, \overline{b}) \longmapsto [(a, b)_k]$$

   is a well-defined symmetric bilinear map.

3. Let p be an odd prime number.

   (a) Show that $(-1, p)_{\mathbb{Q}}$ is a division algebra if and only if $p \equiv 3[4]$.

   (b) Let p, q be two prime numbers congruent to 3 modulo 4. Show that $(-1, pq)_{\mathbb{Q}}$ is split if and only if $p = q$.

   (c) Deduce that $\mathrm{Br}(\mathbb{Q})$ is infinite.

4. Let Q be a division quaternion k-algebra. Compute $\lambda^m Q$ for all $m \geq 1$.

5. Let A be a central simple k-algebra. Recall that the symmetric bilinear form

   $$T_A: \begin{array}{c} A \times A \longrightarrow k \\ (a, a') \longmapsto \mathrm{Trd}_A(aa') \end{array}$$

   is non-singular.

(a) Let $(e_1, \ldots, e_{n^2})$ be a k-basis of A, and let $(e'_1, \ldots, e'_{n^2})$ be the corresponding dual basis with respect to T_A. Show that the Goldman element g_A is

$$g_A = \sum_{i=1}^{n^2} e_i \otimes e'_i.$$

(b) Assume that $\operatorname{char}(k) \neq 2$. Let Q be a quaternion k-algebra, with basis $1, i, j, ij$. Show that we have

$$g_Q = 1 \otimes 1 + i \otimes i^{-1} + j \otimes j^{-1} + ij \otimes (ij)^{-1}.$$

6. Let k be a field of characteristic different from 2. Recover the fact that $(a, b)_k \otimes (a, c) \sim_k (a, bc)_k$ by exhibiting an appropriate bimodule.

Hint: Set $Q_1 = (a, b)_k, Q_2 = (a, c)_k, Q_3 = (a, bc)_k$. One may assume that a is not a square. Set $L = k(\sqrt{a})$, so that Q_i carries the structure of an L-algebra (for which external product law?). Consider the L-algebra $M = Q_1 \otimes_L Q_2$, and define the structure of a left $Q_1 \otimes_k Q_2$-module on M. Now show the existence of a ring morphism $f : Q_3 \longrightarrow M$ such that

$$f(i_3) = i_2 \otimes 1 \text{ and } f(j_3) = j_1 \otimes j_2,$$

and use it to define the structure of a right Q_3-module on M.

CHAPTER VI

Crossed products

In this chapter, we study more closely the central simple k-algebras containing a Galois maximal subfield L. This will lead to the notion of a crossed product. We then prove some general results on such algebras.

VI.1. Definition of crossed products

Let L/k be a Galois extension of degree n. We are going to give a characterization of central simple k-algebras of degree n containing L/k as a maximal subfield. We start with a lemma.

LEMMA VI.1.1. *Let A be a central simple k-algebra of degree n, containing L as a maximal subfield. Assume that we have a family of elements $e_\sigma \in A^\times$ for all $\sigma \in \mathrm{Gal}(L/k)$ such that*

$$\lambda e_\sigma = e_\sigma \sigma^{-1}(\lambda) \text{ for all } \lambda \in L.$$

Then $A = \bigoplus_{\sigma \in \mathrm{Gal}(L/k)} e_\sigma L$.

Proof. First, notice that multiplication in A endows A with a structure of right L-vector space. Since $\dim_k(A) = n^2$ and $[L : k] = n$, we have $\dim_L(A) = n$. Hence, to prove the lemma, it is enough to prove that the elements $e_\sigma, \sigma \in \mathrm{Gal}(L/k)$ are linearly independent over L. Suppose the contrary, and let $X \subset \mathrm{Gal}(L/k)$ be a subset of minimal cardinality such that we have

$$\sum_{\sigma \in X} e_\sigma \lambda_\sigma = 0,$$

for some $\lambda_\sigma \in L^\times$ which are not all zero. Since each e_σ is invertible, we necessarily have $|X| \geq 2$. For $\lambda \in L$, we have

$$0 = \lambda \sum_{\sigma \in X} e_\sigma \lambda_\sigma = \sum_{\sigma \in X} e_\sigma \sigma^{-1}(\lambda)\lambda_\sigma.$$

Let $\sigma_0 \in X$. We have

$$0 = \Big(\sum_{\sigma \in X} e_\sigma \lambda_\sigma\Big)\sigma_0^{-1}(\lambda) = \sum_{\sigma \in X} e_\sigma \sigma_0^{-1}(\lambda)\lambda_\sigma,$$

and therefore

$$\sum_{\sigma \in X} e_\sigma(\sigma^{-1}(\lambda) - \sigma_0^{-1}(\lambda))\lambda_\sigma = 0 \text{ for all } \lambda \in L.$$

We therefore get a dependence relation with fewer terms, since the term corresponding to σ_0 cancels out. By minimality of $|X|$, all the coefficients in the relation above are zero, that is

$$(\sigma^{-1}(\lambda) - \sigma_0^{-1}(\lambda))\lambda_\sigma = 0 \text{ for all } \sigma \in X.$$

Notice now that the minimality of $|X|$ implies that $\lambda_\sigma \neq 0$ for all $\sigma \in X$, so we get

$$\sigma^{-1}(\lambda) = \sigma_0^{-1}(\lambda) \text{ for all } \sigma \in X, \lambda \in L,$$

that is

$$\sigma = \sigma_0 \text{ for all } \sigma \in X.$$

We then get $|X| = 1$, a contradiction. □

We can now characterize the central simple k-algebras containing L/k as a maximal subfield.

PROPOSITION VI.1.2. *Let A be a central simple k-algebra of degree n, and let L/k be a Galois extension of degree n. Then A contains L as a maximal subfield if and only if there exists a family of elements $(e_\sigma)_{\sigma \in \mathrm{Gal}(L/k)}$ and a map*

$$\xi: \begin{array}{c} \mathrm{Gal}(L/k) \times \mathrm{Gal}(L/k) \longrightarrow L^\times \\ (\sigma, \tau) \longmapsto \xi_{\sigma,\tau} \end{array}$$

satisfying the following properties:

(1) $e_\sigma \in A^\times$ *for all* $\sigma \in \mathrm{Gal}(L/k)$, *and* $e_{\mathrm{Id}} = 1_A$;

(2) $A = \displaystyle\bigoplus_{\sigma \in \mathrm{Gal}(L/k)} e_\sigma L$;

(3) $\lambda e_\sigma = e_\sigma \sigma^{-1}(\lambda)$ *for all* $\sigma \in \mathrm{Gal}(L/k), \lambda \in L$;

(4) $e_\sigma e_\tau = e_{\sigma\tau} \xi_{\sigma,\tau}$ *for all* $\sigma, \tau \in \mathrm{Gal}(L/k)$;

(5) *for all* $\sigma, \tau, \rho \in \mathrm{Gal}(L/k)$, *we have*

$$\xi_{\sigma,\mathrm{Id}} = \xi_{\mathrm{Id},\tau} = 1_L$$

and

$$\xi_{\sigma,\tau\rho}\xi_{\tau,\rho} = \xi_{\sigma\tau,\rho}\rho^{-1}(\xi_{\sigma,\tau}).$$

Proof. If $n = 1$, there is nothing to prove since $A = k$ in this case, so we assume that $n \geq 2$. If properties $(1) - (5)$ hold, then $L = Le_{\mathrm{Id}}$ is a subfield of A which is maximal since $[L : k] = n = \deg(A)$. Conversely, assume that A contains L as a maximal subfield. Let ι be the inclusion $L \subset A$, and let $\sigma \in \mathrm{Gal}(L/k)$. By Skolem-Noether's Theorem, the two k-algebra morphisms $\iota \circ \sigma$ and ι differ by an inner automorphism of A. Hence there exists $e_\sigma \in A^\times$ such that

$$e_\sigma \lambda e_\sigma^{-1} = \sigma(\lambda) \text{ for all } \lambda \in L.$$

Replacing λ by $\sigma^{-1}(\lambda)$, we get

$$e_\sigma \sigma^{-1}(\lambda) = \lambda e_\sigma \text{ for all } \lambda \in L.$$

Moreover, we may choose e_{Id} to be equal to 1_A. Hence, properties (1) and (3) hold. We now proceed to show the existence of the map ξ. For all $\sigma, \tau \in \mathrm{Gal}(L/k)$, set $\xi_{\sigma,\tau} = e_{\sigma\tau}^{-1}e_\sigma e_\tau$. Let us prove that $\xi_{\sigma,\tau} \in L^\times$. For all $\lambda \in L$, we have

$$
\begin{aligned}
\operatorname{Int}(e_{\sigma\tau}^{-1}e_\sigma e_\tau)(\lambda) &= (\operatorname{Int}(e_{\sigma\tau})^{-1}\circ\operatorname{Int}(e_\sigma)\circ\operatorname{Int}(e_\tau))(\lambda) \\
&= (\operatorname{Int}(e_{\sigma\tau})^{-1}\circ\operatorname{Int}(e_\sigma))(\tau(\lambda)) \\
&= \operatorname{Int}(e_{\sigma\tau})^{-1}(\sigma\tau(\lambda)) \\
&= \lambda.
\end{aligned}
$$

Therefore, $\xi_{\sigma,\tau}=e_{\sigma\tau}^{-1}e_\sigma e_\tau\in C_A(L)$. Since L is a maximal subfield, we have $C_A(L)=L$ by Remark IV.1.10 (1). Hence $\xi_{\sigma,\tau}\in L$. Since each e_σ is invertible, so is $\xi_{\sigma,\tau}$. Therefore, we have $\xi_{\sigma,\tau}\in L^\times$. By definition, we have

$$e_\sigma e_\tau = e_{\sigma\tau}\xi_{\sigma,\tau} \text{ for all } \sigma,\tau\in\operatorname{Gal}(L/k),$$

and thus ξ satisfies (4). We now prove (5). For all $\sigma,\tau,\rho\in\operatorname{Gal}(L/k)$, we have

$$(e_\sigma e_\tau)e_\rho = e_\sigma(e_\tau e_\rho),$$

since A is an associative ring. On the one hand, we have

$$
\begin{aligned}
(e_\sigma e_\tau)e_\rho &= e_{\sigma\tau}\xi_{\sigma,\tau}e_\rho \\
&= e_{\sigma\tau}e_\rho\rho^{-1}(\xi_{\sigma,\tau}) \\
&= e_{\sigma\tau\rho}\xi_{\sigma\tau,\rho}\rho^{-1}(\xi_{\sigma,\tau}).
\end{aligned}
$$

On the other hand, we have

$$
\begin{aligned}
e_\sigma(e_\tau e_\rho) &= e_\sigma e_{\tau\rho}\xi_{\tau,\rho} \\
&= e_{\sigma\tau\rho}\xi_{\sigma,\tau\rho}\xi_{\tau,\rho}.
\end{aligned}
$$

We then get $e_{\sigma\tau\rho}\xi_{\sigma\tau,\rho}\rho^{-1}(\xi_{\sigma,\tau})=e_{\sigma\tau\rho}\xi_{\sigma,\tau\rho}\xi_{\tau,\rho}$, and since $e_{\sigma\tau\rho}\in A^\times$, we have

$$\xi_{\sigma,\tau\rho}\xi_{\tau,\rho}=\xi_{\sigma\tau,\rho}\rho^{-1}(\xi_{\sigma,\tau}) \text{ for all } \sigma,\tau,\rho\in\operatorname{Gal}(L/k).$$

The equalities $\xi_{\sigma,\operatorname{Id}}=\xi_{\operatorname{Id},\tau}=1$ come from the definition of the map ξ, since $e_{\operatorname{Id}}=1_A$. Finally, (2) follows from the previous lemma. This concludes the proof. $\qquad\square$

EXAMPLE VI.1.3. Assume that $\operatorname{char}(k)\neq2$. Any division quaternion k-algebra Q contains a quadratic subfield L/k, which is necessarily a Galois extension. Hence we can find a decomposition of Q as described in the proposition. For example, if $Q=(a,b)_k$ is a division quaternion algebra, with standard basis $1,i,j,ij$, then $L=k(j)$ is a quadratic subfield of Q (isomorphic to $k(\sqrt{b})$), and $\operatorname{Gal}(L/k)=\{\operatorname{Id}_L,\iota\}$, where

$$
\iota:\quad
\begin{aligned}
L &\longrightarrow L \\
x+yj &\longmapsto x-yj.
\end{aligned}
$$

Moreover, we have $Q=L\oplus iL$ and

$$(x+jy)i = xi+yji = xi-yij = i(x-yj) = i\iota^{-1}(x+yj)$$

for all $x,y\in k$. Since $i^2=a=1{\cdot}a$, we may set

$$e_\iota=i,\xi_{\operatorname{Id},\iota}=\xi_{\iota,\operatorname{Id}}=1 \text{ and } \xi_{\iota,\iota}=a.$$

$\qquad\square$

REMARK VI.1.4. The elements e_σ and $\xi_{\sigma,\tau}$ described in the proposition are not unique. For example, if we choose $z_\sigma\in L^\times$, with $z_{\operatorname{Id}}=1_A$, then the elements

$$f_\sigma=e_\sigma z_\sigma\in A^\times \text{ and } y_{\sigma,\tau}=\xi_{\sigma,\tau}z_\tau\tau^{-1}(z_\sigma)z_{\sigma\tau}^{-1}\in L^\times$$

also satisfy the required conditions. Indeed, (1) and (2) are clear by choice of f_σ. Moreover, $\mathrm{Int}(f_\sigma)$ and $\mathrm{Int}(e_\sigma)$ have the same restriction to L since $z_\sigma \in L$ and L is commutative, so (3) holds as well. Now (4) and (5) follow from a direct computation, using the properties of the e_σ's.   □

Given a central simple k-algebra A containing L as a maximal subfield, one may find a map

$$\xi \colon \begin{array}{c} \mathrm{Gal}(L/k) \times \mathrm{Gal}(L/k) \longrightarrow L^\times \\ (\sigma, \tau) \longmapsto \xi_{\sigma,\tau} \end{array}$$

and an L-basis $(e_\sigma)_{\sigma \in \mathrm{Gal}(L/k)}$ of the right L-vector space A satisfying conditions $(1)-(5)$ of the previous proposition. We are now interested in the following problem: given a map ξ satisfying (5), can we find an L-vector space A of dimension n and an L-basis $(e_\sigma)_{\sigma \in \mathrm{Gal}(L/k)}$ such that A may be endowed with a structure of a central simple k-algebra, containing L as a maximal subfield, and such that conditions $(1)-(5)$ of the proposition hold?

The answer is affirmative, and leads to the notion of crossed product. Before giving the definition, we start by giving a name to the maps ξ satisfying (5).

DEFINITION VI.1.5. Let G be a finite group, and let C be an abelian group **denoted multiplicatively** on which G acts on the right by automorphisms, that is

$$(c_1 c_2)^\sigma = c_1^\sigma c_2^\sigma \text{ for all } \sigma \in G, c_1, c_2 \in C,$$

where c^σ denotes the action of $\sigma \in G$ on the element $c \in C$.

A 2-**cocycle** from G with values in C is a map

$$\begin{array}{c} G \times G \longrightarrow C \\ (\sigma, \tau) \longmapsto \xi_{\sigma,\tau} \end{array}$$

satisfying

$$\xi_{1,\tau} = \xi_{\sigma,1} = 1$$

and

$$\xi_{\sigma,\tau\rho}\xi_{\tau,\rho} = \xi_{\sigma\tau,\rho}\xi_{\sigma,\tau}^\rho \text{ for all } \sigma, \tau, \rho \in G.$$

One may check that the set of 2-cocycles from G with values in C is an abelian group, denoted by $Z^2(G, C)$, where the neutral element is the constant map

$$\begin{array}{c} G \times G \longrightarrow C \\ (\sigma, \tau) \longmapsto 1, \end{array}$$

called the **trivial cocycle**. It is simply denoted by 1.

EXAMPLES VI.1.6.

(1) If L/k is a Galois extension, then $\mathrm{Gal}(L/k)$ acts naturally by automorphisms on $C = L$ or $L^\times$ by

$$\lambda^\sigma = \sigma^{-1}(\lambda) \text{ for all } \sigma \in \mathrm{Gal}(L/k), \lambda \in C.$$

Then a 2-cocycle from $\mathrm{Gal}(L/k)$ with values in $L^\times$ is nothing but a map

$$\mathrm{Gal}(L/k) \times \mathrm{Gal}(L/k) \longrightarrow L^\times$$

satisfying condition (5) of the previous proposition.

(2) If G is a finite group acting by automorphisms on an abelian group C, then for any map $G \longrightarrow C, \sigma \longmapsto z_\sigma$ such that $z_1 = 1$, the map

$$G \times G \longrightarrow C$$

$$(\sigma, \tau) \longmapsto z_\tau z_\sigma^\tau z_{\sigma\tau}^{-1}$$

is a 2-cocycle. The verification is left to the reader as an exercise.

$\square$

Let L/k be a Galois extension of degree n, and let $G = \mathrm{Gal}(L/k)$. If $\lambda \in L$ and $\sigma \in G$, we set as in the previous example

$$\lambda^\sigma = \sigma^{-1}(\lambda).$$

We will keep this notation throughout this chapter and the next ones.

The set

$$\mathrm{Map}(G, L) = \{f : G \longrightarrow L\}$$

of all maps from G to L is a right L-vector space of dimension n, with L-basis $(e_\sigma)_{\sigma \in G}$, where $e_\sigma \in \mathrm{Map}(G, L)$ is defined by

$$e_\sigma(\tau) = \delta_{\sigma,\tau} \text{ for all } \tau \in G.$$

In particular, we have

$$\mathrm{Map}(G, L) = \bigoplus_{\sigma \in G} e_\sigma L.$$

Let $\xi \in Z^2(G, L^\times)$ be a 2-cocycle, and let

$$\mu_\xi : \mathrm{Map}(G, L) \times \mathrm{Map}(G, L) \longrightarrow \mathrm{Map}(G, L)$$

be the map defined by

$$\mu_\xi \Big(\sum_{\sigma \in G} e_\sigma \lambda_\sigma, \sum_{\tau \in G} e_\tau \lambda'_\tau \Big) = \sum_{\sigma,\tau \in G} e_{\sigma\tau} \xi_{\sigma,\tau} \lambda_\sigma^\tau \lambda'_\tau.$$

LEMMA VI.1.7. *With the previous notation, the pair* $(\mathrm{Map}(G, L), \mu_\xi)$ *is a central simple k-algebra of degree n, containing a maximal subfield isomorphic to L.*

Proof. We first have to prove that μ_ξ endows $A = \mathrm{Map}(G, L)$ with a structure of unital associative finite-dimensional k-algebra. It is clear from the definition that A is a right k-vector space of dimension n^2. Notice first that μ_ξ is biadditive. Moreover, for all $\lambda, \lambda' \in L, c, c' \in k$ and $\sigma, \tau \in G$, we have

$$\begin{aligned}
\mu_\xi(e_\sigma \lambda c, e_\tau \lambda' c') &= e_{\sigma\tau} \xi_{\sigma,\tau} (\lambda c)^\tau \lambda' c' \\
&= e_{\sigma\tau} \xi_{\sigma,\tau} \tau^{-1}(\lambda c) \lambda' c' \\
&= e_{\sigma\tau} \xi_{\sigma,\tau} \tau^{-1}(\lambda) c \lambda' c' \\
&= e_{\sigma\tau} \xi_{\sigma,\tau} \lambda^\tau \lambda' c c' \\
&= \mu_\xi(e_\sigma \lambda, e_\tau \lambda') c c',
\end{aligned}$$

since any element of G restricts to the identity on k. Hence μ_ξ is a k-bilinear map. Therefore, for the rest of the proof, we will write $a \cdot a'$ or aa' rather than $\mu_\xi(a, a')$ for the product of two elements $a, a' \in A$.

By definition of μ_ξ, we have

$$e_\sigma \lambda \cdot e_\tau \lambda' = e_{\sigma\tau} \xi_{\sigma,\tau} \lambda^\tau \lambda' \text{ for all } \lambda, \lambda' \in L, \sigma, \tau \in G.$$

To prove that the multiplication is associative, by biadditivity of the multiplication, it is enough to prove

$$(e_\sigma \lambda \cdot e_\tau \lambda') \cdot e_\rho \lambda'' = e_\sigma \lambda \cdot (e_\tau \lambda' \cdot e_\rho \lambda'')$$

for all $\lambda, \lambda', \lambda'' \in L, \sigma, \tau, \rho \in G$. Using the definition, we have

$$
\begin{aligned}
(e_\sigma \lambda \cdot e_\tau \lambda') \cdot e_\rho \lambda'' &= (e_{\sigma\tau} \xi_{\sigma,\tau} \lambda^\tau \lambda') \cdot e_\rho \lambda'' \\
&= e_{\sigma\tau\rho} \xi_{\sigma\tau,\rho} (\xi_{\sigma,\tau} \lambda^\tau \lambda')^\rho \lambda'' \\
&= e_{\sigma\tau\rho} \xi_{\sigma\tau,\rho} \xi_{\sigma,\tau}^\rho \lambda^{\tau\rho} (\lambda')^\rho \lambda''.
\end{aligned}
$$

We also have

$$
\begin{aligned}
e_\sigma \lambda \cdot (e_\tau \lambda' \cdot e_\rho \lambda'') &= e_\sigma \lambda \cdot e_{\tau\rho} \xi_{\tau,\rho} (\lambda')^\rho \lambda'' \\
&= e_{\sigma\tau\rho} \xi_{\sigma,\tau\rho} \lambda^{\tau\rho} \xi_{\tau,\rho} (\lambda')^\rho \lambda'' \\
&= e_{\sigma\tau\rho} \xi_{\sigma,\tau\rho} \xi_{\tau,\rho} \lambda^{\tau\rho} (\lambda')^\rho \lambda''.
\end{aligned}
$$

These two quantities are equal because ξ is a 2-cocycle. Hence, multiplication in A is associative.

Moreover, by definition of the multiplication we have

$$e_\sigma \lambda \cdot e_{\mathrm{Id}} = e_\sigma \xi_{\sigma,\mathrm{Id}} \lambda = e_\sigma \lambda \text{ for all } \sigma \in G, \lambda \in L$$

and similarly

$$e_{\mathrm{Id}} \cdot e_\tau \lambda' = e_\tau \xi_{\mathrm{Id},\tau} \lambda' = e_\tau \lambda' \text{ for all } \tau \in G, \lambda' \in L$$

since ξ is a 2-cocycle. By biadditivity of multiplication, we conclude that e_{Id} is a neutral element for the multiplication. Hence the pair (A, μ_ξ) is an associative unital k-algebra of dimension n^2. Notice finally that we have

$$e_{\mathrm{Id}} \lambda \cdot e_{\mathrm{Id}} \lambda' = e_{\mathrm{Id}} \lambda \lambda' \text{ for all } \lambda, \lambda' \in L.$$

It follows easily that the subspace $e_{\mathrm{Id}} L$ is a maximal subfield of A which is isomorphic to L. It remains to prove that A is a central simple k-algebra. By Theorem IV.1.20 (3), it is enough to prove that $A \otimes_k L \cong_k \mathrm{M}_n(L)$.

By Proposition IV.1.6, we have a unique morphism of L-algebras

$$f : A \otimes_k L \longrightarrow \mathrm{End}_L(A)$$

satisfying

$$f(a \otimes \lambda)(z) = az\lambda \text{ for all } a, z \in A, \lambda \in L.$$

To prove that it is an isomorphism, it is enough to prove that it is injective, since $\dim_L \mathrm{End}_L(A) = n^2 = \dim_L(A \otimes_k L)$.

Let $\alpha \in A \otimes_k L$ such that $f(\alpha) = 0 \in \mathrm{End}_L(A)$. Let $(u_\sigma)_{\sigma \in G}$ be a k-basis of L (we may index this basis by the elements of G since we have $[L : k] = |G|$). Then $(e_\sigma u_\tau)_{\sigma,\tau \in G}$ is a k-basis of A. Therefore, we can write

$$\alpha = \sum_{\sigma,\tau \in G} e_\sigma u_\tau \otimes \lambda_{\sigma,\tau} \text{ for some } \lambda_{\sigma,\tau} \in L.$$

We then have

$$f(\alpha)(e_{\mathrm{Id}}) = 0 = \sum_{\sigma \in G} e_\sigma u_\tau \lambda_{\sigma,\tau}.$$

Since $(e_\sigma)_{\sigma \in G}$ is an L-basis of A, we get $u_\tau \lambda_{\sigma,\tau} = 0$ for all $\sigma, \tau \in G$. Since u_τ is an element of an L-basis, we get

$$\lambda_{\sigma,\tau} = 0 \text{ for all } \sigma, \tau \in G,$$

that is $\alpha = 0$. This concludes the proof. $\qquad\square$

DEFINITION VI.1.8. Let L/k be a Galois extension of group G, and let $\xi \in Z^2(G, L^\times)$ be a 2-cocycle. The central simple k-algebra defined in the previous lemma is denoted by $(\xi, L/k, G)$ and is called a **crossed product** over L/k. This is a central simple k-algebra of degree $|G|$ containing L as a maximal subfield (where L is identified to $e_{\mathrm{Id}}L$), and defined by generators and relations as follows:

$$(\xi, L/k, G) = \bigoplus_{\sigma \in G} e_\sigma L,$$

$$\lambda e_\sigma = e_\sigma \lambda^\sigma = e_\sigma \sigma^{-1}(\lambda),$$

$$e_\sigma e_\tau = e_{\sigma\tau} \xi_{\sigma,\tau}.$$

REMARK VI.1.9. If L/k and L'/k are isomorphic Galois extensions, any crossed product over L/k is isomorphic to a crossed product over L'/k.

Indeed, let L/k and L'/k be two Galois extensions with Galois groups G and G' respectively. Assume that we have an isomorphism of k-algebras

$$f : L \xrightarrow{\sim} L'.$$

In particular, the map

$$\varphi : \begin{aligned} G' &\longrightarrow G \\ \sigma' &\longmapsto f^{-1} \circ \sigma' \circ f \end{aligned}$$

is a group isomorphism. Let $\xi \in Z^2(G, L^\times)$. Then the map

$$\xi' : \begin{aligned} G' \times G' &\longrightarrow (L')^\times \\ (\sigma', \tau') &\longmapsto f(\xi_{\varphi(\sigma'),\varphi(\tau')}) \end{aligned}$$

is a 2-cocycle.

Before checking this fact, notice first that we have

$$f(\lambda^{\varphi(\sigma')}) = f(\lambda)^{\sigma'} \text{ for all } \sigma' \in G', \lambda \in L.$$

Indeed,

$$f(\lambda^{\varphi(\sigma')}) = f(\varphi(\sigma')^{-1}(\lambda)) = f(f^{-1}(\sigma'^{-1}(f(\lambda)))) = f(\lambda)^{\sigma'}.$$

Now let $\sigma', \tau', \rho' \in G'$. Then we have

$$\begin{aligned}
\xi'_{\sigma'\tau',\rho'} \xi'^{\rho'}_{\sigma',\tau'} &= f(\xi_{\varphi(\sigma'\tau'),\varphi(\rho')})(f(\xi_{\varphi(\sigma'),\varphi(\tau')})^{\rho'} \\
&= f(\xi_{\varphi(\sigma'\tau'),\varphi(\rho')}) f(\xi^{\varphi(\rho')}_{\varphi(\sigma'),\varphi(\tau')}) \\
&= f(\xi_{\varphi(\sigma'\tau'),\varphi(\rho')} \xi^{\varphi(\rho')}_{\varphi(\sigma'),\varphi(\tau')}) \\
&= f(\xi_{\varphi(\sigma')\varphi(\tau'),\varphi(\rho')} \xi^{\varphi(\rho')}_{\varphi(\sigma'),\varphi(\tau')}).
\end{aligned}$$

Since ξ is a 2-cocycle, this yields

$$
\begin{aligned}
\xi'_{\sigma'\tau',\rho'}\xi'^{\rho'}_{\sigma',\tau'} &= f(\xi_{\varphi(\sigma'),\varphi(\tau')\varphi(\rho')}\xi_{\varphi(\tau'),\varphi(\rho')}) \\
&= f(\xi_{\varphi(\sigma'),\varphi(\tau'\rho')}\xi_{\varphi(\tau'),\varphi(\rho')}) \\
&= f(\xi_{\varphi(\sigma'),\varphi(\tau'\rho')})f(\xi_{\varphi(\tau'),\varphi(\rho')}) \\
&= \xi'_{\sigma',\tau'\rho'}\xi'_{\tau',\rho'},
\end{aligned}
$$

which is the equality defining a 2-cocycle.

We now claim that we have an isomorphism of k-algebras

$$
(\xi', L'/k, G') \cong_k (\xi, L/k, G).
$$

Indeed, if $(e_\sigma)_{\sigma \in G}$ and $(e'_{\sigma'})_{\sigma' \in G'}$ are the associated bases of $(\xi, L/k, G)$ and $(\xi', L'/k, G')$ respectively, then one may check that the map

$$
(\xi', L'/k, G') \longrightarrow (\xi, L/k, G)
$$
$$
\sum_{\sigma \in G'} e'_{\sigma'}\lambda'_{\sigma'} \longmapsto \sum_{\sigma \in G} e_{\varphi(\sigma')}f^{-1}(\lambda'_{\sigma'})
$$

is an isomorphism of k-algebras. Details are left to the reader as an exercise. □

VI.2. Some properties of crossed products

We continue this chapter by studying the properties of crossed products. We start with a reformulation of Proposition VI.1.2.

PROPOSITION VI.2.1. *Let L/k be a finite Galois extension. Then a central simple k-algebra contains L/k as a maximal subfield if and only if it is isomorphic to a crossed product over L/k.*

COROLLARY VI.2.2. *Let A be a central simple k-algebra, and assume that A is split by a Galois extension L/k. Then A is Brauer equivalent to a crossed product over L/k. In particular, any central simple algebra is Brauer equivalent to a crossed product.*

Proof.   By Proposition IV.1.12, there exists a central simple k-algebra A' Brauer equivalent to A which contains a maximal subfield L' isomorphic to L. By Proposition VI.2.1, A' is isomorphic to a crossed product over L'/k. By Remark VI.1.9, A' is then isomorphic to a crossed product over L/K. Hence A is Brauer equivalent to a crossed product over L/k. The last part follows from the fact that any central simple k-algebra has a Galois splitting field by Corollary IV.1.18; this concludes the proof.   □

Let $(\xi, L/k, G)$ and $(\xi', L/k, G)$ be two crossed products over L/k. Since they contain L as a maximal subfield, they are both split by L, and therefore so is their tensor product. The previous corollary then shows that $(\xi, L/k, G) \otimes_k (\xi', L/k, G)$ is Brauer equivalent to a crossed product over L/k. The next proposition tells us which one.

PROPOSITION VI.2.3. *Let L/k be a Galois extension of degree n with Galois group G, and let $\xi, \xi' \in Z^2(G, L^\times)$ be two 2-cocycles. Then the central simple k-algebra*

$(\xi, L/k, G) \otimes_k (\xi', L/k, G)$ *is Brauer equivalent to* $(\xi\xi', L/k, G)$. *In particular, we have*

$$(1, L/k, G) \cong_k \mathrm{M}_n(k).$$

Proof. Let $A = (\xi, L/k, G)$, $B = (\xi', L/k, G)$ and $C = (\xi\xi', L/k, G)$, and let $(e_\sigma)_{\sigma \in G}, (f_\sigma)_{\sigma \in G}$ and $(g_\sigma)_{\sigma \in G}$ be the corresponding L-bases. Notice that we have

$$\deg(A \otimes_k B) = n^2 \text{ and } \deg(C) = n.$$

By Proposition V.2.3, to prove that C and $A \otimes_k B$ are Brauer equivalent, it is enough to find a $A \otimes_k B - C$-bimodule M of dimension n^3 over k.

Since A and B are right L-vector spaces, we may consider the right L-vector space $M = A \otimes_L B$. By definition of the L-vector space structure, we have

$$(a \otimes b) \cdot \lambda = a\lambda \otimes b = a \otimes b\lambda \text{ for all } a \in A, b \in B, \lambda \in L.$$

Notice that we have $\dim_L(M) = n^2$, and therefore $\dim_k(M) = n^3$. We are going to show that M is the bimodule we are looking for. First, we are going to endow M with a ring structure.

Since $A = \bigoplus_{\sigma \in G} e_\sigma L$ and $B = \bigoplus_{\sigma \in G} f_\sigma L$, we have

$$M = \bigoplus_{\sigma, \tau \in G} e_\sigma L \otimes_L f_\tau L = \bigoplus_{\sigma, \tau \in G} (e_\sigma \otimes f_\tau) L.$$

Hence, there exists a unique biadditive map

$$\varphi : M \times M \longrightarrow M$$

such that

$$\varphi(e_\sigma \lambda \otimes f_\tau \mu, e_{\sigma'} \lambda' \otimes f_{\tau'} \mu') = (e_\sigma \lambda)(e_{\sigma'} \lambda') \otimes (f_\tau \mu)(f_{\tau'} \mu')$$

for all $\sigma, \tau, \sigma', \tau' \in G, \lambda, \lambda', \mu, \mu' \in L$.

Using the biadditivity of φ, one can see easily that we have

$$\varphi(a \otimes b, a' \otimes b') = aa' \otimes bb' \text{ for all } a, a' \in A, b, b' \in B.$$

For all $x, x' \in M$, set

$$x \cdot x' = \varphi(x, x').$$

Let us show that $(M, +, \cdot)$ is a ring. Since φ is biadditive, we have

$$(x_1 + x_2) \cdot x_3 = x_1 \cdot x_3 + x_2 \cdot x_3 \text{ and } x_1 \cdot (x_2 + x_3) = x_1 \cdot x_2 + x_1 \cdot x_3,$$

for all $x_1, x_2, x_3 \in M$. Moreover, we have

$$1_A \otimes 1_B \cdot a' \otimes b' = a' \otimes b' \cdot 1_A \otimes 1_B = a' \otimes b' \text{ for all } a' \in A, b' \in B,$$

and it follows from the biadditivity of φ that we have

$$1_A \otimes 1_B \cdot x = x \cdot 1_A \otimes 1_B = x \text{ for all } x \in M.$$

Finally, we also have easily

$$((a_1 \otimes b_1) \cdot (a_2 \otimes b_2)) \cdot (a_3 \otimes b_3) = (a_1 \otimes b_1) \cdot ((a_2 \otimes b_2) \cdot (a_3 \otimes b_3)),$$

for all $a_1, a_2, a_3 \in A, b_1, b_2, b_3 \in B$, and by biadditivity we get

$$(x_1 \cdot x_2) \cdot x_3 = x_1 \cdot (x_2 \cdot x_3) \text{ for all } x_1, x_2, x_3 \in M.$$

Hence the operation

$$M \times M \longrightarrow M$$
$$(x, x') \longmapsto x \cdot x' = \varphi(x, x')$$

endows M with the structure of a ring with unit element $1_A \otimes 1_B$.

We are now ready to define a structure of a $A \otimes_k B - C$-bimodule on M. The map

$$A \times B \longrightarrow M$$
$$(a, b) \longmapsto a \otimes b$$

is L-bilinear, hence k-bilinear, and then induces a unique k- linear map

$$f_1 : A \otimes_k B \longrightarrow M$$

satisfying

$$f_1(a \otimes b) = a \otimes b \text{ for all } a \in A, b \in B.$$

At this point, the reader should remember that the tensor product on the right hand side is taken over L. One may verify easily that f_1 is a ring morphism. It readily follows that the external product

$$A \otimes_k B \times M \longrightarrow M$$
$$(z, x) \longmapsto z \bullet x = f_1(z) \cdot x$$

endows M with the structure of a left $A \otimes_k B$-module.

Now let $f_2 : C \longrightarrow M$ be the unique L-linear map defined by

$$f_2(g_\sigma) = e_\sigma \otimes f_\sigma \text{ for all } \sigma \in G.$$

We claim that f_2 is a ring morphism. A distributivity argument shows that is enough to check that $f_2(1_C) = 1_M = 1_A \otimes 1_B$ and

$$f_2((g_\sigma \lambda)(g_\tau \lambda')) = f_2(g_\sigma \lambda) f_2(g_\tau \lambda') \text{ for all } \sigma, \tau \in G, \lambda, \lambda' \in L.$$

By definition, we have

$$f_2(1_C) = f_2(g_{\mathrm{Id}}) = e_{\mathrm{Id}} \otimes f_{\mathrm{Id}} = 1_A \otimes 1_B.$$

Moreover, we have

$$
\begin{aligned}
f_2((g_\sigma \lambda)(g_\tau \lambda')) &= f_2(g_\sigma(\lambda g_\tau)\lambda')) \\
&= f_2(g_{\sigma\tau} \xi_{\sigma,\tau} \xi'_{\sigma,\tau} \lambda^\tau \lambda') \\
&= (e_{\sigma\tau} \otimes f_{\sigma\tau}) \cdot \xi_{\sigma,\tau} \xi'_{\sigma,\tau} \lambda^\tau \lambda' \\
&= e_{\sigma\tau} \otimes f_{\sigma\tau} \xi_{\sigma,\tau} \xi'_{\sigma,\tau} \lambda^\tau \lambda'.
\end{aligned}
$$

Now we have

$$
\begin{aligned}
f_2(g_\sigma \lambda) f_2(g_\tau \lambda') &= ((e_\sigma \otimes f_\sigma) \cdot \lambda)((e_\tau \otimes f_\tau) \cdot \lambda') \\
&= (e_\sigma \otimes f_\sigma \lambda)(e_\tau \otimes f_\tau \lambda') \\
&= e_\sigma e_\tau \otimes f_\sigma \lambda f_\tau \lambda' \\
&= e_{\sigma\tau} \xi_{\sigma,\tau} \otimes f_{\sigma\tau} \xi'_{\sigma,\tau} \lambda^\tau \lambda' \\
&= e_{\sigma\tau} \otimes f_{\sigma\tau} \xi'_{\sigma,\tau} \lambda^\tau \lambda' \xi_{\sigma,\tau} \\
&= e_{\sigma\tau} \otimes f_{\sigma\tau} \xi_{\sigma,\tau} \xi'_{\sigma,\tau} \lambda^\tau \lambda' \\
&= f_2((g_\sigma \lambda)(g_\tau \lambda')).
\end{aligned}
$$

Hence f_2 is a ring morphism, and thus the external product

$$M \times C \longrightarrow M$$
$$(x, z') \longmapsto x * z' = x \cdot f_2(z')$$

endows M with the structure of a right C-module.

It remains to check the equality

$$(z \bullet x) * z' = z \bullet (x * z') \text{ for all } z \in A \otimes_k B, x \in M, z' \in C.$$

But we have

$$(z \bullet x) * z' = (f_1(z) \cdot x) \cdot f_2(z') = f_1(z) \cdot (x \cdot f_2(z')) = z \bullet (x * z').$$

Hence M has a structure of $A \otimes_k B - C$-bimodule, and this concludes the proof of the first part.

To prove the last part, applying the first point with $\xi' = 1$ gives

$$[(\xi, L/k, G)] + [(1, L/k, G)] = [(\xi, L/k, G)],$$

that is

$$[(1, L/k, G)] = 0 = [\mathrm{M}_n(k)] \in \mathrm{Br}(k).$$

By Lemma V.1.1 (2), we get $(1, L/k, G) \cong_k \mathrm{M}_n(k)$; this concludes the proof. $\qquad \square$

REMARK VI.2.4. Let us keep the notation of the proof above. One may find a non-zero idempotent $e \in A \otimes_k B$ such that $e(A \otimes_k B)e \cong_k C$, and therefore get a new proof of the previous proposition, in view of Proposition V.2.3 (4). See [**41**], Section 14.3 for instance for a proof using this approach. $\qquad \square$

We now study under which conditions two crossed products over L/k are isomorphic.

LEMMA VI.2.5. *Let L/k be a Galois extension of Galois group G, and let $\xi, \xi' \in Z^2(G, L^\times)$. Then $(\xi, L/k, G)$ and $(\xi', L/k, G)$ are isomorphic if and only if there exists a map*

$$z: \begin{array}{c} G \longrightarrow L^\times \\ \sigma \longmapsto z_\sigma \end{array}$$

such that $z_{\mathrm{Id}} = 1$ and

$$\xi_{\sigma,\tau} = \xi'_{\sigma,\tau} z_\tau z_\sigma^\tau z_{\sigma\tau}^{-1} \text{ for all } \sigma, \tau \in G.$$

Proof. Let $(e_\sigma)_{\sigma \in G}$ and $(e'_\sigma)_{\sigma \in G}$ be the corresponding L-bases of $A = (\xi, L/k, G)$ and $A' = (\xi', L/k, G)$ respectively. Assume that we have a map

$$z: \begin{array}{c} G \longrightarrow L^\times \\ \sigma \longmapsto z_\sigma \end{array}$$

satisfying the conditions of the lemma, and let

$$f : (\xi, L/k, G) \longrightarrow (\xi', L/k, G)$$

be the map defined by

$$f\left(\sum_{\sigma \in G} e_\sigma \lambda_\sigma\right) = \sum_{\sigma \in G} e'_\sigma z_\sigma \lambda_\sigma.$$

Clearly, f is an isomorphism of L-vector spaces. Let us check that f is a k-algebra morphism. The k-linearity being obvious, we just have to check that f is a ring morphism. Recall that the unit elements of $(\xi, L/k, G)$ and $(\xi', L/k, G)$ are e_{Id} and e'_{Id} respectively. Since $z_{\mathrm{Id}} = 1$, we get $f(e_{\mathrm{Id}}) = e'_{\mathrm{Id}}$. Now let us prove that f

preserves products. Since f is additive, a distributivity argument shows that it is enough to check that

$$f(e_\sigma\lambda\cdot e_\tau\lambda') = f(e_\sigma\lambda)f(e_\tau\lambda') \text{ for all } \sigma,\tau \in G, \lambda,\lambda' \in L.$$

Since we have

$$
\begin{aligned}
f(e_\sigma\lambda\cdot e_\tau\lambda') &= f(e_{\sigma\tau}\xi_{\sigma,\tau}\lambda^\tau\lambda')\\
&= e'_{\sigma\tau}z_{\sigma\tau}\xi_{\sigma,\tau}\lambda^\tau\lambda'\\
&= e'_{\sigma\tau}\xi'_{\sigma,\tau}z_\tau z_\sigma^\tau\lambda^\tau\lambda'\\
&= e'_{\sigma\tau}\xi'_{\sigma,\tau}(z_\sigma\lambda)^\tau z_\tau\lambda'\\
&= e'_\sigma z_\sigma\lambda\cdot e'_\tau z_\tau\lambda'\\
&= f(e_\sigma\lambda)f(e_\tau\lambda'),
\end{aligned}
$$

we are done. Conversely, assume that there exists an isomorphism of k-algebras

$$f : A \xrightarrow{\sim} A.$$

We identify L to $e_{\mathrm{Id}}L$ and $e'_{\mathrm{Id}}L$ in A and A' respectively. The map

$$f_{|L} : \begin{aligned} L &\longrightarrow A'\\ \lambda &\longmapsto f(\lambda) \end{aligned}$$

is a k-algebra morphism. By Skolem-Noether's Theorem applied to $f_{|L}$ and the inclusion $L \subset A'$, there exists an (inner) automorphism $\varphi : A' \longrightarrow A'$ such that $\varphi_{|L} = f_{|L}$. Then the map

$$g = \varphi^{-1} \circ f : A \xrightarrow{\sim} A'$$

is an isomorphism of k-algebras satisfying

$$g(\lambda) = \lambda \text{ for all } \lambda \in L,$$

since we have the equalities

$$g_{|L} = \varphi_{|L}^{-1} \circ f_{|L} = f_{|L}^{-1} \circ f_{|L} = \mathrm{Id}_L.$$

Now we have

$$e_\sigma\lambda^\sigma e_\sigma^{-1} = \lambda \text{ for all } \lambda \in L,$$

and therefore applying g to this equation yields

$$g(e_\sigma)\lambda^\sigma g(e_\sigma)^{-1} = \lambda \text{ for all } \lambda \in L.$$

Since we also have

$$e'_\sigma\lambda^\sigma (e'_\sigma)^{-1} = \lambda \text{ for all } \lambda \in L,$$

we get

$$g(e_\sigma)^{-1}e'_\sigma\lambda^\sigma = \lambda^\sigma g(e_\sigma)^{-1}e'_\sigma \text{ for all } \lambda \in L.$$

Since λ^σ runs through L when λ does, it follows that $g(e_\sigma)^{-1}e'_\sigma$ lies in $C_{A'}(L) = L$ (recall that L equals its own centralizer in A' since it is a maximal subfield by Remark IV.1.10 (1)). Since $g(e_\sigma)^{-1}e'_\sigma$ is invertible, we have in fact $g(e_\sigma)^{-1}e'_\sigma \in L^\times$, and thus $z_\sigma = (e'_\sigma)^{-1}g(e_\sigma) \in L^\times$ as well. By definition, we have

$$g(e_\sigma) = e'_\sigma z_\sigma \text{ for all } \sigma \in G.$$

In particular, $z_{\mathrm{Id}} = 1$. Since $\xi_{\sigma,\tau} \in L^\times$, we get $\xi_{\sigma,\tau} = g(\xi_{\sigma,\tau})$ for all $\sigma,\tau \in G$, and therefore

$$\xi_{\sigma,\tau} = g(\xi_{\sigma,\tau}) = g(e_{\sigma\tau}^{-1}e_\sigma e_\tau) = g(e_{\sigma\tau})^{-1}g(e_\sigma)g(e_\tau),$$

that is

$$
\begin{aligned}
\xi_{\sigma,\tau} &= z_{\sigma\tau}^{-1}(e'_{\sigma\tau})^{-1}e'_{\sigma}z_{\sigma}e'_{\tau}z_{\tau} \\
&= z_{\sigma\tau}^{-1}(e'_{\sigma\tau})^{-1}e'_{\sigma}e'_{\tau}z_{\sigma}^{\tau}z_{\tau} \\
&= z_{\sigma\tau}^{-1}\xi'_{\sigma,\tau}z_{\sigma}^{\tau}z_{\tau},
\end{aligned}
$$

and this concludes the proof.   □

As pointed out in Example VI.1.6 (2), if C is an abelian group with a right action of G by automorphisms, the maps

$$
G \times G \longrightarrow C
$$
$$
(\sigma,\tau) \longmapsto z_{\tau}z_{\sigma}^{\tau}z_{\sigma\tau}^{-1},
$$

where $z : G \longrightarrow C$ is a map satisfying $z_{\mathrm{Id}} = 1$, are 2-cocycles. We give them a special name.

DEFINITION VI.2.6. Let G be a finite group acting on the right by automorphisms on an abelian group C. A 2-**coboundary** is a 2-cocycle of the form

$$
G \times G \longrightarrow C
$$
$$
(\sigma,\tau) \longmapsto z_{\tau}z_{\sigma}^{\tau}z_{\sigma\tau}^{-1},
$$

where $z : G \longrightarrow C$ is a map satisfying $z_{\mathrm{Id}} = 1$.

The set of 2-coboundaries form a subgroup $B^2(G,C)$ of $Z^2(G,C)$. The factor group

$$
H^2(G,C) = Z^2(G,C)/B^2(G,C)
$$

is called the **second cohomology group** of G with values in C. The equivalence class of a 2-cocycle ξ is denoted by $[\xi]$. The neutral element of $H^2(G,C)$ is $[1]$. Cocycles in the same equivalence class are said to be **cohomologous**.

Recall now that $\mathrm{Br}(L/k)$ is by definition the kernel of the restriction map

$$
\mathrm{Res}_{L/k}: \quad \begin{aligned} \mathrm{Br}(k) &\longrightarrow \mathrm{Br}(L) \\ [A] &\longmapsto [A \otimes_k L], \end{aligned}
$$

that is the subgroup of $\mathrm{Br}(k)$ consisting of Brauer classes of central simple k-algebras which are split by L.

We can now summarize all the previous results in a condensed form as follows:

THEOREM VI.2.7. *Let L/k be a finite Galois extension of Galois group G. Then the map*

$$
H^2(G,L^{\times}) \longrightarrow \mathrm{Br}(L/k)
$$
$$
[\xi] \longmapsto [(\xi, L/k, G)]
$$

is a well-defined group isomorphism.

Proof. Recall first that a crossed product over L/k is split by L, since it contains L as a maximal subfield (see Proposition IV.1.6). Hence the Brauer class of such a crossed product is an element of $\mathrm{Br}(L/k)$. By Proposition VI.2.3, the map

$$
\varphi: \quad \begin{aligned} Z^2(G,L^{\times}) &\longrightarrow \mathrm{Br}(L/k) \\ \xi &\longmapsto [(\xi, L/k, G)] \end{aligned}
$$

is a group morphism. Lemma VI.2.5 (applied to $\xi' = 1$) and the fact that the algebra $(1, L/k, G)$ is split show that $\ker(\varphi) = B^2(G, L^\times)$. hence we get an injective morphism

$$\overline{\varphi} : \begin{array}{c} H^2(G, L^\times) \hookrightarrow \mathrm{Br}(L/k) \\[1mm] [\xi] \longmapsto [(\xi, L/k, G)]. \end{array}$$

By Corollary VI.2.2, this group morphism is surjective, and this concludes the proof. □

We will investigate the case where L/k is either cyclic or biquadratic in the next chapters. For the moment, to finish this section, we would like to give a few general results on crossed products.

LEMMA VI.2.8. *Let L/k be a finite Galois extension of Galois group G, let $k \subset F \subset L$ and let $H = \mathrm{Gal}(L/F)$. For any $\xi \in Z^2(G, L^\times)$, the map $\mathrm{Res}_H^G(\xi) = \xi_{|H \times H}$ is an element of $Z^2(H, L^\times)$ and $(\xi, L/k, G) \otimes_k F$ is Brauer equivalent to $(\mathrm{Res}_H^G(\xi), L/F, H)$.*

Proof. The first part is obvious.

To prove the second part, let $A = (\xi, L/k, G)$, let $(e_\sigma)_{\sigma \in G}$ be the corresponding L-basis of A and let $B = \bigoplus_{\tau \in H} e_\tau L$. Clearly, B is a subalgebra of A which is isomorphic to $(\mathrm{Res}_H^G(\xi), L/F, H)$. In particular, B is a central simple F-algebra of degree $|H|$ over F. Notice that Corollary III.5.3 implies that $A \otimes_k F$ is Brauer equivalent to $C_A(F)$. Hence, we are going to show that $B = C_A(F)$, which will prove the desired result. If $\lambda \in F$ and $b = \sum_{\tau \in H} e_\tau \lambda_\tau \in B$, we have

$$\lambda b = \sum_{\tau \in H} \lambda e_\tau \lambda_\tau = \sum_{\tau \in H} e_\tau \lambda^\tau \lambda_\tau.$$

Since $H = \mathrm{Gal}(L/F)$, we have $\lambda^\tau = \lambda$ for all $\tau \in H$, and therefore

$$\lambda b = \sum_{\tau \in H} e_\tau \lambda \lambda_\tau = b\lambda,$$

so $b \in C_A(F)$. Thus, we have proved that $B \subset C_A(F)$. By the Centralizer Theorem, we have $\dim_k(C_A(F)) = \dfrac{n^2}{[F : k]}$, and thus

$$\dim_F(C_A(F)) = \frac{n^2}{[F : k]^2}.$$

Now we have

$$|H| = [L : F] = \frac{[L : k]}{[F : k]} = \frac{n}{[F : k]},$$

and therefore

$$\dim_F(C_A(F)) = |H|^2 = \dim_F(B).$$

It follows that $B = C_A(F)$, and this concludes the proof. □

We now examine the opposite case of a field extension F/k, not necessarily of finite degree, such that L and F are both contained in a field E, and satisfying $L \cap F = k$. Recall from Galois theory that in this case, L and F are linearly disjoint over k (since L/k is Galois). Hence any k-basis $(u_1, \ldots, u_n)$ of L is also an F-basis of

LF (the compositum of L and F in E), and therefore any $\sigma \in \mathrm{Gal}(L/k)$ extends in a unique way (by F-linearity) to an element $\tilde{\sigma} \in \mathrm{Gal}(LF/F)$. We then have a canonical group isomorphism

$$f: \begin{array}{c} \mathrm{Gal}(LF/F) \xrightarrow{\sim} \mathrm{Gal}(L/k) \\ \tau \longmapsto \tau_{|L}, \end{array}$$

the inverse map being given by

$$f^{-1}: \begin{array}{c} \mathrm{Gal}(L/k) \xrightarrow{\sim} \mathrm{Gal}(LF/F) \\ \sigma \longmapsto \tilde{\sigma}. \end{array}$$

LEMMA VI.2.9. *Let L/k be a Galois extension of Galois group G, and let F/k be an arbitrary field extension. Assume that L and F are both contained in a same field E, and that $L \cap F = k$.*

For any $\xi \in Z^2(G, L^\times)$, the map

$$\tilde{\xi}: \begin{array}{c} G \times G \longrightarrow (LF)^\times \\ (\tilde{\sigma}, \tilde{\tau}) \longmapsto \xi_{\sigma, \tau} \end{array}$$

is an element of $Z^2(G, (LF)^\times)$, and $(\xi, L/k, G) \otimes_k F$ is isomorphic to $(\tilde{\xi}, LF/F, G)$.

Proof. The first part is clear. Let $(e_\sigma)_{\sigma \in \mathrm{Gal}(L/k)}$ be the L-basis of $(\xi, L/k, G)$ and let $(f_{\tilde{\sigma}})_{\tilde{\sigma} \in \mathrm{Gal}(LF/F)}$ be the L-basis of $(\tilde{\xi}, LF/F, G)$.

Let $\varphi: (\xi, L/k, G) \longrightarrow (\tilde{\xi}, LF/F, G)$ be the unique L-linear map satisfying

$$\varphi(e_\sigma) = f_{\tilde{\sigma}} \text{ for all } \sigma \in \mathrm{Gal}(L/k).$$

One may see that φ is a k-algebra morphism, and that the image of φ commutes with the image of the k-algebra morphism

$$\begin{array}{c} F \longrightarrow (\tilde{\xi}, LF/F, G) \\ \lambda \longmapsto e_{\widetilde{\mathrm{Id}}_L} \lambda. \end{array}$$

Hence we get a k-algebra morphism

$$(\xi, L/k, G) \otimes_k F \longrightarrow (\tilde{\xi}, LF/F, G),$$

which is easily checked to be F-linear, that is an F-algebra morphism. Since $(\xi, L/k, G) \otimes_k F$ and $(\tilde{\xi}, LF/F, G)$ have the same dimension, φ is an isomorphism by Lemma I.2.2. $\qquad\square$

If L/k is a Galois extension of Galois group G, then the previous considerations show that every $\tau \in \mathrm{Gal}(L/L \cap F)$ canonically extends by F-linearity in a unique way to an element $\tilde{\tau} \in \mathrm{Gal}(LF/F)$ and we have a canonical group isomorphism

$$\mathrm{Gal}(L/L \cap F) \simeq \mathrm{Gal}(LF/F).$$

Combining the two previous lemmas, we get the following theorem.

THEOREM VI.2.10 (Restriction theorem). *Let L/k be a Galois field extension of Galois group G, let F/k be any field extension. Assume that L and F are both* *contained in a same field E and let $H = \mathrm{Gal}(L/L \cap F)$. For any $\tau \in H$, let $\tilde{\tau} \in \mathrm{Gal}(LF/F)$ be its unique extension to LF. For any $\xi \in Z^2(G, L^\times)$, the map*

$$\xi' \colon \begin{array}{c} H \times H \longrightarrow (LF)^\times \\ (\tilde{\tau}_1, \tilde{\tau}_2) \longmapsto \xi_{\tau_1, \tau_2} \end{array}$$

is an element of $Z^2(H, (LF)^\times)$, and $(\xi, L/k, G) \otimes_k F$ is Brauer equivalent to $(\xi', LF/F, H)$.

Let L/k be a Galois extension of Galois group G, and let $k \subset F \subset L$. Assume that F/k is Galois. Then $H = \mathrm{Gal}(L/F)$ is a normal subgroup of G. Moreover, the map

$$\begin{array}{c} G \longrightarrow \mathrm{Gal}(F/k) \\ \sigma \longmapsto \sigma_{|F} \end{array}$$

is a surjective group morphism of kernel H, and therefore induces a group isomorphism $G/H \simeq \mathrm{Gal}(F/k)$. We then have the following result:

THEOREM VI.2.11 (Inflation theorem). *Let L/k be a Galois extension of Galois group G and let $k \subset F \subset L$. Assume that F/k is Galois, and let $\overline{G} = \mathrm{Gal}(F/k)$. For any $\xi \in Z^2(\overline{G}, F^\times)$, the map*

$$\mathrm{Inf}_F^L(\xi) \colon \begin{array}{c} G \times G \longrightarrow F^\times \subset L^\times \\ (\sigma, \tau) \longmapsto \xi_{\sigma_{|F}, \tau_{|F}} \end{array}$$

is an element of $Z^2(G, L^\times)$, and $(\xi, F/k, \overline{G})$ is Brauer equivalent to $(\mathrm{Inf}_F^L(\xi), L/k, G)$.

Proof.   Let $r = [L : F] = |H|$. We are going to establish the isomorphism of k-algebras

$$(\mathrm{Inf}_F^L(\xi), L/k, G) \cong_k \mathrm{M}_r((\xi, F/k, \overline{G})),$$

which will yield the desired result by definition of Brauer equivalence.

Let $A = (\mathrm{Inf}_F^L(\xi), L/k, G)$ and let $B = (\xi, F/k, \overline{G})$, and let $(b_\tau)_{\tau \in \overline{G}}$ be the corresponding F-basis of B.

For all $\sigma \in G$ and $M = (m_{ij}) \in \mathrm{M}_r(B)$, set $M^\sigma = (m_{ij}^\sigma)$.

Claim: There exist an F-algebra morphism $f : L \longrightarrow \mathrm{M}_r(F)$ (necessarily injective since L is a field) and a map $g : G \longrightarrow \mathrm{GL}_r(F)$ satisfying

$$g(\sigma\tau) = g(\tau)g(\sigma)^\tau \text{ and } g(\sigma)f(\lambda)^\sigma = f(\lambda^\sigma)g(\sigma),$$

for all $\sigma, \tau \in G, \lambda \in L$.

Assume that the claim has been proved, and let us show how it allows us to conclude. Notice that we have necessarily $g(\mathrm{Id}) = I_r$ (apply the relation above with $\sigma = \tau = \mathrm{Id}$). Moreover $L' = f(L)$ is a Galois subfield of $\mathrm{M}_r(B)$ of Galois group G' isomorphic to G, the isomorphism being induced by

$$\rho \colon \begin{array}{c} G' \longrightarrow G \\ \sigma' \longmapsto f^{-1} \circ \sigma' \circ f. \end{array}$$

Set

$$e_{\sigma'} = b_{\rho(\sigma')_{|F}} g(\rho(\sigma'))^{-1} \in \mathrm{M}_r(B) \text{ for all } \sigma' \in G'.$$

Notice that for every matrix $M = (m_{ij}) \in \mathrm{M}_r(F)$, we have

$$Mb_{\sigma|_F} = (m_{ij}b_{\sigma|_F}) = (b_{\sigma|_F}m_{ij}^{\sigma|_F}),$$

and therefore

$$Mb_{\sigma|_F} = b_{\sigma|_F}M^\sigma.$$

Let $\sigma' \in G'$. Then for all $\lambda \in L$, we have

$$
\begin{aligned}
f(\lambda)e_{\sigma'} &= f(\lambda)b_{\rho(\sigma')|_F}g(\rho(\sigma'))^{-1} \\
&= b_{\rho(\sigma')|_F}f(\lambda)^{\rho(\sigma')}g(\rho(\sigma'))^{-1} \\
&= b_{\rho(\sigma')|_F}g(\rho(\sigma'))^{-1}f(\lambda^{\rho(\sigma')}) \\
&= e_{\sigma'}f(\rho(\sigma')^{-1}(\lambda)) \\
&= e_{\sigma'}(\sigma'^{-1}(f(\lambda)) \\
&= e_{\sigma'}f(\lambda)^{\sigma'}.
\end{aligned}
$$

Hence we have $\lambda'e_{\sigma'} = e_{\sigma'}(\lambda')^{\sigma'}$ for all $\lambda' \in L'$ and all $\sigma' \in G'$. By construction, the elements $e_{\sigma'}$ are invertible in $\mathrm{M}_r(B)$, so by Lemma VI.1.1 we have

$$\mathrm{M}_r(B) = \bigoplus_{\sigma' \in G'} e_{\sigma'}L'.$$

Notice that we have $e_{\mathrm{Id}} = b_{\mathrm{Id}}g(\mathrm{Id})^{-1} = 1_B I_r = 1_{\mathrm{M}_r(B)}$. Let $\sigma', \tau' \in G'$, and write $\sigma = \rho(\sigma'), \tau = \rho(\tau')$.

Then we have

$$
\begin{aligned}
e_{\sigma'}e_{\tau'} &= b_{\sigma|_F}g(\sigma)^{-1}b_{\tau|_F}g(\tau)^{-1} \\
&= b_{\sigma|_F}b_{\tau|_F}(g(\sigma)^{-1})^\tau g(\tau)^{-1} \\
&= b_{\sigma|_F}b_{\tau|_F}g(\sigma\tau)^{-1} \\
&= b_{\sigma\tau|_F}\xi_{\sigma|_F,\tau|_F}g(\sigma\tau)^{-1} \\
&= b_{\sigma\tau|_F}g(\sigma\tau)^{-1}\xi_{\sigma|_F,\tau|_F} \\
&= e_{\sigma'\tau'}\xi_{\rho(\sigma')|_F,\rho(\tau')|_F} \\
&= e_{\sigma'\tau'}\mathrm{Inf}_F^L(\xi)_{\rho(\sigma'),\rho(\tau')} \\
&= e_{\sigma'\tau'}f(\mathrm{Inf}_F^L(\xi)_{\rho(\sigma'),\rho(\tau')}),
\end{aligned}
$$

the last equality coming from the fact that f is F-linear and $\mathrm{Inf}_F^L(\xi)$ takes values in F. Hence $\mathrm{M}_r(B) \cong_k (\xi', L'/F, G')$ where ξ' is the 2-cocycle

$$
\xi': \begin{array}{c} G \times G \longrightarrow L' \\ (\sigma', \tau') \longmapsto f(\mathrm{Inf}_F^L(\xi)_{\rho(\sigma'),\rho(\tau')}). \end{array}
$$

Applying Remark VI.1.9, we get $\mathrm{M}_r(B) \cong_k A$.

Therefore, it remains to construct the maps f and g with the desired properties. Fix an F-basis $(b_1, \ldots, b_r)$ of L. For $\lambda \in L$, we then may write

$$\lambda b_j = \sum_i b_i f_{ij}(\lambda) \text{ for all } j = 1, \ldots, r,$$

for some $f_{ij}(\lambda) \in F$. For $\sigma \in G$, write

$$b_j^\sigma = \sum_i b_i g_{ij}(\sigma) \text{ for all } j = 1, \ldots, r,$$

for some $g_{ij}(\sigma) \in F$. We now set

$$f(\lambda) = (f_{ij}(\lambda)) \text{ and } g(\sigma) = (g_{ij}(\sigma)).$$

The desired relations are then easy to check by straightforward computations. For example, we have

$$
\begin{aligned}
b_j^{\sigma\tau} &= (b_j^\sigma)^\tau \\
&= \sum_i b_i^\tau g_{ij}(\sigma)^\tau \\
&= \sum_i \sum_\ell b_\ell g_{\ell i}(\tau) g_{ij}(\sigma)^\tau \\
&= \sum_\ell b_\ell \Big(\sum_i g_{\ell i}(\tau) g_{ij}(\sigma)^\tau \Big),
\end{aligned}
$$

for all $j = 1, \ldots, r$. Since we also have $b_j^{\sigma\tau} = \sum_\ell b_\ell g_{\ell j}(\sigma\tau)$, we conclude that

$$
g_{\ell j}(\sigma\tau) = \sum_i g_{\ell i}(\tau) g_{ij}(\sigma)^\tau \text{ for all } \ell, j,
$$

that is

$$
g(\sigma\tau) = g(\tau) g(\sigma)^\tau \text{ for all } \sigma, \tau \in G.
$$

The other relations may be checked similarly, and are left to the reader. The fact that $g(\sigma)$ is invertible follows from the fact that we necessarily have $g(\mathrm{Id}) = I_r$ and that we have therefore

$$
g(\sigma) g(\sigma^{-1})^\sigma = I_r \text{ for all } \sigma \in G.
$$

This concludes the proof.   □

VI.3. Shaping and crossed products based codes

Let L/k be a Galois extension of degree n with Galois group G, and then let $\xi \in Z^2(G, L^\times)$ be a 2-cocycle. We consider the crossed product $A = (\xi, L/k, G) = \bigoplus_{\sigma \in G} e_\sigma L$ over L/k, where the generators e_σ are subject to the relations

$$
\lambda e_\sigma = e_\sigma \lambda^\sigma, e_\sigma e_\tau = e_{\sigma\tau} \xi_{\sigma,\tau} \text{ for all } \sigma, \tau \in G, \lambda \in L,
$$

where we have set $\lambda^\sigma = \sigma^{-1}(\lambda)$. Since L is a subfield of A, we may consider the k-algebra morphism

$$
\varphi_{A,L} : A \hookrightarrow \mathrm{M}_n(L)
$$

described in Remark I.2.12. Recall that $\varphi_{A,L}$ maps an element $a \in A$ onto M_a, the matrix of left multiplication by a in a chosen L-basis of the right L-vector space A. We will now compute M_a with respect to the basis $(e_\sigma)_{\sigma \in G}$.

LEMMA VI.3.1. *Let $A = (\xi, L/K, G)$, and let $a = \sum_{\sigma \in G} e_\sigma a_\sigma$. Then the matrix M_a of left multiplication by a, relative to the L-basis $(e_\sigma)_{\sigma \in G}$ is*

$$
M_a = (\xi_{\sigma\tau^{-1}, \tau} a_{\sigma\tau^{-1}}^\tau)_{\sigma, \tau}.
$$

Proof. For all $\tau \in G$, we have

$$
\begin{aligned}
ae_\tau &= \sum_{\sigma \in G} e_\sigma a_\sigma e_\tau \\
&= \sum_{\sigma \in G} e_\sigma e_\tau a_\sigma^\tau \\
&= \sum_{\sigma \in G} e_{\sigma\tau} \xi_{\sigma,\tau} a_\sigma^\tau \\
&= \sum_{\sigma \in G} e_\sigma \xi_{\sigma\tau^{-1},\tau} a_{\sigma\tau^{-1}}^\tau,
\end{aligned}
$$

the last equality being obtained by performing the change of variables $\sigma \leftrightarrow \sigma\tau^{-1}$. This concludes the proof. $\qquad\square$

EXAMPLE VI.3.2. Assume that $\mathrm{char}(k) \neq 2$. Let $Q = (a,b)_k$ be a division quaternion k-algebra, with standard basis $1, i, j, ij$.

If $L = k(j)$, then the Galois group G of L/k is generated by

$$
\sigma \colon \begin{array}{c} L \longrightarrow L \\ x + yj \longmapsto x - yj, \end{array}
$$

and we may see Q as a G-crossed product (see Example VI.1.3), with L-basis

$$
e_{\mathrm{Id}} = 1, \ e_\sigma = i,
$$

the corresponding cocycle ξ being given by

$$
\xi_{\mathrm{Id},\sigma} = \xi_{\sigma,\mathrm{Id}} = 1, \ \xi_{\sigma,\sigma} = a.
$$

Thus the matrix of left multiplication by $q = (x + yj) + i(z + tj)$ is given by

$$
M_q = \begin{pmatrix} x + yj & a(z - tj) \\ z + tj & x - yj \end{pmatrix} = \begin{pmatrix} x + y\sqrt{b} & a(z - t\sqrt{b}) \\ z + t\sqrt{b} & x - y\sqrt{b} \end{pmatrix}.
$$

Therefore, we recover the quaternion algebras codes described in Section II.3. $\quad\square$

REMARK VI.3.3. In general, there is no obvious way to decide whether a crossed product is a division algebra. We will see later on criteria for cyclic algebras and central simple algebras of degree 4. $\qquad\square$

We will now introduce a new code design, called *shaping*. Recall from Chapter I that the channel is given by

$$
\mathbf{Y} = \mathbf{H}\mathbf{X} + \mathbf{V}.
$$

If $\mathbf{X} \in \mathrm{M}_n(\mathbb{C})$ is a complex matrix with columns $\mathbf{X}_1, \ldots, \mathbf{X}_n$, we set

$$
vec(\mathbf{X}) = \begin{pmatrix} \mathbf{X}_1 \\ \vdots \\ \mathbf{X}_n \end{pmatrix} \in \mathbb{C}^{n^2}.
$$

An equivalent channel equation is obtained by serializing the columns of each matrix. The previous matrix equation is then equivalent to the system of n equations

$$
\mathbf{Y}_i = \mathbf{H}\mathbf{X}_i + \mathbf{V}_i, i = 1, \ldots, n.
$$

This yields

$$vec(\mathbf{Y}) = \begin{pmatrix} \mathbf{H} & \mathbf{0} & \dots & \mathbf{0} \\ \mathbf{0} & \mathbf{H} & \dots & \mathbf{0} \\ \vdots & \vdots & \ddots & \vdots \\ \mathbf{0} & \mathbf{0} & \dots & \mathbf{H} \end{pmatrix} vec(\mathbf{X}) + vec(\mathbf{V}).$$

Now let $\mathbf{s}$ denote a column vector containing all the information symbols. If the codeword $\mathbf{X}$ is a linear mapping of the information symbols, then there exists a matrix Φ such that

$$vec(\mathbf{X}) = \Phi\mathbf{s}.$$

Thus we can rewrite the equivalent channel as

$$vec(\mathbf{Y}) = \underbrace{\begin{pmatrix} \mathbf{H} & \mathbf{0} & \dots & \mathbf{0} \\ \mathbf{0} & \mathbf{H} & \dots & \mathbf{0} \\ \vdots & \vdots & \ddots & \vdots \\ \mathbf{0} & \mathbf{0} & \dots & \mathbf{H} \end{pmatrix}}_{\hat{\mathbf{H}}} \Phi\mathbf{s} + vec(\mathbf{V}),$$

or equivalently

$$\hat{\mathbf{y}} = \hat{\mathbf{H}}\Phi\mathbf{s} + \hat{\mathbf{v}}$$

by setting $\hat{\mathbf{y}} = vec(\mathbf{Y})$ and $\hat{\mathbf{v}} = vec(\mathbf{V})$. Now $\Phi\mathbf{s}$ can be understood as a lattice point, where Φ is the lattice generator matrix, and the lattice is defined over the alphabet of $\mathbf{s}$ (typically $\mathcal{O}_k$ for division algebra based codes). Since $\mathbf{s}$ takes a finite number of values, we are considering a finite subset of a lattice, whose shape will be dictated by its fundamental parallelotope. Since the norm of each lattice point determines the amount of energy needed to send this point, the average energy required for the finite lattice constellation follows from its shape, and thus depends on the chosen lattice. The most energy efficient shaping is of course the sphere, which is however hard to obtain. The next best choice is thus a cubic shaping, obtained through the cubic lattice, that is, asking Φ to be a unitary matrix.

EXAMPLE VI.3.4. Consider the code $\mathcal{C}$ given in Example VI.3.2. A codeword $\mathbf{X} \in \mathcal{C}$ is of the form

$$\mathbf{X} = \begin{pmatrix} x + y\sqrt{b} & a(z - t\sqrt{b}) \\ z + t\sqrt{b} & x - y\sqrt{b} \end{pmatrix}.$$

By vectorizing $\mathbf{X}$, we have that

$$vec(\mathbf{X}) = \Phi\mathbf{s}$$

with $\mathbf{s} = \begin{pmatrix} x \\ y \\ z \\ t \end{pmatrix}$ and

$$\Phi = \begin{pmatrix} 1 & \sqrt{b} & 0 & 0 \\ 0 & 0 & 1 & \sqrt{b} \\ 0 & 0 & a & -a\sqrt{b} \\ 1 & -\sqrt{b} & 0 & 0 \end{pmatrix}.$$

Furthermore, we can rewrite

$$\Phi = P\Phi'$$

where P is a permutation matrix and Φ' is given by

$$\Phi' = \begin{pmatrix} 1 & \sqrt{b} & 0 & 0 \\ 1 & -\sqrt{b} & 0 & 0 \\ 0 & 0 & 1 & \sqrt{b} \\ 0 & 0 & a & -a\sqrt{b} \end{pmatrix}.$$

Since permutation matrices are unitary, clearly Φ is unitary if and only if Φ' is. However it is easier to determine under which conditions Φ' is unitary, namely

$$B = \begin{pmatrix} 1 & \sqrt{b} \\ 1 & -\sqrt{b} \end{pmatrix}, \begin{pmatrix} 1 & 0 \\ 0 & a \end{pmatrix} B$$

have to be unitary. Equivalently, B has to be unitary and $|a|^2 = 1$. $\square$

Let us now go back to the case where we have codes $\mathcal{C} \subset \mathcal{C}_{A,I}$, where $A = (\xi, L/k, G)$. The field L can be seen as an n-dimensional k-vector space. We will assume that $k/\mathbb{Q}$ is a quadratic imaginary extension, and that $\mathcal{O}_k$ is a principal ideal domain (recall from Remark IV.3.9 that these conditions are justified by the coding theoretic setting).

We now derive the shaping constraint we are looking for. Since any ideal of $\mathcal{O}_L$ is a free $\mathcal{O}_k$-module of rank $n = |G|$, we will index any $\mathcal{O}_k$-basis of I with the elements of G.

Let $A = (\xi, L/k, G)$ and let I be an ideal of $\mathcal{O}_L$.

If we choose an $\mathcal{O}_k$-basis $(\omega_\sigma)_{\sigma \in G}$ of I, we will encode (for now) n^2 information symbols $(a_{\sigma,\tau})_{\sigma,\tau \in G}$ into the matrix $M_a \in \mathcal{C}_{A,I}$, where

$$a = \sum_{\sigma \in G} e_\sigma \Big(\sum_{\tau \in G} a_{\sigma,\tau} \omega_\tau \Big).$$

PROPOSITION VI.3.5. *With this way of encoding, the energy constraint is satisfied if and only if the following two conditions are fulfilled:*

(1) $|\xi_{\sigma,\tau}|^2 = 1$ *for all* $\sigma, \tau \in G$;

(2) *the matrix* $W = (\omega_\tau^\sigma)_{\sigma,\tau}$ *is unitary.*

Proof. For $\sigma \in G$, set $a_\sigma = \sum_{\tau \in G} a_{\sigma,\tau} \omega_\tau$. We would like to have

$$\sum_{\sigma,\tau \in G} |a_{\sigma,\tau}|^2 = \sum_{\sigma,\tau \in G} |\xi_{\sigma\tau^{-1},\tau} a_{\sigma\tau^{-1}}^\tau|^2 = \sum_{\sigma,\tau \in G} |\xi_{\sigma,\tau} a_\sigma^\tau|^2$$

for all $a_{\sigma,\tau} \in \mathcal{O}_k$. For $\sigma \in G$, we consider the two column vectors of L^n

$$X_\sigma = (\xi_{\sigma,\rho} a_\sigma^\rho)_{\rho \in G}, A_\sigma = (a_{\sigma,\rho})_{\rho \in G}.$$

Let D_σ be the diagonal matrix of $\mathrm{M}_n(L)$ whose non-zero entry at column ρ is $\xi_{\sigma,\rho}$. Since $a_{\sigma,\tau} \in k$, we have

$$\xi_{\sigma,\tau} a_\sigma^\tau = \sum_{\rho \in G} \xi_{\sigma,\tau} a_{\sigma,\rho}^\tau \omega_\rho^\tau = \sum_{\rho \in G} \xi_{\sigma,\tau} a_{\sigma,\rho} \omega_\rho^\tau.$$

Now if $W = (\omega_\rho^\tau)_{\tau,\rho}$, then $D_\sigma W = (\xi_{\sigma,\tau} \omega_\rho^\tau)_{\tau,\rho}$, and therefore we get

$$X_\sigma = D_\sigma W A_\sigma \text{ for all } \sigma \in G.$$

Let $\mathbf{x}$ and $\mathbf{a}$ be the block column vectors defined by

$$\mathbf{x} = \begin{pmatrix} \vdots \\ X_\sigma \\ \vdots \end{pmatrix}, \mathbf{a} = \begin{pmatrix} \vdots \\ A_\sigma \\ \vdots \end{pmatrix},$$

and let $M \in \mathrm{M}_{n^2}(L)$ be the block diagonal matrix

$$M = \begin{pmatrix} \ddots & & \\ & D_\sigma W & \\ & & \ddots \end{pmatrix}.$$

Then we have $\mathbf{x} = M\mathbf{a}$. Since $\mathbf{x}$ contains all the entries of M_a and $\mathbf{a}$ contains all the information symbols, fulfilling the energy constraint is equivalent to asking for M to be unitary. It is equivalent to saying that $D_\sigma W$ is unitary for all $\sigma \in G$. Since $D_{\mathrm{Id}} = I_n$, it is equivalent to asking for W to be unitary and for D_σ to be unitary for all $\sigma \in G$. In view of the definition of D_σ, this is equivalent to conditions (1) and (2). $\qquad \square$

Finding an $\mathcal{O}_k$-basis of I satisfying condition (2) is not easy. In order to simplify the problem, we will make the extra assumption that complex conjugation induces a $\mathbb{Q}$-automorphism on L, which commutes with every element of $\mathrm{Gal}(L/k)$.

In this case, it is easy to check that $\overline{W}^t W = (\mathrm{Tr}_{L/k}(\overline{\omega}_\sigma \omega_\tau))_{\sigma,\tau}$. Hence, we may find an $\mathcal{O}_k$-basis of I for which encoding satisfy the shaping constraint if and only if the hermitian complex $\mathcal{O}_k$-lattice (I, h_1) defined by

$$h_1 : \begin{array}{l} I \times I \longrightarrow \mathcal{O}_k \\ (x, y) \longmapsto \mathrm{Tr}_{L/k}(\overline{x}y) \end{array}$$

is isomorphic to the cubic lattice $\mathcal{O}_k^n$. In particular, we should have $\det(I, h_1) = 1$. By Proposition C.2.4, we would have

$$1 = N_{L/\mathbb{Q}}(I)d_{L/k},$$

where $d_{L/k} = \det(\mathcal{O}_L, h_1)$ is the relative discriminant of L/k (see definition C.2.3). In particular, we would get $d_{L/k} = 1$, since $d_{L/k}$ and $N_{L/\mathbb{Q}}(I)$ are integers. But Corollary C.2.11 then shows on the other hand that $d_{L/k} \geq n^n > 1$ as soon as $n \geq 2$.

This proves that (I, h_1) is never isomorphic to the cubic lattice, so we now modify the encoding process as follows. Let $\lambda \in L^\times$ satisfying the following conditions:

(a) $\overline{\lambda} = \lambda$;

(b) λ^σ is a positive real number for all $\sigma \in G$;

(c) $\mathrm{Tr}_{L/k}(\lambda \overline{x} y) \in \mathcal{O}_k$ for all $x, y \in I$.

Notice that in this case, $N_{L/k}(\lambda)$ and $\mathrm{Tr}_{L/k}(\lambda)$ are positive real numbers.

Let $D_\lambda \in \mathrm{M}_n(\mathbb{R})$ be the diagonal matrix whose diagonal entries are the real numbers $\sqrt{\lambda^\sigma}, \sigma \in G$. If $(\omega_\sigma)_{\sigma \in G}$ is an $\mathcal{O}_k$-basis of I, we will encode n^2 information symbols

$(a_{\sigma,\tau})_{\sigma,\tau \in G}$ into the matrix

$$\mathbf{X}_a = M_a D_\lambda = (\sqrt{\lambda^\tau} \xi_{\sigma\tau^{-1},\tau} a^\tau_{\sigma\tau^{-1}})_{\sigma,\tau},$$

where $a_\sigma = \sum_{\tau \in G} a_{\sigma,\tau} \omega_\tau$ for all $\sigma \in G$.

The reader will check that this new way of encoding simply replaces W by $W_\lambda = D_\lambda W$ in the proof of the previous proposition. Now $\overline{W}^t_\lambda W_\lambda = (\mathrm{Tr}_{L/k}(\lambda \overline{\omega}_\sigma \omega_\tau))_{\sigma,\tau}$. Hence we may find an $\mathcal{O}_k$-basis of I for which encoding fulfills the shaping constraint if and only if the hermitian $\mathcal{O}_k$-lattice

$$I \times I \longrightarrow \mathcal{O}_k$$
$$(x,y) \longmapsto \mathrm{Tr}_{L/k}(\lambda \overline{x} y)$$

is isomorphic to the cubic lattice $\mathcal{O}^n_k$. We then set

$$\mathcal{C}_{A,\lambda,I} = \{\mathbf{X}_a = M_a D_\lambda \mid a \in \Lambda_{A,I}\},$$

where $\Lambda_{A,I} = \bigoplus_{\sigma \in G} e_\sigma I$. Clearly, we have

$$\delta_{min}(\mathcal{C}_{A,\lambda,I}) = N_{L/k}(\lambda) \delta_{min}(\mathcal{C}_{A,I}).$$

Corollary IV.3.8 then shows that this quantity bounded by a positive constant whenever A is a division k-algebra.

At this point, we would like to summarize what we have done so far.

Assume that $k/\mathbb{Q}$ is a quadratic imaginary extension, and that $\mathcal{O}_k$ is a principal ideal domain. Let L/k be a Galois extension with Galois group G satisfying the following conditions:

(1) complex conjugation induces a $\mathbb{Q}$-automorphism of L which commutes with every element of G;

(2) there exists $\lambda \in L^\times$ and I an ideal of $\mathcal{O}_L$ satisfying the following conditions:

(a) $\overline{\lambda} = \lambda$, that is $\lambda \in \mathbb{R}$;

(b) λ^σ is a positive real number for all $\sigma \in G$;

(c) $\mathrm{Tr}_{L/k}(\lambda \overline{x} y) \in \mathcal{O}_k$ for all $x, y \in I$;

(d) the hermitian $\mathcal{O}_k$-lattice

$$h_\lambda : \begin{array}{c} I \times I \longrightarrow \mathcal{O}_k \\ (x,y) \longmapsto \mathrm{Tr}_{L/k}(\lambda \overline{x} y) \end{array}$$

is isomorphic to the cubic lattice $\mathcal{O}^n_k$.

Then for any orthonormal $\mathcal{O}_k$-basis $(\omega_\sigma)_{\sigma \in G}$ of (I, h_λ), and for any crossed product division k-algebra $A = (\xi, L/k, G)$ such that

$$|\xi_{\sigma,\tau}|^2 = 1 \text{ for all } \sigma, \tau \in G,$$

the encoding map

$$\mathcal{O}^n_k \longrightarrow \mathcal{C}_{A,\lambda,I}$$
$$(a_{\sigma,\tau})_{\sigma,\tau \in G} \longmapsto (\sqrt{\lambda^\tau} \xi_{\sigma\tau^{-1},\tau} a^\tau_{\sigma\tau^{-1}})_{\sigma,\tau},$$

where $a_\sigma = \sum_{\tau \in G} a_{\sigma,\tau} \omega_\tau$ for all $\sigma \in G$ is fulfilling the shaping constraint. Moreover, $\delta_{min}(\mathcal{C}_{A,\lambda,I})$ is bounded by a positive constant.

We would now like to apply the results proved in Appendix C to give an estimation of the minimum determinant of a crossed product based code, that is we have to estimate

$$\delta_{min}(\mathcal{C}_{A,\lambda,I}) = \inf_{a \neq 0} |\det(\mathbf{X}_a)|^2.$$

We will let the reader refer to Appendix C for the definitions and results concerning hermitian lattices and complex ideal lattices (I, h_λ). Let us introduce some notation first. The set

$$\mathcal{E}_\xi^{(\tau)} = \{c \in \mathcal{O}_k \mid c\xi_{\sigma\tau^{-1},\tau} \in \mathcal{O}_L \text{ for all } \sigma \in G\}$$

is an ideal of $\mathcal{O}_k$. We will denote by $\Delta_\xi^{(\tau)}$ the norm of this ideal. Equivalently, we have

$$\Delta_\xi^{(\tau)} = N_{k/\mathbb{Q}}(c_\xi^{(\tau)}) = |c_\xi^{(\tau)}|^2,$$

for any generator $c_\xi^{(\tau)}$ of $\mathcal{E}_\xi^{(\tau)}$. Notice that by the definition of a cocycle, we have $\Delta_\xi^{(\text{Id})} = 1$.

By definition, we have

$$\Delta_\xi^{(\tau)} = 1 \iff \xi_{\sigma\tau^{-1},\tau} \in \mathcal{O}_L \text{ for all } \sigma \in G.$$

Now set

$$\Delta_\xi = \prod_{\sigma \in G} \Delta_\xi^{(\sigma)}.$$

We then have

$$\Delta_\xi = 1 \iff \xi_{\sigma,\tau} \in \mathcal{O}_L \text{ for all } \sigma, \tau \in G.$$

Finally, we set

$$N_{min}(I) = \min_{x \in I \setminus \{0\}} |N_{L/\mathbb{Q}}(x)|.$$

Notice that $N_{min}(I) = N_{L/\mathbb{Q}}(I)$ if I is a principal ideal. In general, equality does not hold.

Recall from Appendix C that the **relative discriminant** of L/k, denoted by $d_{L/k}$, is the determinant of $(\mathcal{O}_L, h_1)$. In other words,

$$d_{L/k} = \det(\text{Tr}_{L/k}(\overline{w}_i w_j)) \in \mathbb{Z}$$

for any $\mathcal{O}_k$-basis $(w_1, \ldots, w_n)$ of $\mathcal{O}_L$. We then have the following result:

PROPOSITION VI.3.6. *Assume that $A = (\xi, L/k, G)$ is a central division k-algebra. With the previous notation, we have*

$$\frac{1}{\Delta_\xi d_{L/k}} \leq \delta_{min}(\mathcal{C}_{A,\lambda,I}) \leq N_{L/k}(\lambda) N_{min}(I).$$

If moreover $\Delta_\xi = 1$ and I is principal, we have

$$\delta_{min}(\mathcal{C}_{A,\lambda,I}) = \frac{1}{d_{L/k}}.$$

Proof. Let $x \in I \setminus \{0\}$ with minimal absolute norm. Then we have

$$\mathbf{X}_{e_{\mathrm{Id}}x} = M_{e_{\mathrm{Id}}x}D_\lambda = xM_{e_{\mathrm{Id}}}D_\lambda = xD_\lambda.$$

It follows that

$$\det(\mathbf{X}_{e_{\mathrm{Id}}x}) = \prod_{\tau \in G} x^\tau \sqrt{\lambda^\tau} = N_{L/k}(x)\sqrt{N_{L/k}(\lambda)},$$

and therefore

$$|\det(\mathbf{X}_{e_{\mathrm{Id}}x})|^2 = N_{L/\mathbb{Q}}(x)N_{L/k}(\lambda) = N_{min}(I)N_{L/k}(\lambda),$$

since we have

$$|N_{L/k}(x)|^2 = N_{k/\mathbb{Q}}(N_{L/k}(x)) = N_{L/\mathbb{Q}}(x).$$

The upper bound then follows from the definition of $\delta_{min}(\mathcal{C}_{A,\lambda,I})$.

Now let $\mathbf{X}_a = M_a D_\lambda \in \mathcal{C}_{A,\lambda,I}$. To establish the lower bound, notice that

$$|\det(\mathbf{X}_a)|^2 = |\det(M_a)|^2 N_{L/k}(\lambda).$$

Recall that we have $\det(M_a) = \mathrm{Nrd}_A(a) \in k$. Now let $c_\xi^{(\tau)}$ be a generator of $\mathcal{E}_\xi^{(\tau)}$, let C be the invertible diagonal matrix whose diagonal entry at column τ is $c_\xi^{(\tau)}$, and let $M_a' = M_a C$. By definition of $c_\xi^{(\tau)}$, we have $M_a' = (c_\xi^{(\tau)}\xi_{\sigma\tau^{-1},\tau}a_{\sigma\tau^{-1}}^\tau)_{\sigma,\tau} \in \mathrm{M}_n(\mathcal{O}_L)$. Thus $\det(M_a') \in \mathcal{O}_L \cap k = \mathcal{O}_k$ and $|\det(M_a')|^2 \in \mathbb{Z}$.

Since each coefficient in the τ^{theo}-column of M_a' lies in I^τ, the definition of the determinant and the previous observation show that we have

$$\det(M_a') \in (\prod_{\tau \in G} I^\tau) \cap \mathcal{O}_k = \mathcal{N}_{L/k}(I).$$

It follows that

$$|\det(M_a')|^2 \in \overline{\mathcal{N}_{L/k}(I)}\mathcal{N}_{L/k}(I) \cap \mathbb{Z} = \mathcal{N}_{L/\mathbb{Q}}(I) = N_{L/\mathbb{Q}}(I)\mathbb{Z}.$$

Now we have

$$|\det(\mathbf{X}_a)|^2 = |\det(M_a'C^{-1}D_\lambda)|^2 = \frac{1}{\Delta_\xi}N_{L/k}(\lambda)|\det(M_a')|^2,$$

and thus

$$|\det(\mathbf{X}_a)|^2 \in \frac{1}{\Delta_\xi}N_{L/k}(\lambda)N_{L/\mathbb{Q}}(I)\mathbb{Z}.$$

Since $\det(\mathbf{X}_a) \neq 0$ if $a \neq 0$, we get

$$|\det(\mathbf{X}_a)|^2 \geq \frac{1}{\Delta_\xi}N_{L/k}(\lambda)N_{L/\mathbb{Q}}(I) \text{ for all } a \in \Lambda_{A,I}, a \neq 0.$$

Using Proposition C.2.4 and the definition of the minimum determinant, we get the desired lower bound.

Finally, if I is a principal ideal, then $N_{min}(I) = N_{L/\mathbb{Q}}(I)$. Using Proposition C.2.4 again, we see that the two bounds are equal whenever $\Delta_\xi = 1$ and I is principal. $\qquad\square$

REMARK VI.3.7. Notice that if $x \in I \setminus \{0\}$ is an element with minimal norm, the first isomorphism theorem applied to the surjective morphism $\mathcal{O}_L/x\mathcal{O}_L \longrightarrow \mathcal{O}_L/I$ shows that $N_{min}(I) = N_{L/\mathbb{Q}}(I)[I : x\mathcal{O}_L]$. Hence the equation in the previous proposition may be rewritten as

$$\frac{1}{\Delta_\xi d_{L/k}} \leq \delta_{min}(\mathcal{C}_{A,\lambda,I}) \leq \frac{[I : x\mathcal{O}_L]}{d_{L/k}}.$$

This shows that maximizing $\delta_{min}(\mathcal{C}_{A,\lambda,I})$ is essentially equivalent to minimizing $d_{L/k}$. The lower bound also shows that it is in our interest to choose the cocycle values to be algebraic integers whenever it is possible. $\qquad\square$

EXERCISES

1. Let k be a field of characteristic different from 2.

   Let $Q_i = (a_i, b_i)_k, i = 1, 2$ be two quaternion k-algebras, and assume that $a_i \notin k^{\times 2}$. Prove that $Q_1 \otimes Q_2$ has a maximal subfield isomorphic to $L = k(\sqrt{a_1}, \sqrt{a_2})$, and compute a 2-cocycle ξ such that

   $$Q_1 \otimes_k Q_2 \cong_k (\xi, L/k, \mathbb{Z}/2\mathbb{Z} \times \mathbb{Z}/2\mathbb{Z}).$$

2. Let G be a finite group of order n.

   (a) Let C be a multiplicative abelian group on which G acts on the right. Show that $H^2(G, C)$ is killed by n.
   Hint: If $\xi : G \times G \longrightarrow C$ is a 2-cocycle, consider the map

   $$z: \quad \begin{array}{c} G \longrightarrow C \\ \sigma \longmapsto \prod_{\rho \in G} \xi_{\sigma,\rho}^{\rho^{-1}}. \end{array}$$

   (b) Recover the fact that the Brauer group is a torsion group.

3. Let F/k be a field extension. We say that k is **algebraically closed in** F if $F \cap \overline{k} = k$, that is if every element of F which is algebraic over k is an element of k.

   Let A be a central simple k-algebra. The main goal of this exercise is to show the existence of a field extension F/k which splits A and such that k is algebraically closed in F.

   (a) Reduce to the case of crossed products.

   Let $A = (\xi, L/k, G)$, and let $E = L(X_\sigma, \sigma \in G)$, where $X_\sigma, \sigma \in G$ are algebraically independent indeterminates over L.

   (b) Show that there is a unique right action of G by k-algebra automorphisms on E extending the natural right action of G on L such that

   $$X_\tau^\sigma = \xi_{\sigma,\tau} X_\sigma^{-1} X_{\sigma\tau} \text{ for all } \sigma, \tau \in G.$$

   (c) Let $F = E^G$. Show that k is algebraically closed in F.

   Hint: Show that $E \cap \overline{k} = L$.

(d) Show that F/k splits A.

Hint: Show that $L \otimes_k F \cong_k LF = L(X_\sigma, \sigma \in G)$.

We now give an application of the previous result. Recall first that if $b : V \times V \longrightarrow$ is a non-singular bilinear form, the **determinant** of b is the square-class of $\det((b(e_i, e_j))_{i,j})$ in $k^\times/k^{\times 2}$, where $(e_1, \ldots, e_n)$ is any k-basis of V.

Let A be a central simple k-algebra. Recall that the symmetric bilinear form

$$T_A : \begin{array}{c} A \times A \longrightarrow k \\ (a, a') \longmapsto \mathrm{Trd}_A(aa') \end{array}$$

is non-singular.

(e) Show that $\det(T_{\mathrm{M}_n(k)}) = (-1)^{\frac{n(n-1)}{2}}$ for all $n \geq 1$.

Hint: Write the matrix of $T_{\mathrm{M}_n(k)}$ in the basis

$$E_{11}, \ldots, E_{nn}, E_{ij}, E_{ji}, i < j.$$

(f) Deduce from the previous results that $\det(T_A) = (-1)^{\frac{n(n-1)}{2}}$ for every central simple k-algebra A of degree n.

Hint: Observe that if k is algebraically closed in F, the map

$$k^\times/k^{\times 2} \longrightarrow F^\times/F^{\times 2}$$

is injective.

CHAPTER VII

Cyclic algebras

In this chapter, we investigate the structure of crossed products over a cyclic extension. This will lead to the notion of a cyclic algebra. When the base field contains a primitive root of 1, these algebras have a nice description and are called symbol algebras in this particular situation. We then describe the structure of central simple algebras over local fields and number fields.

VII.1. Cyclic algebras

In order to have a more explicit description of crossed products over a cyclic extension L/k, we need to compute $H^2(\mathrm{Gal}(L/k), L^\times)$. Since it does not cost any extra work, we will compute $H^2(G, C)$ for any cyclic group G and any abelian group C (written multiplicatively) on which G acts on the right by automorphisms. We start by exhibiting examples of 2-cocycles.

LEMMA VII.1.1. *Let G be a cyclic group of order n, acting on the right on an abelian group C by automorphisms. Let $\sigma \in G$ be a generator of G, and let $c \in C^G$ be an element of C fixed by G.*

For all $0 \leq i, j \leq n - 1$, set

$$\xi^{\sigma,c}_{\sigma^i,\sigma^j} = \begin{cases} 1 & \text{if } i + j < n \\ c & \text{if } i + j \geq n \end{cases}$$

Then the map $\xi^{\sigma,c} : G \times G \longrightarrow C$ is a 2-cocycle.

Proof. To simplify notation, we will write ξ instead of $\xi^{\sigma,c}$. Notice that, since G acts by automorphisms on C, we have $g \cdot 1 = 1$ for all $g \in G$. Hence $\xi_{\sigma^i,\sigma^j} \in C^G$ for all i, j. Therefore, the relations to check may be rewritten as

$$\xi_{1,\sigma^j} = \xi_{\sigma^i,1} = 1$$

and

$$\xi_{\sigma^i,\sigma^{j+\ell}}\xi_{\sigma^j,\sigma^\ell} = \xi_{\sigma^{i+j},\sigma^\ell}\xi_{\sigma^i,\sigma^j} \text{ for all } 0 \leq i, j, \ell \leq n - 1.$$

The first relation is clearly satisfied by definition of ξ, so let us check the other one.

If $i + j < n$ and $j + \ell < n$, then we have

$$\xi_{\sigma^{i+j},\sigma^\ell} = \xi_{\sigma^i,\sigma^{j+\ell}}$$

by definition of ξ. Moreover, we have

$$\xi_{\sigma^i,\sigma^j} = \xi_{\sigma^j,\sigma^\ell} = 1$$

in this case, and therefore the desired equality holds.

If $i + j \geq n$ and $j + \ell < n$, we have

$$\xi_{\sigma^i, \sigma^j} = c \text{ and } \xi_{\sigma^j, \sigma^\ell} = 1.$$

Notice that we have $i + j + \ell \geq i + j \geq n$, and thus

$$\xi_{\sigma^i, \sigma^{j+\ell}} = c.$$

Finally, we have $0 \leq i + j - n < n$, so we get

$$\xi_{\sigma^{i+j}, \sigma^\ell} = \xi_{\sigma^{i+j-n}, \sigma^\ell}.$$

Since $i + j - n + \ell = i - n + j + \ell < i < n$, we have

$$\xi_{\sigma^{i+j}, \sigma^\ell} = 1.$$

hence the relation is once again satisfied.

If $i + j < n$ and $j + \ell \geq n$, similar arguments show the desired equality holds.

If $i + j \geq n$ and $j + \ell \geq n$, we have

$$\xi_{\sigma^i, \sigma^j} = \xi_{\sigma^j, \sigma^\ell} = c.$$

Moreover we have

$$\xi_{\sigma^{i+j}, \sigma^\ell} = \xi_{\sigma^{i+j-n}, \sigma^\ell} \text{ and } \xi_{\sigma^i, \sigma^{j+\ell}} = \xi_{\sigma^i, \sigma^{j+\ell-n}}.$$

By definition of ξ, these two quantities are equal. This concludes the proof.   □

EXAMPLE VII.1.2. Let L/k be a cyclic extension of degree n with Galois group G generated by σ.

Let $a \in k^\times = (L^\times)^G$, and let $(a, L/k, \sigma) = (\xi^{\sigma,a}, L/k, G)$ the corresponding crossed product. Set $e = e_\sigma$. An easy induction argument and the definition of $\xi^{\sigma,a}$ shows that we have

$$e_\sigma^i = e_{\sigma^i} \text{ for all } i = 0, \ldots, n - 1.$$

Therefore, we have

$$(a, L/k, \sigma) = \bigoplus_{i=0}^{n-1} e^i L.$$

Moreover, we have

$$\lambda e = e \lambda^\sigma \text{ for all } \lambda \in L.$$

Finally, we have

$$e^n = e^{n-1} e = e_{\sigma^{n-1}} e_\sigma = e_{\mathrm{Id}} a = a.$$

These relations completely describe the k-algebra $(a, L/k, \sigma)$.   □

DEFINITION VII.1.3. The k-algebra $(a, L/k, \sigma)$ is called a **cyclic algebra**. It is the k-algebra $(a, L/k, \sigma) = \bigoplus_{i=0}^{n-1} e^i L$ generated by one element e subject to the relations

$$e^n = a, \lambda e = e \lambda^\sigma \text{ for all } \lambda \in L.$$

REMARK VII.1.4. Assume that $\mathrm{char}(k) \neq 2$. Let $L = k(\sqrt{b})$ be a quadratic extension, and let σ be the unique non-trivial k-automorphism of L. One can easily see that $(a, L/k, \sigma) \cong_k (a, b)_k$. Hence cyclic algebras may be viewed as a generalization of quaternion algebras. We will see in the sequel that cyclic algebras share properties similar to those of quaternion algebras.   □

We now give an explicit description of $H^2(G, C)$.

THEOREM VII.1.5. *Let G be a cyclic group of order n, acting on the right on an abelian group C by automorphisms. Let $\sigma \in G$ be a generator of G. For all $c \in C$, set*

$$N(c) = cc^{\sigma} \cdots c^{\sigma^{n-1}} \in C^G.$$

Then the map

$$\overline{\varphi}: \quad \begin{array}{c} C^G/N(C) \longrightarrow H^2(G, C) \\ \overline{c} \longmapsto [\xi^{\sigma,c}] \end{array}$$

is a well-defined group isomorphism.

Proof. Consider the map

$$\varphi: \quad \begin{array}{c} C^G \longrightarrow H^2(G, C) \\ c \longmapsto [\xi^{\sigma,c}]. \end{array}$$

It easily follows from the definition that we have

$$\xi^{\sigma,c}\xi^{\sigma,c'} = \xi^{\sigma,cc'} \text{ for all } c, c' \in C^G.$$

Hence the map φ is a group morphism.

Let us check that φ is surjective. Let $\xi \in Z^2(G, C)$, and set

$$c = \prod_{m=0}^{n-1} \xi_{\sigma,\sigma^m}.$$

We claim that $c \in C^G$. To prove this, it is enough to check that $c^{\sigma} = c$. Since ξ is a 2-cocycle, we have

$$c^{\sigma} = \prod_{m=0}^{n-1} \xi_{\sigma,\sigma^m}^{\sigma} = \prod_{m=0}^{n-1} \xi_{\sigma,\sigma^{m+1}} \xi_{\sigma^m,\sigma} \xi_{\sigma^{m+1},\sigma}^{-1}.$$

Therefore we get

$$c^{\sigma} = \Big(\prod_{m=1}^{n} \xi_{\sigma,\sigma^m} \Big) \Big(\prod_{m=0}^{n-1} \xi_{\sigma^m,\sigma} \Big) \Big(\prod_{m=1}^{n} \xi_{\sigma^m,\sigma} \Big)^{-1}.$$

Hence we get

$$c^{\sigma} = \Big(\prod_{m=1}^{n} \xi_{\sigma,\sigma^m} \Big) \xi_{1,\sigma} \xi_{\sigma^n,\sigma}^{-1} = (c\xi_{\sigma,1}^{-1}\xi_{\sigma,\sigma^n}) \xi_{1,\sigma} \xi_{\sigma^n,\sigma}^{-1}.$$

Since we have $\sigma^n = 1$, we get $c^{\sigma} = c$ as claimed. We are now going to prove that ξ is cohomologous to $\xi^{\sigma,c}$. For $i = 0, \ldots, n-1$, we set

$$z_{\sigma^i} = \prod_{m=0}^{i-1} \xi_{\sigma,\sigma^m}.$$

Notice that we have $z_1 = 1$. Let us compute

$$b(\sigma^i, \sigma^j) = z_{\sigma^j} z_{\sigma^i}^{\sigma^j} z_{\sigma^{i+j}}^{-1},$$

for all $0 \leq i, j \leq n-1$. If $i + j < n$, we have

$$b(\sigma^i, \sigma^j) = \Big(\prod_{m=0}^{j-1} \xi_{\sigma,\sigma^m} \Big) \Big(\prod_{m=0}^{i-1} \xi_{\sigma,\sigma^m}^{\sigma^j} \Big) \Big(\prod_{m=0}^{i+j-1} \xi_{\sigma,\sigma^m} \Big)^{-1}.$$

Since ξ is a 2-cocycle, we get

$$b(\sigma^i,\sigma^j) = \Big(\prod_{m=0}^{j-1}\xi_{\sigma,\sigma^m}\Big)\Big(\prod_{m=0}^{i-1}\xi_{\sigma,\sigma^{m+j}}\xi_{\sigma^m,\sigma^j}\xi_{\sigma^{m+1},\sigma^j}^{-1}\Big)\Big(\prod_{m=0}^{i+j-1}\xi_{\sigma,\sigma^m}\Big)^{-1}.$$

Notice that we have

$$\Big(\prod_{m=0}^{i-1}\xi_{\sigma,\sigma^{m+j}}\xi_{\sigma^m,\sigma^j}\xi_{\sigma^{m+1},\sigma^j}^{-1}\Big) = \Big(\prod_{m=0}^{i-1}\xi_{\sigma,\sigma^{m+j}}\Big)\xi_{\sigma^0,\sigma^j}\xi_{\sigma^i,\sigma^j}^{-1}$$

$$= \Big(\prod_{m=j}^{i+j-1}\xi_{\sigma,\sigma^m}\Big)\xi_{\sigma^i,\sigma^j}^{-1},$$

since $\xi_{1,\sigma^j}=1$. Hence

$$b(\sigma^i,\sigma^j) = \Big(\prod_{m=0}^{j-1}\xi_{\sigma,\sigma^m}\Big)\Big(\prod_{m=j}^{i+j-1}\xi_{\sigma,\sigma^m}\Big)\xi_{\sigma^i,\sigma^j}^{-1}\Big(\prod_{m=0}^{i+j-1}\xi_{\sigma,\sigma^m}\Big)^{-1} = \xi_{\sigma^i,\sigma^j}^{-1}.$$

If $i+j\geq n$, we have

$$b(\sigma^i,\sigma^j) = z_{\sigma^j}z_{\sigma^i}^{\sigma^j}z_{\sigma^{i+j}}^{-1} = z_{\sigma^j}z_{\sigma^i}^{\sigma^j}z_{\sigma^{i+j-n}}^{-1},$$

and therefore the computations above show that we have

$$b(\sigma^i,\sigma^j) = \Big(\prod_{m=0}^{j-1}\xi_{\sigma,\sigma^m}\Big)\Big(\prod_{m=j}^{i+j-1}\xi_{\sigma,\sigma^m}\Big)\xi_{\sigma^i,\sigma^j}^{-1}\Big(\prod_{m=0}^{i+j-n-1}\xi_{\sigma,\sigma^m}\Big)^{-1}$$

$$= \Big(\prod_{m=i+j-n}^{i+j-1}\xi_{\sigma,\sigma^m}\Big)\xi_{\sigma^i,\sigma^j}^{-1}.$$

Thus, we get

$$b(\sigma^i,\sigma^j) = \Big(\prod_{m=0}^{n-1}\xi_{\sigma,\sigma^{m+i+j-n}}\Big)\xi_{\sigma^i,\sigma^j}^{-1} = \Big(\prod_{m=0}^{n-1}\xi_{\sigma,\sigma^m}\Big)\xi_{\sigma^i,\sigma^j}^{-1} = c\xi_{\sigma^i,\sigma^j}^{-1},$$

the second equality coming from the fact that multiplication by σ^{i+j-n} induces a permutation of the elements of G. Therefore, in any case, we get

$$b(\sigma^i,\sigma^j) = \xi_{\sigma^i,\sigma^j}^{\sigma,c}\xi_{\sigma^i,\sigma^j}^{-1} \text{ for all } 0\leq i,j\leq n-1,$$

that is

$$\xi_{\sigma^i,\sigma^j}^{\sigma,c} = \xi_{\sigma^i,\sigma^j}z_{\sigma^j}z_{\sigma^i}^{\sigma^j}z_{\sigma^{i+j}}^{-1} \text{ for all } 0\leq i,j\leq n-1.$$

In other words, ξ and $\xi^{\sigma,c}$ are cohomologous and therefore $[\xi]=[\xi^{\sigma,c}]$. This proves the surjectivity of φ. We are now going to show the equality

$$\ker(\varphi) = N(C).$$

The desired result will follow from the first isomorphism theorem.

Let $c\in N(C)$, and let $x\in C$ such that $c=N(x)$. We are going to show that $\xi^{\sigma,c}$ is cohomologous to the trivial cocycle. We set

$$z_{\sigma^i} = \prod_{m=0}^{i-1}x^{\sigma^m} \text{ for all } i=0,\ldots,n-1.$$

Notice that $z_1 = 1$. If $i + j < n$, we have

$$z_{\sigma^j} z_{\sigma^i}^{\sigma^j} z_{\sigma^{i+j}}^{-1} = \Big(\prod_{m=0}^{j-1} x^{\sigma^m}\Big)\Big(\prod_{m=0}^{i-1} x^{\sigma^{m+j}}\Big)\Big(\prod_{m=0}^{i+j-1} x^{\sigma^m}\Big)^{-1}.$$

Hence we have

$$z_{\sigma^j} z_{\sigma^i}^{\sigma^j} z_{\sigma^{i+j}}^{-1} = \Big(\prod_{m=0}^{j-1} x^{\sigma^m}\Big)\Big(\prod_{m=j}^{i+j-1} x^{\sigma^m}\Big)\Big(\prod_{m=0}^{i+j-1} x^{\sigma^m}\Big)^{-1} = 1.$$

If $i + j \geq n$, we have $z_{\sigma^{i+j}} = z_{\sigma^{i+j-n}}$, and therefore

$$\begin{aligned}
z_{\sigma^j} z_{\sigma^i}^{\sigma^j} z_{\sigma^{i+j}}^{-1} &= \Big(\prod_{m=0}^{j-1} x^{\sigma^m}\Big)\Big(\prod_{m=0}^{i-1} x^{\sigma^{m+j}}\Big)\Big(\prod_{m=0}^{i+j-n-1} x^{\sigma^m}\Big)^{-1} \\
&= \Big(\prod_{m=0}^{j-1} x^{\sigma^m}\Big)\Big(\prod_{m=0}^{i+j-1} x^{\sigma^j}\Big)\Big(\prod_{m=0}^{i+j-n-1} x^{\sigma^m}\Big)^{-1}.
\end{aligned}$$

Thus

$$z_{\sigma^j} z_{\sigma^i}^{\sigma^j} z_{\sigma^{i+j}}^{-1} = \prod_{m=i+j-n}^{i+j-1} x^{\sigma^j} = \Big(\prod_{m=0}^{n-1} x^{\sigma^j}\Big)^{\sigma^{i+j-n}}.$$

Consequently, we have

$$z_{\sigma^j} z_{\sigma^i}^{\sigma^j} z_{\sigma^{i+j}}^{-1} = c^{\sigma^{i+j-n}} = c,$$

since $c \in N(C) \subset C^G$. Hence, in any case we have

$$\xi^{\sigma,c}_{\sigma^i,\sigma^j} = z_{\sigma^j} z_{\sigma^i}^{\sigma^j} z_{\sigma^{i+j}}^{-1} \text{ for all } 0 \leq i, j \leq n - 1,$$

meaning that $\xi^{\sigma,c}$ is cohomologous to the trivial cocycle. In other words, $\varphi(c) = [1]$ is $c \in N(C)$, that is

$$N(C) \subset \ker(\varphi).$$

Conversely, assume that $c \in \ker(\varphi)$. By assumption, we have $[\xi^{\sigma,c}] = 0$, and thus there exists a map $z : G \longrightarrow C$ satisfying $z_1 = 1$ such that

$$\xi^{\sigma,c}_{\sigma^i,\sigma^j} = z_{\sigma^j} z_{\sigma^i}^{\sigma^j} z_{\sigma^{i+j}}^{-1} \text{ for all } 0 \leq i, j \leq n - 1.$$

Let $x = z_\sigma$. For $j = 0, \ldots, n - 2$, we have

$$\xi^{\sigma,c}_{\sigma,\sigma^j} = z_{\sigma^j} z_\sigma^{\sigma^j} z_{\sigma^{j+1}}^{-1},$$

that is

$$z_{\sigma^{j+1}} = z_{\sigma^j} x^{\sigma^j} \text{ for all } j = 0, \ldots, n - 2.$$

This yields to

$$z_{\sigma^j} = \prod_{m=0}^{j-1} x^{\sigma^m} \text{ for all } j = 0, \ldots, n - 1.$$

We then have

$$c = \xi_{\sigma,\sigma^{n-1}} = z_{\sigma^{n-1}} z_\sigma^{\sigma^{n-1}} z_{\sigma^n}^{-1} = z_{\sigma^{n-1}} z_\sigma^{\sigma^{n-1}} z_1^{-1} = z_{\sigma^{n-1}} z_\sigma^{\sigma^{n-1}}.$$

Therefore, we get

$$c = \Big(\prod_{m=0}^{n-2} x^{\sigma^m}\Big) x^{\sigma^{n-1}} = \prod_{m=0}^{n-1} x^{\sigma^m} = N(x).$$

This concludes the proof of the theorem. $\qquad\square$

REMARK VII.1.6. This result may be obtained easily using homological algebra techniques. See for example [**44**] for more details. □

COROLLARY VII.1.7. *Let L/k be a cyclic extension, with Galois group G generated by σ. The map*

$$\Theta \colon \begin{array}{c} k^\times / N_{L/k}(L^\times) \longrightarrow \mathrm{Br}(L/k) \\ \overline{a} \longmapsto [(a, L/k, \sigma)] \end{array}$$

is a well-defined group isomorphism.

Proof. Use the previous theorem with $C = L^\times$ together with Theorem VI.2.7. □

REMARK VII.1.8. We can use the previous result to give another proof of Frobenius' Theorem. Let D be a central division $\mathbb{R}$-algebra. If $\deg(D) > 1$, then D is split by a finite non-trivial extension of $\mathbb{R}$ by Corollary IV.1.9, that is D is split by $\mathbb{C}$. Hence $[D] \in \mathrm{Br}(\mathbb{C}/\mathbb{R})$. The previous result immediately shows that

$$\mathrm{Br}(\mathbb{C}/\mathbb{R}) \cong_K \mathbb{R}^\times / N_{\mathbb{C}/\mathbb{R}}(\mathbb{C}^\times) \cong_K \mathbb{Z}/2\mathbb{Z},$$

the non-trivial class being given by the image of -1, namely $[\mathbb{H}]$. Hence $[D] = [\mathbb{H}]$, and since D and $\mathbb{H}$ are both division algebras, we conclude that $D \cong_\mathbb{R} \mathbb{H}$. □

The following properties of cyclic algebras will be very useful in the sequel.

PROPOSITION VII.1.9. *Let L/k be a cyclic extension of degree n, and let σ be a generator of $\mathrm{Gal}(L/k)$. Then the following properties hold:*

(1) *for all $a, b \in L^\times$, $(a, L/k, \sigma) \otimes_k (b, L/k, \sigma)$ is Brauer equivalent to $(ab, L/k, \sigma)$;*

(2) *for all $a, b \in L^\times$, $(a, L/k, \sigma) \simeq (b, L/k, \sigma)$ if and only if we have $ba^{-1} \in N_{L/k}(L^\times)$. In particular, the k-algebra $(a, L/k, \sigma)$ is split if and only if $a \in N_{L/k}(L^\times)$;*

(3) *let F/k be a field extension, and assume that L and F are both contained in the same field E. Then $(a, L/k, \sigma) \otimes_k F$ is Brauer equivalent to $(a, LF/F, \tilde{\tau})$, where $\tau = \sigma^{[L \cap F:k]}$ and $\tilde{\tau}$ is the unique extension of τ to LF;*

(4) *for every $m \geq 1$ prime to n, we have*

$$(a^m, L/k, \sigma^m) \cong_K (a, L/k, \sigma);$$

(5) *let $m \geq 1$ be an integer prime to n, and let ℓ be the unique integer satisfying $0 \leq \ell \leq n-1$ and $\ell m \equiv 1[n]$. Then we have*

$$m[(a, L/k, \sigma)] = [(a, L/k, \sigma^\ell)];$$

(6) *let $m \mid n$, and let $F = L^{\langle \sigma^m \rangle}$. Then we have*

$$m[(a, L/k, \sigma)] = [(a, F/k, \sigma_{|F})].$$

Proof. Points (1) and (2) just come from Theorem VII.1.5. To prove (3), notice first that $L/L \cap F$ is a cyclic extension, whose Galois group is a subgroup of $\mathrm{Gal}(L/k)$. Hence we have $\mathrm{Gal}(L/L \cap F) = \langle \sigma^m \rangle$ for some $m \geq 1$. Since we have

$$[L : L \cap F] = |\mathrm{Gal}(L/L \cap F)| = \frac{n}{m},$$

we get

$$[L \cap F : k] = \frac{n}{[L : L \cap F]} = m.$$

Hence $H = \mathrm{Gal}(L/L \cap F) = \langle \tau \rangle$, and $\mathrm{Gal}(LF/F) = \langle \tilde{\tau} \rangle$. Notice that $\tilde{\tau}$ has order $\dfrac{n}{m}$. It is immediate to check that the map

$$\xi': \begin{array}{c} H \times H \longrightarrow (LF)^\times \\ (\widetilde{\tau^i}, \widetilde{\tau^j}) \longmapsto \xi_{\tau^i, \tau^j} \end{array}$$

is nothing but $\xi^{a, \tilde{\tau}}$. Now we conclude using the Restriction theorem.

Let us prove (4). Let e and f be the generators of $(a, L/k, \sigma)$ and $(a^m, L/k, \sigma^m)$ respectively, and let

$$\varphi : (a^m, L/k, \sigma^m) \longrightarrow (a, L/k, \sigma)$$

be the unique L-linear map defined by

$$\varphi(f^i) = e^{im} \text{ for all } i = 0, \ldots, n-1.$$

It is left to the reader to check that φ is a k-algebra morphism. An application of Lemma I.2.2 then gives the desired conclusion. Notice that the fact that m is prime to n is implicitly used to define $(a^m, L/k, \sigma^m)$, since σ^m is a generator of $\mathrm{Gal}(L/k) = \langle \sigma \rangle$ in this case.

To prove (5), notice that from (1) we have

$$m[(a, L/k, \sigma)] = [(a^m, L/k, \sigma)].$$

Since $\sigma = \sigma^{\ell m} = (\sigma^\ell)^m$, applying (4) gives the desired result.

It remains to prove (6). Since $m[(a, L/k, \sigma)] = [(a^m, L/k, \sigma)]$, we need to prove that $(a^m, L/k, \sigma)$ and $(a, F/k, \sigma_{|F})$ are Brauer equivalent. Let $\overline{G} = \mathrm{Gal}(F/k) = \langle \tau \rangle$, where $\tau = \sigma_{|F}$ has order $d = \dfrac{n}{m}$. By the Inflation Theorem, we have

$$[(a, F/k, \tau)] = [(\xi^{\tau, a}, F/k, \overline{G})] = [(\mathrm{Inf}_F^L(\xi^{\tau, a}), L/k, G)].$$

On the other hand, we have

$$[(a^m, L/k, \sigma)] = [(\xi^{\sigma, a^m}, L/k, G)].$$

Hence, we need to prove that $(\mathrm{Inf}_F^L(\xi^{\tau, a}), L/k, G)$ and $(\xi^{\sigma, a^m}, L/k, G)$ are Brauer equivalent. This is equivalent to showing that $\mathrm{Inf}_F^L(\xi^{\tau, a})$ and ξ^{σ, a^m} are cohomologous by Theorem VI.2.7.

For $0 \leq i, j \leq n-1$, write the Euclidean division of i and j by d. We have

$$i = r + \ell d \text{ and } j = s + \ell' d$$

for some unique integers r, s satisfying $0 \leq r, s \leq d-1$. We then have

$$\mathrm{Inf}_F^L(\xi^{\tau, a})_{\sigma^i, \sigma^j} = \xi^{\tau, a}_{\tau^i, \tau^j} = \left\{ \begin{array}{ll} 1 & \text{if } r + s < d \\ a & \text{if } r + s \geq d \end{array} \right.$$

Now for $i = 0, \ldots, n-1$, set $z_{\sigma^i} = a^\ell$, where ℓ is the quotient of the Euclidean division of i by d.

We claim that we have

$$\mathrm{Inf}_F^L(\xi^{\tau, a})_{\sigma^i, \sigma^j} (z_{\sigma^j} z_{\sigma^i}^{\sigma^j} z_{\sigma^{i+j}}^{-1}) = \xi^{a^m, \sigma}_{\sigma^i, \sigma^j}$$

for all $0 \leq i, j \leq n-1$. As above, write

$$i = r + \ell d \text{ and } j = s + \ell' d$$

for some unique integers r, s satisfying $0 \leq r, s \leq d-1$. We consider four cases.

(*a*) Assume that $i + j < n$ and $r + s < d$. We have $i + j = r + s + (\ell + \ell')d$. Since $0 \leq r + s < d$, $\ell + \ell'$ is the quotient of the division of $i + j$ by d, and therefore $z_{\sigma^{i+j}} = a^{\ell+\ell'}$. Thus we have

$$z_{\sigma^j} z_{\sigma^i}^{\sigma^j} z_{\sigma^{i+j}}^{-1} = 1.$$

Now since $r + s < d$, we have $\mathrm{Inf}_F^L(\xi^{\tau,a})_{\sigma^i,\sigma^j} = 1$, and since $i + j < n$ we have $\xi_{\sigma^i,\sigma^j}^{a^m,\sigma} = 1$ as well. Hence the desired equality is satisfied in this case.

(*b*) Assume that $i + j < n$ and $r + s \geq d$. Since $d \leq r + s < 2d$, we have $0 \leq r+s-d < d$, and therefore $i+j = r+s-d+(\ell+\ell'+1)d$. Since $0 \leq r+s-d < d$, $\ell+\ell'+1$ is the quotient of the division of $i+j$ by d in this case. Hence $z_{\sigma^{i+j}} = a^{\ell+\ell'+1}$. Thus we get

$$z_{\sigma^j} z_{\sigma^i}^{\sigma^j} z_{\sigma^{i+j}}^{-1} = a^{-1}.$$

Since $r+s \geq d$, we have $\mathrm{Inf}_F^L(\xi^{\tau,a})_{\sigma^i,\sigma^j} = a$, and since $i+j < n$ we have $\xi_{\sigma^i,\sigma^j}^{a^m,\sigma} = 1$. Thus the equality is also satisfied in this case.

(*c*) Assume that $i + j \geq n$ and $r + s < d$. Then $0 \leq i + j - n \leq n - 1$, and $z_{\sigma^{i+j}} = z_{\sigma^{i+j-n}}$. We have $i + j - n = i + j - md = r + s + (\ell + \ell' - m)d$, so $z_{\sigma^{i+j}} = a^{\ell+\ell'-m}$. Thus we have

$$z_{\sigma^j} z_{\sigma^i}^{\sigma^j} z_{\sigma^{i+j}}^{-1} = a^m.$$

Now we have $\mathrm{Inf}_F^L(\xi^{\tau,a})_{\sigma^i,\sigma^j} = 1$ since $r + s < d$, and $\xi_{\sigma^i,\sigma^j}^{a^m,\sigma} = a^m$ since $i + j \geq n$. The required equality follows.

(*d*) Assume that $i+j \geq n$ and $r+s \geq d$. Then $0 \leq i+j-n \leq n-1$, $z_{\sigma^{i+j}} = z_{\sigma^{i+j-n}}$ and we have $i + j - n = r + s - d + (\ell + \ell' + 1 - m)d$. Thus $z_{\sigma^{i+j}} = a^{\ell+\ell'+1-m}$, and we have

$$z_{\sigma^j} z_{\sigma^i}^{\sigma^j} z_{\sigma^{i+j}}^{-1} = a^{m-1}.$$

Now we have $\mathrm{Inf}_F^L(\xi^{\tau,a})_{\sigma^i,\sigma^j} = a$ since $r + s \geq d$, and $\xi_{\sigma^i,\sigma^j}^{a^m,\sigma} = a^m$ since $i + j \geq n$. Then we have the desired equality, and this concludes the proof. $\square$

Using the second point of the previous proposition, we can derive a way to compute the exponent of a cyclic algebra.

COROLLARY VII.1.10. *Let* $A = (a, L/k, \sigma)$ *be a cyclic* k-*algebra of degree* n. *Then the following properties hold:*

(1) *the exponent of* A *is the smallest integer* $m \geq 1$ *such that* $m \mid n$ *and* $a^m \in N_{L/k}(L^\times)$;

(2) *the exponent of* A *is the smallest integer* $m \geq 1$ *such that* $m \mid n$ *and* $a \in N_{K/k}(L^\times)$, *where* K *is the unique subfield of* L *of degree* $\dfrac{n}{m}$.

Proof. By definition, the exponent of A is the smallest integer $m \geq 1$ such that $m[A] = 0 \in \mathrm{Br}(k)$. By Theorem V.3.5, the exponent of A divides the index of A, hence the degree of A. Therefore, m divides n. By Proposition VII.1.9 (1) (applied several times), we have $m[A] = (a^m, L/k, \sigma)$. Point (1) now follows from the second point of the same proposition. Using Proposition VII.1.9 (6), we see that (2) is just a reformulation of (1). $\square$

COROLLARY VII.1.11. *Let* L/k *be a cyclic extension of degree* n, *and let* $a \in k^\times$. *Then we have the following properties:*

(1) *for every $m \geq 1$ prime to n, the cyclic algebras $(a^m, L/k, \sigma)$ and $(a, L/k, \sigma)$ have the same index;*

(2) *if σ and σ' are two generators of $\mathrm{Gal}(L/k)$, the cyclic algebras $(a, L/k, \sigma)$ and $(a, L/k, \tau)$ have the same index.*

Proof. Let $A = (a, L/k, \sigma)$. By Proposition VII.1.9 (1) applied several times, $A^{\otimes m}$ is Brauer equivalent to $(a^m, L/k, \sigma)$. Since m is prime to n, it is also prime to $\mathrm{ind}(A)$. The first part then follows from point (11) of Theorem V.3.1. To prove the second part, notice that $\tau = \sigma^\ell$ for some $\ell \geq 1$ which is prime to n. Now, by Proposition VII.1.9 (5), we see that $[(a, L/k, \tau)] = m[A]$, where m is prime to n. Now, use the first part to conclude. $\qquad\square$

We now give two sufficient conditions for a cyclic algebra to be a division algebra.

COROLLARY VII.1.12. *For every cyclic k-algebra $A = (a, L/k, \sigma)$ of degree n, the following properties hold:*

(1) *if $a^m \notin N_{L/k}(L^\times)$ for all $m \neq n$ dividing n, then A is a division k-algebra;*

(2) *if $a \notin N_{K/k}(K^\times)$ for all subfields $K \neq k$ of L, then A is a division k-algebra.*

Proof. The assumptions imply that $\exp(A) = n$. Since $\exp(A)$ divides $\mathrm{ind}(A)$, we get $\mathrm{ind}(A) = n = \deg(A)$, proving that A is a division algebra. $\qquad\square$

VII.2. Central simple algebras over local fields

In this paragraph and the following ones, we give without proof some important results on the structure of central simple algebras over local fields and number fields. The proofs may be found in [**11**] for example. We refer to Appendix B for the missing definitions and results in number theory.

In this section, (k, υ) is a local field.

Any central simple algebra A of degree n over a local field (k, υ) is isomorphic to a cyclic algebra. More precisely, we can write

$$A \cong_k (a, L_n/k, \mathrm{Frob}(L_n/k)),$$

where L_n/k is the unique unramified cyclic extension of degree n of k (see Proposition B.1.19), and $\mathrm{Frob}(L/k)$ is the Frobenius map. We then set

$$\mathrm{inv}(A) = \frac{n_\upsilon(a)}{n} \in \mathbb{Q}/\mathbb{Z},$$

where $n_\upsilon(a) \in \mathbb{Z}$ is the unique integer satisfying

$$a = u\pi^{n_\upsilon(a)},$$

where u is a unit of $\mathcal{O}_\upsilon$ and π is a local parameter.

One can show that $\mathrm{inv}(A)$ is well-defined, and that it only depends on the class of A in the Brauer group. For any $[A] \in \mathrm{Br}(k)$, we set $\mathrm{inv}([A]) = \mathrm{inv}(A)$. This makes sense by the previous considerations.

We then have the following structure theorem:

THEOREM VII.2.1. *We have the group isomorphism*

$$\mathrm{inv} : \mathrm{Br}(k) \xrightarrow{\sim} \mathbb{Q}/\mathbb{Z}.$$

Moreover, for every central simple k-algebra A, we have

$$\mathrm{ind}(A) = \exp(A).$$

REMARK VII.2.2. In particular, we get that $\mathrm{ind}(A)$ is the order of $\mathrm{inv}([A])$ in $\mathbb{Q}/\mathbb{Z}$. $\qquad\qquad\qquad\qquad\qquad\qquad\qquad\qquad\qquad\qquad\square$

If $A = (a, L/k, \sigma)$, where L/k is not necessarily unramified, it may be difficult to compute the invariant of A, since it implies that we have to write A in the form $(a', L_n/k, \mathrm{Frob}(L_n/k))$, which can be tricky. Therefore, it might be more fruitful to consider another approach. Since the index and the exponent of a central simple k-algebra A coincide, we have to compute the smallest integer $m \mid n$ such that $m[A] = 0 \in \mathrm{Br}(k)$. But we have

$$m[A] = [(a, M/k, \sigma_{|M})],$$

where $M = L^{\langle \sigma^m \rangle}$ by Proposition VII.1.9 (6). Using Proposition VII.1.9 (2), we see that $m[A] = 0$ if and only if a is a norm of M/k.

Thus the problem boils down to the following question: given a cyclic extension of local fields M/k, how does one decide if a given element of k is a norm of M/k?

The answer is given by the Hasse symbol. We will not define it in full generality, but only in the tame case.

We will assume until the end of this section that $\mu_n \subset k$, where n is prime to the characteristic of the residue field $\kappa(v)$. One can show in this case that the natural projection

$$\mathcal{O}_k \longrightarrow \kappa(v)$$

identifies μ_n with a subgroup of $\kappa(v)^\times$. If particular, if $\kappa(v) \simeq \mathbb{F}_q$, then $n \mid q - 1$.

Therefore the following statement makes sense:

PROPOSITION VII.2.3. *Keeping the previous notation, let $a, b \in k^\times$. Then*

$$(a,b)_{n,v} = \overline{(-1)^{n_v(a)n_v(b)}a^{n_v(b)}b^{-n_v(a)}}^{\frac{q-1}{n}} \in \kappa(v)^\times$$

is a nth-root of 1. Moreover, for all $a, a', b,' \in k^\times$, the following properties hold:

(1) $(aa', b)_{v,n} = (a, b)_{v,n}(a', b)_{v,n}$;

(2) $(a, bb')_{v,n} = (a, b)_{v,n}(a, b')_{v,n}$;

(3) $(a, b)_{v,n}(b, a)_{v,n} = 1$;

(4) $(a, b)_{v,n} = 1$ *if and only if a is a norm in $k(\sqrt[n]{b})/k$.*

DEFINITION VII.2.4. The element $(a, b)_{v,n}$ is called the (tame) **Hasse symbol** of a and b.

If $\mu_n \subset k$, then every cyclic extension of k of degree dividing n has the form $k(\sqrt[n]{b})/k$. Therefore, the Hasse symbol allows us to compute the index of a given cyclic k-algebra in the tame case.

PROPOSITION VII.2.5. *Assume $\mu_n \subset k$, where n is prime to the characteristic of the residue field $\kappa(v)$, and let $A = (a, k(\sqrt[n]{b})/k, \sigma)$ be a cyclic k-algebra of degree n. Then the index of A is the order of $(a, b)_{n,v}$ in $\kappa(v)^{\times}$.*

Proof. For $m \geq 1$, we have $m[A] = (a^m, k(\sqrt[n]{b})/k, \sigma)$ by Proposition VII.1.9 (1). By Proposition VII.1.9 (1) and the previous proposition, we have $m[A] = 0$ if and only if $(a^m, b)_{n,v} = 1$. The properties of the Hasse symbol imply that we have

$$m[A] = 0 \iff (a, b)_{n,v}^m = 1 \text{ for all } m \geq 1.$$

Hence $\exp(A)$ is the order of $(a, b)_{n,v}$. Since $\mathrm{ind}(A) = \exp(A)$ by Theorem VII.2.1, the result follows. $\square$

VII.3. Central simple algebras over number fields

In all this section, k will denote a number field. In this case, we have the following important theorem.

THEOREM VII.3.1 (Brauer-Hasse-Noether). *Let A be a central simple k-algebra. Then the following properties hold:*

(1) *the set of places v of k such that $A \otimes_k k_v$ is not split is finite. Moreover, A is split if and only if $A \otimes_k k_v$ is split for all places v of k except maybe one.*

(2) *A is isomorphic to a cyclic k-algebra;*

(3) *$\mathrm{ind}(A) = \exp(A)$. Moreover, $\mathrm{ind}(A)$ is the least common multiple of the indices of $A \otimes_k k_v$, when v describes the set of places of k.*

REMARK VII.3.2. The first point of this theorem says that if $A \otimes_k k_v$ is split for all places $v \neq v_0$, then A is split. It follows that $A \otimes_k k_{v_0}$ is split as well. $\square$

COROLLARY VII.3.3. *Let A be a central simple k-algebra of degree p^m, where p is a prime number. Then A is a division k-algebra if and only if there exists a place v of k such that $A \otimes_k k_v$ is a division algebra.*

Proof. If such a place v exists, then A is a division algebra since the index decreases by scalar extension. Conversely, if no such place exists, then the last point of the previous theorem shows that the index of A is a strict divisor of p^m, meaning that A is not a division algebra. $\square$

Let us now examine the case of a cyclic k-algebra in more details (everything boils down to this case, since every central simple k-algebra is isomorphic to such an algebra).

Let $A = (a, L/k, \sigma)$ and let v be a finite place of k. Then $\mathrm{ind}(A \otimes_k k_v)$ may be computed easily in some cases.

PROPOSITION VII.3.4. *Let $A = (a, L/k, \sigma)$, let $v_{\mathfrak{p}}$ be a finite place of k corresponding to a prime ideal $\mathfrak{p}$ of $\mathcal{O}_k$ which does not ramify in L, and let $\mathfrak{P} \mid \mathfrak{p}$. Then $\mathrm{ind}(A \otimes_k k_{\mathfrak{p}})$ is the order of $\dfrac{n_{\mathfrak{p}}(a)}{f_{\mathfrak{p}}}$ in $\mathbb{Q}/\mathbb{Z}$.*

Proof. By Proposition VII.1.9 (3), we have

$$[A \otimes k_{v_{\mathfrak{p}}}] = [(a, L_{\mathfrak{P}}/k_{\mathfrak{p}}, \tau)],$$

where $\mathfrak{P}$ is any prime above $\mathfrak{p}$, and τ is a suitable generator of the group $\mathrm{Gal}(L_\mathfrak{P}/k_\mathfrak{p})$. Since $L_\mathfrak{P}/k_\mathfrak{p}$ is unramified by Proposition B.3.6, $\mathrm{Gal}(L_\mathfrak{P}/k_\mathfrak{p})$ is generated by $\mathrm{Frob}(L_\mathfrak{P}/k_\mathfrak{p})$. By Corollary VII.1.11, the central simple $k_\mathfrak{p}$-algebras $(a, L_\mathfrak{P}/k_\mathfrak{p}, \tau)$ and $(a, L_\mathfrak{P}/k_\mathfrak{p}, \mathrm{Frob}(L_\mathfrak{P}/k_\mathfrak{p}))$ have the same index, so one may assume that $\tau = \mathrm{Frob}(L_\mathfrak{P}/k_\mathfrak{p})$. In this case, we get

$$\mathrm{inv}_\mathfrak{p}([A]) = \frac{n_{v_\mathfrak{p}(a)}}{e_\mathfrak{p} f_\mathfrak{p}} = \frac{n_\mathfrak{p}(a)}{f_\mathfrak{p}}.$$

Since $\mathrm{inv}_\mathfrak{p}$ induces a group isomorphism between $\mathbb{Q}/\mathbb{Z}$ and $\mathrm{Br}(k_\mathfrak{p})$, $\mathrm{ind}(A \otimes_k k_\mathfrak{p}) = \exp(A \otimes_k k_\mathfrak{p})$ is the order of $\mathrm{inv}_\mathfrak{p}([A])$ in $\mathbb{Q}/\mathbb{Z}$. This concludes the proof. $\qquad\square$

When $\mathfrak{p}$ ramifies in L, as already explained earlier, it may be difficult to find the index of $A \otimes_k k_\mathfrak{p}$ by computing its invariant. Of course, we can use Hilbert symbols in the case where $\mu_n \subset k$ and the characteristic of $\kappa(\mathfrak{p})$ is prime to n to compute the index. However, we would like to give some results which do not need these extra assumptions.

We would like first to reduce the case of prime degree cyclic algebras. This is given by the following result.

PROPOSITION VII.3.5. *Let k be an number field, and let L/k be a cyclic extension of degree n, generated by σ. For every $d \geq 1, d \mid n$, let $L^{(d)}$ be the unique subfield of L of degree d. Finally, let $A = (a, L/k, \sigma)$ be a cyclic algebra. Then A is a division algebra if and only if for every prime divisor p of n, the cyclic algebra $(a, L^{(p)}/k, \sigma_{|_{L^{(p)}}})$ is not split.*

Proof. If L/k is cyclic of degree $n = p_1^{m_1} \cdots p_r^{m_r}$, then we have

$$L = L^{(p_1^{m_1})} \cdots L^{(p_r^{m_r})}.$$

Let $u_1, \ldots, u_r \in \mathbb{Z}$ satisfying $u_1 \dfrac{n}{p_1^{m_1}} + \ldots u_r \dfrac{n}{p_r^{m_r}} = 1$. The reader will check as an exercise (using the properties of cyclic algebras) that we have a decomposition

$$(a, L/k, \sigma) \cong_k (a^{u_1}, L^{(p_1^{m_1})}/k, \sigma_{|_{L^{(p_1^{m_1})}}}) \otimes_k \cdots \otimes_k (a^{u_r}, L^{(p_r^{m_r})}/k, \sigma_{|_{L^{(p_r^{m_r})}}}).$$

Notice that $(a^{u_i}, L^{(p_i^{m_i})}/k, \sigma_{|_{L^{(p_i^{m_i})}}})$ has degree $p_i^{m_i}$, so the decomposition above is the primary decomposition of A given by Theorem V.3.8. The same theorem says that A is a division algebra if and only if each cyclic k-algebra $(a^{u_i}, L^{(p_i^{m_i})}/k, \sigma_{|_{L^{(p_i^{m_i})}}})$ is.

Now for each i, we have an equation of the form

$$\frac{n}{p_i^{m_i}} u_i + p_i^{m_i} v_i = 1,$$

for some $v_i \in \mathbb{Z}$. Then u_i is prime to p_i, so $(a^{u_i}, L^{(p_i^{m_i})}/k, \sigma_{|_{L^{(p_i^{m_i})}}})$ is a division algebra if and only if $A^{(i)} = (a, L^{(p_i^{m_i})}/k, \sigma_{|_{L^{(p_i^{m_i})}}})$ is, by Corollary VII.1.11 (1).

Assume that $A^{(i)}$ is a division algebra. Since $\exp(A^{(i)}) = \mathrm{ind}(A^{(i)})$, we get $\exp(A^{(i)}) = p_i^{m_i}$, and therefore $p_i^{m_i-1}[A] \neq 0$. Conversely, if $A^{(i)}$ is not a division algebra, then $\mathrm{ind}(A^{(i)})$ is a proper divisor of $p_i^{m_i}$, that is $\mathrm{ind}(A^{(i)}) \mid p_i^{m_i-1}$. In this case we have $\exp(A^{(i)}) \mid p_i^{m_i-1}$, and therefore $p_i^{m_i-1}[A^{(i)}] = 0$. Hence we have proved that $A^{(i)}$ is a division algebra if and only if $p_i^{m_i-1}[A^{(i)}] \neq 0$. Now by Proposition VII.1.9 (6), we have $p_i^{m_i-1}[A] = [(a, L^{(p_i)}/k, \sigma_{|_{L^{(p_i)}}})]$. The result follows immediately. $\qquad\square$

REMARK VII.3.6. By Corollary VII.1.11 (2), we may replace $\sigma_{|_{L^{(p)}}}$ by any other generator of the Galois group $\mathrm{Gal}(L^{(p)}/k)$. $\qquad\square$

To sum up, the question of deciding if a cyclic algebra is a division algebra boils down to the following question: given a cyclic algebra A of prime degree p, how can we decide concretely if it is split or not? In the next section, we examine this question in detail.

VII.4. Cyclic algebras of prime degree over number fields

Let k be a number field. As explained in the previous section, it is crucial to find an explicit way to decide whether a given cyclic k-algebra A of prime degree is split or not. By the Brauer-Hasse-Noether's theorem, it is enough to look at it over all the completions of k. If v is a complex place, $A \otimes_k k_v$ is split, since $k_v \cong \mathbb{C}$. We now deal with the case of real places.

LEMMA VII.4.1. Let $A = (a, L/k, \sigma)$ be a cyclic k-algebra of prime degree p, and let v_τ be a real place of k corresponding to a real embedding $\tau : k \hookrightarrow \mathbb{R}$.

(1) If $p > 2$, then $A \otimes_k k_v$ is split;
(2) if $p = 2$, write $L = k(\sqrt{b})$. Then $A \otimes_k k_v$ is not split if and only if $\tau(a) < 0$ and $\tau(b) < 0$ as elements of $k_v \cong_k \mathbb{R}$.

Proof. Assume that $p > 2$. Since $\exp(A) \mid \deg(A)$, we get $p[A] = 0$, and therefore $p[A \otimes_k k_v] = 0$. On the other hand $2[A \otimes_k k_v] = 0$, since $k_v \cong_k \mathbb{R}$ and $\mathrm{Br}(\mathbb{R}) \cong_k \mathbb{Z}/2\mathbb{Z}$. Since p is odd, we get $[A \otimes_k k_v] = 0$, meaning that $A \otimes_k k_v$ is split.

Assume now that $p = 2$, and write $L = k(\sqrt{b})$. Then $A \cong_k (a, b)_k$. Moreover, the external product

$$k \times k_v \longrightarrow k_v$$
$$(\lambda, x) \longmapsto \lambda * x = \tau(\lambda)x$$

endows k_v with the structure of a k-vector space, so we can form the tensor product $A \otimes_k k_v$ with respect to this structure.

Let $1, i, j, ij$ be the standard basis of A, and let

$$i' = i \otimes 1, j' = j \otimes 1 \in A \otimes_k k_v.$$

Clearly $1, i', j', i'j'$ is a k_v-basis of $A \otimes_k k_v$. Finally, let $1, e, f, ef$ be the standard basis of $(\tau(a), \tau(b))_{k_v}$. Notice that we have

$$i'^2 = i^2 \otimes 1 = a \otimes 1 = 1 \otimes \tau(a) = \tau(a) \cdot 1_{A \otimes_k k_v},$$

the last equality coming from the definition of the structure of k_v-vector space on $A \otimes_k k_v$. Similarly, we have

$$j'^2 = \tau(b) \cdot 1_{A \otimes_k k_v}.$$

It is then easy to verify that map

$$f: \begin{array}{c} (a,b)_k \otimes k_v \longrightarrow (a,b)_{k_v} \\ x + yi' + zj' + ti'j' \longmapsto x + ye + zf + tef \end{array}$$

is an isomorphism of k_v-algebras. Now if $\tau(a) > 0$ as an element of k_v, then $\tau(a) = \lambda^2$, and therefore $(\tau(a), \tau(b))_{k_v} \cong_k (1, \tau(b))_{k_v}$ is split. Similarly, if $\tau(b) > 0$, $(\tau(a), \tau(b))_{k_v}$ is split. Finally, if $\tau(a)$ and $\tau(b)$ are both negative, then $-\tau(a)$ and $-\tau(b)$ are squares and therefore $(\tau(a), \tau(b))_{k_v} \cong_k (-1, -1)_{k_v} \cong_{k_v} \mathbb{H}$, which is a division algebra.   □

EXAMPLE VII.4.2. Let $k = \mathbb{Q}(\sqrt{2})$, $L = k(i)$ and consider the cyclic k-algebra $A = (1 + \sqrt{2}, L/k, \sigma)$, where σ is the non-trivial k-automorphism of L. We then have

$$A \cong_k (1 + \sqrt{2}, -1)_k.$$

Let v_0 be the trivial real place corresponding to the inclusion $k \subset \mathbb{R}$, and let v_1 be the real place corresponding to

$$\tau: \begin{array}{c} k \hookrightarrow \mathbb{R} \\ x + y\sqrt{2} \longmapsto x - y\sqrt{2}. \end{array}$$

Then $A \otimes_k k_{v_0}$ splits since $1 + \sqrt{2} > 0$, but $A \otimes_k k_{v_1}$ is a division algebra since $\tau(1 + \sqrt{2}) = 1 - \sqrt{2} < 0$ and $\tau(-1) = -1 < 0$.   □

We now deal with the case of finite places. Since L/k has degree p, given prime ideal $\mathfrak{p}$ of $\mathcal{O}_k$, they are only three possibilities: $\mathfrak{p}$ totally splits, $\mathfrak{p}$ is inert or $\mathfrak{p}$ totally ramifies in L (this comes from the formula $efg = p$ and the fact that p is prime).

The following results show that we have a complete answer in all but finitely many cases, which are the cases of wild ramification. Until the end of the section, $A = (a, L/k, \sigma)$ is a central simple k-algebra of prime degree p, $\mathfrak{p}$ is a prime ideal of $\mathcal{O}_k$ and $\mathfrak{P}$ denotes a prime ideal of $\mathcal{O}_L$ lying above $\mathfrak{p}$.

LEMMA VII.4.3. If $\mathfrak{p}$ is a prime ideal of $\mathcal{O}_k$ which totally splits in L, then $A \otimes_k k_{\mathfrak{p}}$ is split.

Proof.   Since $\mathfrak{p}$ totally splits, we have $e_{\mathfrak{p}} = f_{\mathfrak{p}} = 1$, and therefore $L_{\mathfrak{P}} = k_{\mathfrak{p}}$ by Proposition B.3.5. Since we have

$$[A \otimes_k k_{\mathfrak{p}}] = [(a, L_{\mathfrak{P}}/k_{\mathfrak{p}}, \tau)],$$

for a suitable τ by Proposition VII.1.9 (3), the result follows.   □

LEMMA VII.4.4. If $\mathfrak{p}$ is a prime ideal of $\mathcal{O}_k$ which is inert in L, then $A \otimes_k k_{\mathfrak{p}}$ is split if and only if $p \mid n_{\mathfrak{p}}(a)$.

Proof.   Since $\mathfrak{p}$ is inert, we have $e_{\mathfrak{p}} = 1$ and $f_{\mathfrak{p}} = p$, and therefore $L_{\mathfrak{P}}/k_{\mathfrak{p}}$ is an unramified extension of degree p by Proposition B.3.6. Since $A \otimes_k k_{\mathfrak{p}}$ has degree p over $k_{\mathfrak{P}}$ and $A \otimes_k k_{\mathfrak{p}}$ is Brauer equivalent to $(a, L_{\mathfrak{P}}/k_{\mathfrak{p}}, \tau)$ for some suitable τ by Proposition VII.1.9 (3), we have

$$A \otimes_k k_{\mathfrak{p}} \cong_{k_{\mathfrak{p}}} (a, L_{\mathfrak{P}}/k_{\mathfrak{p}}, \tau).$$

Proposition VII.3.4 then shows that the index of $A \otimes_k k_{\mathfrak{p}}$ is the order of $\dfrac{n_{\mathfrak{p}}(a)}{p}$ in $\mathbb{Q}/\mathbb{Z}$. This order is 1 if and only if $p \mid n_{\mathfrak{p}}(a)$. $\square$

EXAMPLE VII.4.5. Let A be the cyclic algebra of degree 5 given by

$$A = \left(\frac{3+2i}{2+3i}, L/k, \sigma\right) = ((3+2i)(2+3i)^4, L/\mathbb{Q}(i), \sigma),$$

where $k = \mathbb{Q}(i), L = k(\zeta_{11} + \zeta_{11}^{-1})$ and $\sigma : L \longrightarrow L$ is the unique $\mathbb{Q}(i)$-automorphism of L satisfying $\sigma(\zeta_{11} + \zeta_{11}^{-1}) = \zeta_{11}^2 + \zeta_{11}^{-2}$.

One may show that the prime ideal $\mathfrak{p} = (3+2i)$ of $\mathcal{O}_k$ is inert in L, and thus $A \otimes_k k_{\mathfrak{p}}$ is not split by the previous lemma. Hence A is not split, and therefore is a division algebra since A has prime degree. $\square$

LEMMA VII.4.6. *Let $\mathfrak{p}$ is a prime ideal of $\mathcal{O}_k$ which totally ramifies in L. Let $\ell = |\kappa(\mathfrak{p})|$, and write*

$$a = u\pi_{\mathfrak{p}}^{n_{\mathfrak{p}}(a)} \in k_{\mathfrak{p}},$$

where $\pi_{\mathfrak{p}}$ is a local parameter and $u \in \mathcal{O}_{\mathfrak{p}}^{\times}$.

If $\mathfrak{p}$ does not lie above p (that is $p \nmid \ell$), then $A \otimes_k k_{\mathfrak{p}}$ is split if and only if

$$\overline{(-1)^{n_{\mathfrak{p}}(a)}u^{\frac{\ell-1}{p}}} = 1 \in \mathbb{F}_{\ell}.$$

Proof. Notice that $\kappa(\upsilon) \cong_k \mathbb{F}_{\ell}$, by definition of ℓ. Since $\mathfrak{p}$ totally ramifies, the extension $L_{\mathfrak{P}}/k_{\mathfrak{p}}$ is a totally ramified Galois extension of degree p by Proposition B.3.6, so $\mu_p \subset k_{\mathfrak{p}}$ and $L_{\mathfrak{P}} = k_{\mathfrak{p}}(\sqrt[p]{\pi_{\mathfrak{p}}})$ by Proposition B.1.21, since $p \nmid \ell$.

We then have

$$A \otimes_k k_{\mathfrak{p}} \cong_{k_{\mathfrak{p}}} (a, k_{\mathfrak{p}}(\sqrt[p]{\pi_{\mathfrak{p}}})/k_{\mathfrak{p}}, \tau),$$

for a suitable τ. Hence $A \otimes k_{\mathfrak{p}}$ splits if and only if $(a, \pi_{\mathfrak{p}})_{p, \upsilon_p} = 1 \in \mathbb{F}_{\ell}$ by Proposition VII.2.5. Now apply the formula defining the Hasse symbol to conclude. $\square$

We now apply the results above to a particular case, which will be useful in the sequel.

LEMMA VII.4.7. *Let k be a totally imaginary number field (i.e. k has no real embeddings), and let $A = (a, L/k, \sigma)$ be a cyclic k-algebra of degree p, where $a \in \mathcal{O}_k^{\times}$. Assume that there is exactly one prime ideal of $\mathcal{O}_k$ lying above p. Then A is split if and only if $A \otimes_k k_{\mathfrak{p}}$ is split for all prime ideals $\mathfrak{p}$ not lying above p which totally ramify in L.*

Proof. The direct implication is obvious. To prove the converse implication, it is enough to show that $A \otimes_k k_{\upsilon} = 0$ for all places of k, except maybe one, by Brauer-Hasse-Noether's theorem. By assumption, k has no real places, and $A \otimes_k k_{\upsilon}$ automatically splits if υ is a complex place. Hence it remains to deal with the finite places $\upsilon_{\mathfrak{p}}$. Assume first that $\mathfrak{p}$ is not lying above p. If $\mathfrak{p}$ totally splits in L, then $A \otimes_k k_{\mathfrak{p}}$ splits by Lemma VII.4.3. If $\mathfrak{p}$ is inert in L, then $A \otimes_k k_{\mathfrak{p}}$ also splits by Lemma VII.4.4, since $n_{\mathfrak{p}}(a) = 0$ by assumption. If $\mathfrak{p}$ totally ramifies in L, then $A \otimes_k k_{\mathfrak{p}}$ is split by assumption. It remains to consider the case of a prime ideal $\mathfrak{p}$ lying above p. By assumption on p, such a $\mathfrak{p}$ is unique. Hence by the first point of Brauer-Hasse-Noether's Theorem, A is split. This concludes the proof. $\square$

By Proposition B.2.29, given a Kummer extension $L = k(\sqrt[p]{d})$ of prime degree p, where $d \in \mathcal{O}_k$, the only prime ideals not lying above p which totally ramify in L are the prime ideals $\mathfrak{p}$ such that $n_\mathfrak{p}(d)$ is prime to p. Therefore, the result of Lemma VII.4.7 may be rewritten as follows.

PROPOSITION VII.4.8. *Let k be a totally imaginary number field, and let p be a prime number. Assume that $\mu_p \subset k$ and that there is exactly one prime ideal of $\mathcal{O}_k$ lying above p. Let $A = (a, k(\sqrt[p]{d})/k, \sigma)$ be a cyclic k-algebra of degree p, where $a \in \mathcal{O}_k^\times$ and $d \in \mathcal{O}_k$. Then A is split if and only if $A \otimes_k k_\mathfrak{p}$ is split for all prime ideals $\mathfrak{p}$ not lying above p such that $n_\mathfrak{p}(d)$ is prime to p.*

VII.5. Examples

In this section, we apply the previous results to give necessary and sufficient conditions for certain cyclic algebras of degree $n = 2, 3, 4$ and 6 to be division algebras. We will also explain how to construct division cyclic algebras of prescribed degrees.

Let us first reformulate Proposition VII.3.5 in the case of Kummer extensions.

PROPOSITION VII.5.1. *Let k be a number field. Assume that $\mu_n \subset k$, and let $A = (a, k(\sqrt[n]{d})/k, \sigma)$ be a cyclic algebra of degree n. Then A is a division algebra if and only if for all prime divisors p of n the cyclic algebra $(a, k(\sqrt[p]{d})/k, \sigma_{k(\sqrt[p]{d})/k})$ is not split.*

The cyclic extensions L/k we are going to use in the sequel will not be necessarily given under the form $k(\sqrt[n]{d})/k$. The next lemma shows how to compute a suitable $d \in k$ explicitly for any base field k.

LEMMA VII.5.2. *Let k be a field. Assume that $\mathrm{char}(k) \nmid n$, and that $\mu_n \subset k$. Let L/k be a cyclic extension of degree n, whose Galois group is generated by σ. Finally, let $\zeta_n \in k$ be a primitive n^{th}-root of 1. Pick any $z \in L$ such that*

$$\theta = \sum_{m=0}^{n-1} \zeta_n^m \sigma^m(z) \neq 0.$$

Then $d = \theta^n$ is an element of $k^\times$ such that $L = k(\sqrt[n]{d})$.

Proof.   To prove that $d \in k$, Galois theory shows it is enough to prove that $\sigma(d) = d$. We have

$$\zeta_n \sigma(\theta) = \sum_{m=0}^{n-1} \zeta_n^{m+1} \sigma^{m+1}(z) = \sum_{m=1}^{n} \zeta_n^m \sigma^m(z) = \theta.$$

Raising to the nth-power yields $\sigma(d) = d$. Moreover, $\sigma(\theta) = \zeta_n^{-1}\theta$, and therefore $\sigma^m(\theta) = \zeta_n^{-m}\theta$ for $m = 0, \ldots, n-1$. Since ζ_n is a primitive n^{th}-root of 1, it follows that the conjugates of θ are all distinct. Therefore θ is a primitive element of L. By construction, $\theta^n = d$, so we get $L = k(\sqrt[n]{d})$, as claimed. $\square$

We now apply Example B.2.14 to $d = -1$ and -3 to describe the ramification of prime numbers in k.

Assume first that $k = \mathbb{Q}(i)$. Then $\mathcal{O}_k = \mathbb{Z}[i]$ is a principal ideal domain, and $\mathcal{O}_k^\times = \{\pm 1, \pm i\}$. Moreover, $2\mathcal{O}_k = (1+i)^2$, so 2 totally ramifies in k. If $\ell \neq 2$, then ℓ totally splits (respectively ℓ is inert) in k if and only of the polynomial $X^2 + 1 \in \mathbb{F}_\ell[X]$ splits (respectively is irreducible), that is if and only if $-1 \in \mathbb{F}_\ell^{\times 2}$ (respectively $-1 \notin \mathbb{F}_\ell^{\times 2}$).

Therefore, if $\ell \neq 2$, then ℓ totally splits in k if and only if $\ell \equiv 1[4]$, and ℓ is inert if and only if $\ell \equiv 3[4]$.

Moreover, any prime element of $\mathcal{O}_k$ is associate to one of the following elements:

(1) $1 + i$;

(2) ℓ, where ℓ is a prime number satisfying $\ell \equiv 3[4]$;

(3) $a + bi$, where $\ell = a^2 + b^2$ is a prime number satisfying $\ell \equiv 1[4]$.

Notice that, since $\mathcal{O}_k$ is a principal ideal domain, every prime ideal $\mathfrak{p}$ is generated by a prime element $\pi \in \mathcal{O}_k$.

If now $k = \mathbb{Q}(j)$, $\mathcal{O}_k = \mathbb{Z}[j]$ is also a principal ideal domain and $\mathcal{O}_k^\times = \{\pm 1, \pm j, \pm j^2\}$. Moreover, $3\mathcal{O}_k = (1 - j)^2 \mathcal{O}_k$, so 3 totally ramifies in k. If $\ell \neq 2, 3$, then ℓ totally splits (respectively ℓ is inert) in k if and only if the polynomial $X^2 + X + 1 \in \mathbb{F}_\ell[X]$ splits (respectively is irreducible), that is if and only if $-3 \in \mathbb{F}_\ell^{\times 2}$ (respectively $-3 \notin \mathbb{F}_\ell^{\times 2}$).

Notice that for $\ell \neq 2, 3$, we have

$$\left(\frac{-3}{\ell}\right) = \left(\frac{-1}{\ell}\right)\left(\frac{3}{\ell}\right) = (-1)^{\frac{\ell-1}{2}} \cdot (-1)^{\frac{\ell-1}{2}\frac{3-1}{2}}\left(\frac{\ell}{3}\right) = \left(\frac{\ell}{3}\right).$$

We deduce that, if $\ell \neq 2, 3$, $-3 \in \mathbb{F}_\ell^{\times 2}$ if and only if $\ell \equiv 1[3]$, and $-3 \notin \mathbb{F}_\ell^{\times 2}$ if and only if $\ell \equiv 2[3]$.

If $\ell = 2$, the polynomial $X^2 + X + 1$ is irreducible, so 2 is inert.

Therefore, if $\ell \neq 3$, then ℓ totally splits in k if and only if $\ell \equiv 1[3]$, and ℓ is inert if and only if $\ell \equiv 2[3]$.

Moreover, any prime element of $\mathcal{O}_k$ is associate to one of the following elements:

(1) $1 - j$;

(2) ℓ, where ℓ is a prime number satisfying $\ell \equiv 2[3]$;

(3) $a + bj$, where $\ell = a^2 + b^2 - ab$ is a prime number satisfying $\ell \equiv 1[3]$.

We now state and prove the first result of this section.

PROPOSITION VII.5.3. *Let $k = \mathbb{Q}(i)$, and let $d \in \mathcal{O}_k \setminus \{0\}$. Then the quaternion algebra $Q = (i, d)_k$ is a division algebra if and only if there exists a prime element $\pi = a + bi$ such that $n_\pi(d)$ is odd and $\ell = a^2 + b^2$ is a prime number satisfying $\ell \equiv 5[8]$.*

Proof. If d is a square in k, then Q is split since $A \cong_k (i, 1)_k$ and $n_\pi(d)$ is even for any prime element π. Without loss of generality, we then may assume that $L = k(\sqrt{d})$ is a quadratic extension of k. In this case, applying Proposition VII.4.8 with $p = 2$, we get that Q is not split if and only if $Q \otimes_k k_\pi$ is not split for some prime element $\pi \neq 1 + i$ such that $n_\pi(d)$ is odd. Let π be such a prime element.

Assume first that $\pi = \ell$, where $\ell \equiv 3[4]$. Then the corresponding residue field is $\mathbb{F}_{\ell^2}$. Since $\dfrac{\ell^2 - 1}{2}$ is a multiple of 4, we have

$$\bar{i}^{\frac{\ell^2-1}{2}} = 1 \in \mathbb{F}_{\ell^2},$$

since $i^4 = 1$. Therefore $A \otimes_k k_\pi$ is split in this case by Lemma VII.4.6.

Assume now that $\pi = a + bi$, where $\ell = a^2 + b^2$ is a prime number which is congruent to 1 modulo 4. In this case, the corresponding residue field is $\mathbb{F}_\ell$ and we have $\bar{i}^{\frac{\ell-1}{2}} = 1 \in \mathbb{F}_\ell$ if and only if $\bar{i} \in \mathbb{F}_\ell^{\times 2}$, that is if and only if $\mu_8 \subset \mathbb{F}_\ell^\times$. This is equivalent to saying that $8 \mid \ell - 1$. By Lemma VII.4.6, we get that $Q \otimes_k k_\pi$ is split if and only if $\ell \equiv 1[8]$ in this case.

Hence Q is not split if and only if there exists a prime element $\pi = a + bi$ such that $n_\pi(d)$ is odd and $\ell = a^2 + b^2$ is a prime number satisfying $\ell \equiv 5[8]$. This concludes the proof, since Q is a division algebra if and only if it is not split. □

EXAMPLE VII.5.4. Let $k = \mathbb{Q}(i)$, and let $A = (i, k(\zeta_5 + \zeta_5^{-1})/k, \sigma)$, where $\sigma \in \mathrm{Gal}(k(\zeta_5 + \zeta_5^{-1})/k)$ is defined by

$$\sigma(\zeta_5 + \zeta_5^{-1}) = \zeta_5^2 + \zeta_5^{-2}.$$

It is well-known that $\cos\left(\dfrac{2\pi}{5}\right) = \dfrac{-1 + \sqrt{5}}{4}$, so $L = \mathbb{Q}(i)(\sqrt{5})$. Applying Proposition VII.5.3 with $\pi = 1 + 2i$ shows that A is a division algebra. □

Proposition VII.5.1 and Proposition VII.5.3 yield immediately the following result.

COROLLARY VII.5.5. Let $k = \mathbb{Q}(i)$, and let $A = (i, k(\sqrt[4]{d})/k, \sigma)$ be a cyclic k-algebra of degree 4. Then A is a division algebra if and only if there exists a prime element $\pi = a + bi$ such that $n_\pi(d)$ is odd and $\ell = a^2 + b^2$ is a prime number satisfying $\ell \equiv 5[8]$.

EXAMPLE VII.5.6. Let $k = \mathbb{Q}(i)$, let $L = k(\zeta_{15} + \zeta_{15}^{-1})$, and consider the cyclic k-algebra $A = (i, L/k, \sigma)$, where $\sigma \in \mathrm{Gal}(k(\zeta_{15} + \zeta_{15}^{-1})/k)$ is defined by

$$\sigma(\zeta_{15} + \zeta_{15}^{-1}) = \zeta_{15}^2 + \zeta_{15}^{-2}.$$

Using Lemma VII.5.2, one can check that we have $L = k(\sqrt[4]{d})$, with $d = 3^2(1 + 2i)(1 - 2i)^3$. Applying Corollary VII.5.5 with $\pi = 1 + 2i$ shows that A is a division algebra. □

PROPOSITION VII.5.7. Let $k = \mathbb{Q}(j)$, and let $d \in \mathcal{O}_k$. Then the quaternion algebra $Q = (-j, d)_k$ is a division algebra if and only there exists a prime element π of $\mathcal{O}_k$ such that $n_\pi(d)$ is odd and satisfying one of the two following conditions:

(1) $\pi = 1 - j$;
(2) $\pi = a + bj$, where $\ell = a^2 + b^2 - ab$ is a prime number satisfying $\ell \equiv 7[12]$.

Proof. Once again, we may assume that d is not a square in k. Notice now that 2 is inert k. In particular, there is only one prime ideal of $\mathcal{O}_k$ lying above 2. Applying Proposition VII.4.8 with $p = 2$, we get that Q is not split if and only if $Q \otimes_k k_\pi$ is not split for some prime element $\pi \neq 2$ such that $n_\pi(d)$ is odd. Let π be such a prime element.

Assume first that $\pi = 1 - j$. Then the residue field is $\mathbb{F}_3$. Since $\overline{-j} = \overline{-1} \neq \overline{1}$ in $\mathbb{F}_3$, $Q \otimes_k k_\pi$ is not split in this case by Lemma VII.4.6.

Assume now that $\pi = \ell, \ell \equiv 2[3], \ell \neq 2$. In this case, the corresponding residue field is $\mathbb{F}_{\ell^2}$. By assumption, we have $\ell = 2 + 3(2m + 1)$ for some integer $m \geq 0$. Thus, $\ell \equiv 5[6]$ and therefore, we get

$$\frac{\ell^2 - 1}{2} \equiv 0[6].$$

Since $(-j)^6 = 1$, we get

$$\overline{-j}^{\frac{\ell^2-1}{2}} = 1 \in \mathbb{F}_{\ell^2},$$

so $Q \otimes_k k_\pi$ splits in this case.

Finally, assume that $\pi = a + bj$, where $\ell = a^2 + b^2 - ab, \ell \neq 2$ is a prime number satisfying $\ell \equiv 1[3]$. In this case, the corresponding residue field is $\mathbb{F}_\ell$.

We then have $\overline{-j}^{\frac{\ell-1}{2}} = 1 \in \mathbb{F}_\ell$ if and only if $\overline{-j} \in \mathbb{F}_\ell^{\times 2}$, that is $\mu_{12} \subset \mathbb{F}_\ell^\times$. This is equivalent to saying that $\ell \equiv 1[12]$. This concludes the proof. $\qquad\square$

PROPOSITION VII.5.8. *Let* $k = \mathbb{Q}(j)$, *and consider the cyclic k-algebra* $A = (j, k(\sqrt[3]{d})/k, \sigma)$ *of degree 3, where $d \in \mathcal{O}_k$. Then A is a division algebra if and only if there exists a prime element π of $\mathcal{O}_k$ such that $n_\pi(d)$ is prime to 3 and satisfying one of the following conditions:*

(1) $\pi = \ell$, *where ℓ is a prime number satisfying $\ell \equiv 2$ or $5[9]$;*
(2) $\pi = a + bj$, *where $\ell = a^2 + b^2 - ab$ is a prime number satisfying $\ell \equiv 4$ or $7[9]$.*

Proof. Applying Proposition VII.4.8 with $p = 3$, we get that A is not split if and only if $A \otimes_k k_\pi$ is not split for some prime element $\pi \neq 1 - j$ such that $n_\pi(d)$ is prime to 3.

Assume first that $\pi = \ell$, where $\ell \equiv 2[3]$. Then the corresponding residue field is $\mathbb{F}_{\ell^2}$.

If $p \equiv 8[9]$, $\dfrac{p^2 - 1}{3}$ is a multiple of 3, so we have

$$\overline{j}^{\frac{p^2-1}{3}} = 1 \in \mathbb{F}_{p^2},$$

since $j^3 = 1$. Therefore $A \otimes_k k_\pi$ is split in this case by Lemma VII.4.6.

If $p \equiv 2$ or $5[9]$, $\dfrac{\ell^2 - 1}{3} \equiv \pm 1[3]$, so we have

$$\overline{j}^{\frac{\ell^2-1}{2}} = \overline{j}^{\pm 1} \in \mathbb{F}_{\ell^2}.$$

Since $X^2 + X + 1$ is irreducible in $\mathbb{F}_\ell[X]$, we get that $\mathbb{F}_{\ell^2} = \mathbb{F}_\ell(\overline{j})$. In particular, we cannot have $\overline{j}^{\pm 1} = 1 \in \mathbb{F}_{\ell^2}$. By Lemma VII.4.6, $A \otimes_k k_\pi$ is not split.

Assume now that $\pi = a + bj$, where $\ell = a^2 + b^2 - ab$ is a prime number satisfying $\ell \equiv 1[3]$. In this case, the corresponding residue field is $\mathbb{F}_\ell$ and we have $\overline{j}^{\frac{\ell-1}{3}} = 1 \in \mathbb{F}_\ell$ if and only if $\overline{j} \in \mathbb{F}_\ell^{\times 3}$, that is if and only if $\mu_9 \subset \mathbb{F}_\ell^\times$. This is equivalent to saying that $9 \mid \ell - 1$. By Lemma VII.4.6, we get that $A \otimes_k k_\pi$ is split if and only if $\ell \equiv 1[9]$ in this case.

Therefore, $A = (j, k(\sqrt[3]{d})/k, \sigma))$ is not split if and only if there exists a prime element π of $\mathcal{O}_k$ such that $n_\pi(d)$ is prime to 3 satisfying one of the following conditions:

(1) $\pi = \ell$, where ℓ is a prime number satisfying $\ell \equiv 2$ or $5[9]$;

(2) $\pi = a + bj$, where $\ell = a^2 + b^2 - ab$ is a prime number satisfying $\ell \equiv 4$ or $7[9]$.

This concludes the proof. $\qquad\qquad\qquad\qquad\qquad\qquad\qquad\qquad\qquad\qquad$ $\square$

EXAMPLE VII.5.9. Let $k = \mathbb{Q}(j)$, let $L = k(\zeta_7 + \zeta_7^{-1})$ and consider the cyclic k-algebra $A = (j, L/k, \sigma)$, where $\sigma \in \mathrm{Gal}(k(\zeta_7 + \zeta_7^{-1})/k)$ is defined by

$$\sigma(\zeta_7 + \zeta_7^{-1}) = \zeta_7^2 + \zeta_7^{-2}.$$

Using once again Lemma VII.5.2, one can check that $L = k(\sqrt[3]{d})$, with $d = -(2 + 3j)^2(1 + 3j)$. Now applying Proposition VII.5.8 with $\pi = 1 + 3j$ shows that A is a division algebra. $\qquad\qquad\qquad\qquad\qquad\qquad\qquad\qquad\qquad\qquad$ $\square$

COROLLARY VII.5.10. *Let* $k = \mathbb{Q}(j)$, *and consider a cyclic algebra* $A = (-j, k(\sqrt[6]{d})/k, \sigma)$ *of degree 6, where* $d \in \mathcal{O}_k$. *Then* A *is a division algebra if and there exist (not necessarily distinct) prime elements* π *and* π' *of* $\mathcal{O}_k$ *such that* $n_\pi(d)$ *is prime to 3,* $n_{\pi'}(d)$ *is odd and satisfying the following conditions:*

(1) $\pi = 1 - j$ *or* $\pi = a + bj$, *where* $\ell = a^2 + b^2 - ab$ *is a prime number satisfying* $\ell \equiv 7[12]$;

(2) $\pi' = \ell$, *where* ℓ *is a prime number which such that* $\ell \equiv 2$ *or* $5[9]$ *or* $\pi' = a + bj$, *where* $\ell = a^2 + b^2 - ab$ *is a prime number satisfying* $\ell \equiv 4$ *or* $7[9]$.

Proof. By Proposition VII.5.1, A is a division algebra if and only if $(-j, d)_k$ and $(-j, k(\sqrt[3]{d})/k, \sigma)$ are. Since -1 is a cube in k, the latter algebra is isomorphic to $(j, k(\sqrt[3]{d})/k, \sigma)$. Now apply Proposition VII.5.7 and Proposition VII.5.8 to conclude. $\qquad\qquad\qquad\qquad\qquad\qquad\qquad\qquad\qquad\qquad$ $\square$

EXAMPLE VII.5.11. Let $k = \mathbb{Q}(j)$, and let $A = (j, L/k, \sigma)$, where $L = k(\zeta_{28} + \zeta_{28}^{-1})$ and $\sigma \in \mathrm{Gal}(k(\zeta_{28} + \zeta_{28}^{-1})/k)$ is defined by

$$\sigma(\zeta_{28} + \zeta_{28}^{-1}) = \zeta_{28}^2 + \zeta_{28}^{-2}.$$

One can check that $L = k(\sqrt[6]{d})$, with $d = (1 - 2j)(3 + 2j)^5$. Applying Corollary VII.5.10 with $\pi = \pi' = 1 - 2j$, we get that A is a division algebra. $\qquad\qquad$ $\square$

We would like now to apply the results of the previous sections to explain how to construct explicitly cyclic division algebras of prescribed degree n. One may even fix the cyclic extension L/k.

Let L/k be a cyclic extension of degree $n = p_1^{m_1} \cdots p_r^{m_r}, m_i \geq 1$ $(n \geq 2)$, and let σ be any generator of $\mathrm{Gal}(L/k)$. By Bézout's Theorem, we may write

$$u_i \frac{n}{p_i^{m_i}} + p_i^{m_i} v_i = 1, u_i, v_i \in \mathbb{Z}.$$

Let $L^{(p_i)}$ the unique subfield of L of degree p_i of L. Since $L^{(p_i)}/k$ is cyclic, one may find a prime ideal $\mathfrak{p}_i$ of $\mathcal{O}_k$ which is inert in $L^{(p_i)}$ by Proposition B.2.17. Pick $a_i \in \mathfrak{p}_i \setminus \mathfrak{p}_i^2$, and set

$$a = a_1^{u_1 \frac{n}{p_1^{m_1}}} \cdots a_r^{u_r \frac{n}{p_r^{m_r}}}.$$

PROPOSITION VII.5.12. *Keeping the notation above, the cyclic algebra* $(a, L/K, \sigma)$ *is a division k-algebra of degree n.*

Proof. By construction of a, we have $(a_i) \subset \mathfrak{p}_i$ and $(a_i) \not\subset \mathfrak{p}_i^2$, or equivalently $\mathfrak{p}_i \mid (a_i)$ and $\mathfrak{p}_i^2 \nmid (a_i)$. In other words,

$$n_{\mathfrak{p}_i}(a_i) = 1 \ \text{ for all } \ i = 1, \dots, r.$$

Since $u_i \dfrac{n}{p_i^{m_i}} = 1 - p_i^{m_i} v_i$, we get

$$n_{\mathfrak{p}_i}(a_i^{u_i \frac{n}{p_j^{m_i}}}) \equiv 1[p_i].$$

Notice now that for all $j \neq i$, $\dfrac{n}{p_j^{m_j}}$ is a multiple of $p_i^{m_i}$. In particular,

$$n_{\mathfrak{p}_i}(a_j^{u_j \frac{n}{p_j^{m_j}}}) \equiv 0[p_i] \ \text{ for all } \ j \neq i.$$

It follows that we have

$$n_{\mathfrak{p}_i}(a) \equiv 1[p_i] \ \text{ for all } \ i = 1, \dots, r.$$

In particular, we have $p_i \nmid n_{\mathfrak{p}_i}(a)$ for all $i = 1, \dots, r$. Lemma VII.4.4 then shows that $(a, L^{(p_i)}, \sigma_{|_{L^{(p_i)}}}) \otimes_k k_{\mathfrak{p}_i}$ is not split for all $i = 1, \dots, r$. Therefore, $(a, L^{(p_i)}/k, \sigma_{|_{L^{(p_i)}}})$ is not split for all $i = 1, \dots, r$. Proposition VII.3.5 then yields the conclusion. $\square$

REMARK VII.5.13. If $\mathcal{O}_k$ is a principal ideal domain, one may simply take $a_i = \pi_i$, where π_i is a generator of $\mathfrak{p}_i$. $\square$

We now give a concrete example when $k = \mathbb{Q}$.

EXAMPLE VII.5.14. Let p be a prime number, and let $m \geq 1$ be an integer. If p is odd, let L_{p^m} be the unique subfield of $\mathbb{Q}(\zeta_{p^{m+1}})$ of degree p^m over $\mathbb{Q}$. We also set $L_{2^m} = \mathbb{Q}(\zeta_{2^{m+2}} + \zeta_{2^{m+2}}^{-1})$. In all cases, $L_{p^m}/\mathbb{Q}$ is cyclic of degree p^m. We will denote by L_p the unique subfield of L_{p^m} of degree p over $\mathbb{Q}$. Notice that $L_2 = \mathbb{Q}(\sqrt{2})$.

Let $n = p_1^{m_1} \cdots p_r^{m_r} \geq 2$, and let $L_n/\mathbb{Q}$ be the compositum (in $\mathbb{C}$) of the extensions $L_{p_i^{m_i}}/\mathbb{Q}, i = 1, \dots, r$. This is a cyclic Galois extension of degree n. Notice that the unique subfield of L_n of degree p_i is L_{p_i} by construction.

We now explain how to find a prime number ℓ which is inert in the extension $L_p/\mathbb{Q}$, for a given prime number p.

If $p = 2$, then $L_2 = \mathbb{Q}(\sqrt{2})$, and we may take $\ell = 3$. Assume now that p is odd. First of all, there exists a prime number ℓ such that $\bar{\ell}$ has order p in $(\mathbb{Z}/p^2\mathbb{Z})^{\times}$.

Indeed, $(\mathbb{Z}/p^2\mathbb{Z})^{\times}$ is cyclic of order $p(p-1)$, so there exists $x \in \mathbb{Z}$ such that $\bar{x}$ has order p in$(\mathbb{Z}/p^2\mathbb{Z})^{\times}$. Now since x is prime to p^2, by Dirichlet's theorem, there exists a prime number ℓ such that $\ell \equiv x[p^2]$.

Claim: ℓ is inert in $L_p/\mathbb{Q}$.

Since $\ell \neq p$ and $\bar{\ell}$ has order p in $(\mathbb{Z}/p^2\mathbb{Z})^{\times}$, Theorem B.2.15 shows thatℓ is not ramified in $\mathbb{Q}(\zeta_{p^2})/\mathbb{Q}$ and has residual degree p. It follows that $\ell\mathbb{Z}[\zeta_{p^2}]$ is the product of $p-1$ distinct primes ideals of $\mathbb{Z}[\zeta_{p^2}]$. In particular, $\ell\mathcal{O}_{L_p}$ is the product of at most $p-1$ distinct primes ideals of $\mathcal{O}_{L_p}$. Notice that ℓ cannot ramify in $L_p/\mathbb{Q}$, since otherwise it would ramify in $\mathbb{Q}(\zeta_{p^2})/\mathbb{Q}$. Since $L_p/\mathbb{Q}$ has prime degree,

ℓ is therefore either inert or totally split. Assume that ℓ totally splits in $L_p/\mathbb{Q}$. In this case, $\ell\mathcal{O}_{L_p}$ would be a product of p distinct prime ideals of $\mathcal{O}_{L_p}$, which is a contradiction. Hence, ℓ is inert in $L_p/\mathbb{Q}$.

Finally, set

$$a = \ell_1^{u_1 \frac{n}{p_1^{m_1}}} \cdots \ell_r^{u_r \frac{n}{p_r^{m_r}}},$$

where ℓ_i is a prime number such that $\overline{\ell_i}$ has order p_i^2 in $(\mathbb{Z}/p_i^2\mathbb{Z})^\times$ if p_i is odd, and $\ell_i = 3$ is p_i is even. Then $(a, L_n/\mathbb{Q}, \sigma)$ is a division $\mathbb{Q}$-algebra of degree n, for any generator σ of $\mathrm{Gal}(L_n/\mathbb{Q})$.   □

VII.6. Cyclic algebras and perfect codes

We would like now to give some constructions of codes based on a cyclic algebra $A = (a, L/k, \sigma)$ of degree n (see Definition VII.1.3).

A code $\mathcal{C}$ based on a cyclic algebra A will be fully diverse if A is a division algebra, which can be determined by one of the criteria given in Corollary VII.1.12, namely, if $a^m \notin N_{L/k}(L^\times)$ for all $m \neq n$ dividing n, then A is a division k-algebra.

One way of constructing cyclic division algebras over fields of the form $k(t)$, where $t \in \mathbb{C}$ is some transcendental complex number has been proposed in [48].

Let L/k be a cyclic extension, whose Galois group is generated by σ, and let $t \in \mathbb{C}$ be a transcendental complex number. Since t is transcendental, elementary Galois theory shows that $L(t)/k(t)$ is also cyclic, generated by the unique $k(t)$-automorphism $\overline{\sigma} : L(t) \longrightarrow L(t)$ satisfying

$$\overline{\sigma}(t) = t \text{ and } \overline{\sigma}_{|L} = \sigma.$$

We then have the following proposition.

PROPOSITION VII.6.1. [48] *The algebra* $(t, L(t)/k(t), \overline{\sigma})$ *is a division algebra.*

Proof. Let $m \neq n$ be a divisor of n. Suppose that $t^m = N_{L(t)/k(t)}\left(\frac{f}{g}\right)$ for some $f, g \in L[t]$. We then have

$$t^m N_{L(t)/k(t)}(g) = N_{L(t)/k(t)}(f).$$

Since t is a transcendental complex number and any element of L is algebraic over $\mathbb{Q}$, we have an isomorphism of L-algebras $L[t] \cong_L L[X]$. One may then consider the degree of an element of $L[t]$ as a polynomial in t. It is clear from the definition of $\overline{\sigma}$ that the degree of $\overline{\sigma}(f)$ is the same as the degree of f. Since

$$N_{L(t)/k(t)}(f) = f\overline{\sigma}(f) \cdots \overline{\sigma}^{n-1}(f),$$

it follows that we have

$$\deg(N_{L(t)/k(t)}(f)) = n \deg(f),$$

and similarly

$$\deg(N_{L(t)/k(t)}(g)) = n \deg(g).$$

Therefore, we have $m + n \deg(g) = n \deg(f)$. In particular, m is a multiple of n. This yields a contradiction since $1 \leq m < n$. Using Corollary VII.1.12 (1), we obtain the desired conclusion.   □

We have seen in Section IV.3 that once a code $\mathcal{C}$ is fully diverse, the next coding criterion is the minimum determinant, namely, one has to maximize

$$\delta_{min}(\mathcal{C}) = \inf_{a \in A \setminus \{0\}} |\mathrm{Nrd}_A(a)|^2$$

where Nrd_A denotes the reduced norm of A. Using that $\mathrm{Nrd}_A(a) \in k(t)$ for all $a \in A$ (by Lemma IV.2.1), we have that

$$\delta_{min}(\mathcal{C}) = 0$$

since an element of the form $|x + yt|^2$, $x, y \in k$, can be made arbitrarily small. We now use the results of the previous chapters to give constructions of codes based on cyclic division algebras with a non-zero minimum determinant, and satisfying the shaping constraint.

Since A is a particular crossed product, the left multiplication matrix M_x of $x = x_0 + ex_1 + \cdots + e^{n-1}x_{n-1}$, $x_i \in L$, $i = 0, \ldots, n-1$ with respect to the L-basis $(1, e, \ldots, e^{n-1})$ is given by

$$\begin{pmatrix} x_0 & a\sigma(x_{n-1}) & a\sigma^2(x_{n-2}) & \ldots & a\sigma^{n-1}(x_1) \\ x_1 & \sigma(x_0) & a\sigma^2(x_{n-1}) & \ldots & a\sigma^{n-1}(x_2) \\ \vdots & & \vdots & & \vdots \\ x_{n-2} & \sigma(x_{n-3}) & \sigma^2(x_{n-4}) & \ldots & a\sigma^{n-1}(x_{n-1}) \\ x_{n-1} & \sigma(x_{n-2}) & \sigma^2(x_{n-3}) & \ldots & \sigma^{n-1}(x_0) \end{pmatrix},$$

by Lemma VI.3.1 and by the description of the cocycle defining A (see Example VII.1.2).

Assume now that $k/\mathbb{Q}$ is an imaginary quadratic extension, and that $\mathcal{O}_k$ is a principal ideal domain. Let L/k be a cyclic number field extension, whose Galois group is generated by σ, and let I be an ideal of $\mathcal{O}_L$. The results obtained in Section VI.3 in the case of a cyclic algebra then translate as follows. Assume that the cyclic extension L/k and the cyclic k-algebra $A = (a, L/k, \sigma)$ satisfy the following conditions:

(1) complex conjugation induces a $\mathbb{Q}$-automorphism of L which commutes with σ;
(2) there exists $\lambda \in L^\times$ and an ideal I of $\mathcal{O}_L$ satisfying:
(a) $\lambda \in \mathbb{R}$;
(b) λ^{σ^m} is a positive real number for $m = 0, \ldots, n-1$;
(c) $\mathrm{Tr}_{L/k}(\lambda \overline{x} y) \in \mathcal{O}_k$ for all $x, y \in I$;
(d) the hermitian $\mathcal{O}_k$-lattice

$$h_\lambda : \begin{array}{c} I \times I \longrightarrow \mathcal{O}_k \\ (x, y) \longmapsto \mathrm{Tr}_{L/k}(\lambda \overline{x} y) \end{array}$$

is isomorphic to the cubic lattice $\mathcal{O}_k^n$.
(3) $|a|^2 = 1$, and A is a division algebra.

Then for any orthonormal $\mathcal{O}_k$-basis $(\omega_\sigma)_{\sigma \in G}$ of (I, h_λ), the encoding map

$$\mathcal{O}_k^n \longrightarrow \mathcal{C}_{A,\lambda,I}$$

$$(x_{\sigma,\tau})_{\sigma,\tau \in G} \longmapsto (\sqrt{\lambda^\tau} \xi_{\sigma\tau^{-1},\tau} x^\tau_{\sigma\tau^{-1}})_{\sigma,\tau},$$

where $x_\sigma = \sum\limits_{\tau \in G} x_{\sigma,\tau} \omega_\tau$ for all $\sigma \in G$ fulfills the shaping constraint.

REMARK VII.6.2. Codes satisfying all the above properties, that is, full diversity, a minimum determinant bounded away from zero, and with an orthonormal lattice structure have been called **perfect** codes [**40**]. $\square$

Since $\mathcal{O}_k$ is a principal ideal domain, we may write $a = \dfrac{a_1}{a_2}$, where $a_1, a_2 \in \mathcal{O}_k$ are coprime elements. It is then clear that $\Delta_{\xi^{\sigma,a}} = |a_2|^{2n-2}$, where $\xi^{\sigma,a}$ is the cocycle defining A, since this cocycle can take only take 1 or a for values (see Section VI.3 for a definition of Δ_ξ).

Notice now that we have

$$|a_2|^{2n-2} = 1 \iff |a_2|^2 = 1 \iff a_2 \in \mathcal{O}_k^\times,$$

the last equivalence following from the fact that $k/\mathbb{Q}$ is an imaginary quadratic extension. Therefore, $|a_2|^{2n-2} = 1$ if and only if $a \in \mathcal{O}_k$, and since $|a|^2 = 1$, we finally get that

$$\Delta_{\xi^{\sigma,a}} = 1 \iff a \in \mathcal{O}_k^\times.$$

Thus, if A is a cyclic k-algebra (where $k/\mathbb{Q}$ is an imaginary quadratic extension such that $\mathcal{O}_k$ is a principal ideal domain), Proposition VI.3.6 translates as follows:

PROPOSITION VII.6.3. *Assume that the cyclic algebra $A = (a, L/k, \sigma)$ is a division k-algebra. Then we have*

$$\frac{1}{|a_2|^{2n-2}d_{L/k}} \le \delta_{min}(\mathcal{C}_{A,\lambda,I}) \le N_{L/k}(\lambda) N_{min}(I).$$

If moreover $a \in \mathcal{O}_k^\times$ and I is principal, we have

$$\delta_{min}(\mathcal{C}_{A,\lambda,I}) = \frac{1}{d_{L/k}^n}.$$

It is then clear from this proposition that the choice of a influences the performance of the code. In particular, if $a \in \mathcal{O}_k^\times$, we see from Proposition VII.6.3 that the minimum determinant will be better.

Recall now from Section IV.3 that for any fully diverse code $\mathcal{C}$ based on a division algebra, the probability error of the code satisfies

$$\mathbb{P}(\mathbf{X} \to \hat{\mathbf{X}}) \le \frac{\kappa}{\delta_{min}(\mathcal{C})^n},$$

where κ is a constant depending on the channel. In particular, for any perfect code $\mathcal{C} \subset \mathcal{C}_{A,\lambda,I}$ based on a cyclic division k-algebra $A = (a, L/k, \sigma)$, we have

$$\mathbb{P}(\mathbf{X} \to \hat{\mathbf{X}}) \le \frac{\kappa}{\delta_{min}(\mathcal{C})^n} \le \frac{\kappa}{\delta_{min}(\mathcal{C}_{A,\lambda,I})^n} \le \kappa(|a_2|^{2n-2}d_{L/k})^n.$$

In the following, we give examples of perfect codes in dimensions 2,3,4 and 6, [**40**]. We then show that perfect codes with $a \in \mathcal{O}_k^\times$ do not exist in other dimensions. In order for perfect codes to exist in other dimensions, we thus need to choose $a \in k^\times$. An example of such code will be given in dimension 5.

Let us start with the case of dimension 2. Let p be a prime number satisfying $p \equiv 5[8]$, so we may write

$$p = u^2 + v^2, u, v \in \mathbb{Z}.$$

Let $L = \mathbb{Q}(i)(\sqrt{p})$. Recall from Chapter IV Exercise 5 that $\mathcal{O}_L = \mathbb{Z}[i, \theta]$, where $\theta = \dfrac{1 + \sqrt{p}}{2}$. Let I be the ideal of $\mathcal{O}_L$ generated by $\dfrac{p-1}{2} + \theta$ and $u + iv$. Finally, set $\lambda = \dfrac{1}{p}$. Direct computations show that we have

$$\mathrm{Tr}_{L/\mathbb{Q}(i)}(\frac{1}{p}\overline{x}y) \in \mathbb{Z}[i] \text{ for all } x, y \in I,$$

so we may consider the complex ideal $\mathbb{Z}[i]$-lattice (I, h_λ). Computing the representative matrix of (I, h_λ) in a $\mathbb{Z}[i]$-basis of I or using the results of Appendix C, shows that (I, h_λ) is a positive definite $\mathbb{Z}[i]$-lattice of determinant 1. By [24], (I, h_λ) is isomorphic to the cubic lattice. Therefore, we may find an orthonormal basis (w_1, w_2) of (I, h_λ). Since $Q = (i, p)_{\mathbb{Q}(i)}$ is a division algebra by Chapter IV Exercise 5, we may construct a code $\mathcal{C}_{\lambda,Q,I}$ satisfying the shaping constraint. Moreover, according to the same exercise and using the equality

$$\delta_{min}(\mathcal{C}_{A,\lambda,I}) = N_{L/\mathbb{Q}(i)}(\lambda)\delta_{min}(\mathcal{C}_{A,I}),$$

we get that

$$\delta_{min}(\mathcal{C}_{A,\lambda,I}) = \frac{1}{p}.$$

More details will be given in Exercise 4.

REMARK VII.6.4. Notice that we also have $d_{L/\mathbb{Q}(i)} = p$, as we may see by computing the determinant of a representative matrix of $(\mathcal{O}_L, h_1)$, for example. Therefore, we have

$$\delta_{min}(\mathcal{C}_{A,\lambda,I}) = \frac{1}{d_{L/\mathbb{Q}(i)}}.$$

This is not surprising, since one may show using class field theory that I is a principal ideal.   □

We then have the theoretical existence of an infinite family of perfect codes in dimension 2. We now give a concrete example.

EXAMPLE VII.6.5. We describe here the so-called Golden code [6], which corresponds to the case $p = 5$.

Set

$$L = \mathbb{Q}(i, \sqrt{5}), \theta = \frac{1 + \sqrt{5}}{2}, \alpha = 1 + i - i\theta \text{ and } \lambda = \frac{1}{5}.$$

It is easy to check that $\alpha \in I$ and $N_{L/\mathbb{Q}}(\alpha) = p = N_{L/\mathbb{Q}}(I)$. Therefore, we have $I = \alpha\mathcal{O}_L$. One may verify that an orthonormal $\mathbb{Z}[i]$-basis for the complex ideal lattice (I, h_λ) is $(\alpha, \alpha\theta)$. We then get the following perfect code based on $Q = (i, 5)_{\mathbb{Q}(i)}(= (i, L/\mathbb{Q}(i), \sigma))$:

$$\mathcal{C}_{Q,\lambda,I} = \left\{ \mathbf{X} = \frac{1}{\sqrt{5}} \begin{pmatrix} \alpha(a + b\theta) & i\sigma(\alpha)(c + d\sigma(\theta)) \\ \alpha(c + d\theta) & \sigma(\alpha)(a + b\sigma(\theta)) \end{pmatrix}, \ a, b, c, d \in \mathbb{Z}[i]. \right\}$$

satisfying

$$\delta_{min}(\mathcal{C}_{Q,\lambda,I}) = \frac{1}{5}.$$

We will show in the next section that the Golden code has optimal performances among all the perfect codes for two antennas if $k = \mathbb{Q}(i)$. This was originally proved in [**33**]. $\square$

We now give examples of perfect codes in higher dimensions. All the codes below have been presented in [**40**].

EXAMPLE VII.6.6. Let $L = \mathbb{Q}(j, \zeta_7 + \zeta_7^{-1})$. The extension $L/\mathbb{Q}(j)$ has degree 3 and cyclic Galois group generated by

$$\sigma: \begin{array}{c} L \longrightarrow L \\ \zeta_7 + \zeta_7^{-1} \longmapsto \zeta_7^2 + \zeta_7^{-2}. \end{array}$$

Consider the cyclic algebra $A = (j, L/\mathbb{Q}(j), \sigma)$ of degree 3, which is a division algebra (see Example VII.5.9). Set

$$\theta = \zeta_7 + \zeta_7^{-1}, \alpha = (1+j) + \theta, I = \alpha\mathcal{O}_L \text{ and } \lambda = \frac{1}{7}.$$

Then $((1+j)+\theta, (-1-2j)+j\theta^2, (-1-2j)+(1+j)\theta+(1+j)\theta^2)$ is an orthonormal $\mathbb{Z}[j]$-basis of (I, h_λ), and we get a perfect code $\mathcal{C}_{A,\lambda,I}$ in dimension 3. One can compute that $d_{L/\mathbb{Q}(j)} = 49$, so we have

$$\delta_{min}(\mathcal{C}_{A,\lambda,I}) = \frac{1}{d_{L/\mathbb{Q}(i)}} = \frac{1}{49}.$$

$\square$

EXAMPLE VII.6.7. Let $L = \mathbb{Q}(i, \zeta_{15} + \zeta_{15}^{-1})$. The extension $L/\mathbb{Q}(i)$ has degree 4 and cyclic Galois group generated by

$$\sigma: \begin{array}{c} L \longrightarrow L \\ \zeta_{15} + \zeta_{15}^{-1} \longmapsto \zeta_{15}^2 + \zeta_{15}^{-2}. \end{array}$$

We consider the corresponding cyclic algebra $A = (i, L/\mathbb{Q}(i), \sigma)$ of degree 4 . It is a division algebra by Example VII.5.6. Set

$$\theta = \zeta_{15} + \zeta_{15}^{-1}, \alpha = ((1-3i) + i\theta^2), I = \alpha\mathcal{O}_L \text{ and } \lambda = \frac{1}{15}.$$

Then

$$((1-3i)+i\theta^2, (1-3i)\theta+i\theta^3, -i+(-3+4i)\theta+(1-i)\theta^3, (-1+i)-3\theta+\theta^2+\theta^3)$$

is an orthonormal basis of (I, h_λ).

Since I is a principal ideal, we get a perfect code $\mathcal{C}_{A,\lambda,I}$ in dimension 4 satisfying

$$\delta_{min}(\mathcal{C}_{A,\lambda,I}) = \frac{1}{1125}.$$

$\square$

Recall that for any fully diverse code $\mathcal{C}$ based on a cyclic division algebra, the probability error of the code satisfies

$$\mathbb{P}(\mathbf{X} \to \hat{\mathbf{X}}) \leq \kappa(|a_2|^{2n-2} d_{L/k})^n.$$

In particular, in the case of the previous code above, we get

$$\mathbb{P}(\mathbf{X} \to \hat{\mathbf{X}}) \leq \kappa \cdot 1125^n.$$

EXAMPLE VII.6.8. As in the 3 antennas case, the base field is $k = \mathbb{Q}(j)$. Let $\theta = \zeta_{28} + \zeta_{28}^{-1}$ and $L = \mathbb{Q}(j, \theta)$ be the compositum of k and $\mathbb{Q}(\theta)$. The extension L/k is cyclic of degree 6 with generator

$$\sigma : \begin{array}{c} L \longrightarrow L \\ \zeta_{28} + \zeta_{28}^{-1} \longmapsto \zeta_{28}^2 + \zeta_{28}^{-2} \end{array}$$

We consider the cyclic algebra $A = (-j, L/k, \sigma)$ of degree 6. This is a division algebra, as shown in Example VII.5.11. One may show that we have

$$7\mathcal{O}_L = \mathfrak{P}^7 \overline{\mathfrak{P}}^7.$$

Set $I = \mathfrak{P}^7$ and $\lambda = \dfrac{1}{14}$.

One may show that (I, h_λ) has an orthonormal basis, so we obtain a perfect code $\mathcal{C}_{A,\lambda,I}$ in dimension 6. Computations show that we have $d_{L/k} = 2^6 \cdot 7^5$. However, the ideal I is not principal anymore, so we only have the estimation

$$\delta_{min}(\mathcal{C}_{A,\lambda,I}) \geq \frac{1}{2^6 \cdot 7^5}.$$

$\square$

We now prove the following theorem.

THEOREM VII.6.9. ([9]) *Perfect codes based on a cyclic division algebra* $A = (a, L/k, \sigma)$ *of degree* $n \geq 2$, *where* $a \in \mathcal{O}_k^\times$, *only exist for* $n = 2, 3, 4$ *or* 6.

Proof. Since $k/\mathbb{Q}$ is an imaginary quadratic extension, an element $a \in \mathcal{O}_k^\times$ is an mth root of unity, where $m = 1, 2, 3, 4$ or 6. Proposition VII.1.9 applied several times shows that, in the Brauer group $\mathrm{Br}(k)$, we have

$$\begin{aligned} m[A] &= [(a^m, L/k, \sigma)] \\ &= [(1, L/k, \sigma)] = 0 \end{aligned}$$

since $(1, L/k, \sigma)$ is split by Proposition VII.1.9 again. Hence $\exp(A)|m$ by definition. Since k is a number field, by Theorem VII.3.1, we have $n = \mathrm{ind}(A) = \exp(A) \mid m$. Hence $n = 2, 3, 4, 6$, and this concludes the proof. $\square$

We end this section by giving an example of code in dimension 5, where $a \notin \mathcal{O}_k$ [13].

EXAMPLE VII.6.10. For the 5 antennas case, we present the construction of [**13**], where the base field is $k = \mathbb{Q}(i)$. Let $\theta = \zeta_{11} + \zeta_{11}^{-1}$ and let $L = \mathbb{Q}(i, \theta)$ be the compositum of K and $\mathbb{Q}(\theta)$. The extension L/k is cyclic of degree 5 with generator

$$\sigma : \begin{array}{c} L \longrightarrow L \\ \zeta_{11} + \zeta_{11}^{-1} \longmapsto \zeta_{11}^2 + \zeta_{11}^{-2}. \end{array}$$

We consider the cyclic algebra $A = (a, L/k, \sigma)$ of degree 5 with

$$a = \frac{3 + 2i}{2 + 3i}.$$

By Example VII.4.5, A is a division algebra. Note here that a is not a root of unity, but has modulus 1. This way of finding a suitable non-norm element a of modulus 1 has been used more generally in [**13**] to find codes in arbitrary dimensions. One may show that one may find a perfect code based on A (see [**4, 13**] for more details), and that $d_{L/k} = 11^4$. However, since $a_2 = 2 + 3i$, the lower bound for the minimum determinant is $\dfrac{1}{11^4 \cdot 13^4}$.   □

VII.7. Optimality of some perfect codes

Recall from the previous section that for any perfect code $\mathcal{C} \subset \mathcal{C}_{A,\lambda,I}$ based on a cyclic division k-algebra $A = (a, L/k, \sigma)$, we have

$$\mathbb{P}(\mathbf{X} \to \hat{\mathbf{X}}) \leq \kappa (|a_2|^{2n-2} d_{L/k})^n.$$

DEFINITION VII.7.1. We say that a perfect code based on a cyclic division k-algebra $A = (a, L/k, \sigma)$ is **optimal** if for any other perfect code based on a cyclic division k'-algebra $A' = (a', L'/k', \sigma')$, we have

$$|a_2'|^{2n-2} d_{L'/k'} \geq |a_2|^{2n-2} d_{L/k'},$$

where k and k' are either $\mathbb{Q}(i)$ or $\mathbb{Q}(j)$. Notice that in this definition, k and k' may be different.

We would like to show that the perfect codes presented in the previous section are optimal in the previous sense. The proofs presented here for the cases $n = 4$ and $n = 6$ are the ones presented in [**10**](see also [**50**]). We start with some general considerations.

LEMMA VII.7.2. *Let $L = k(\sqrt[n]{d})$ be a Kummer extension of k of degree n. Then complex conjugation induces a $\mathbb{Q}$-automorphism of L which commutes with $\mathrm{Gal}(L/k)$ if and only if $d\bar{d} \in k^{\times n}$.*

Proof. Let $\zeta_n \in k$ be a primitive n^{th}-root of 1. Then a generator σ of $\mathrm{Gal}(L/k)$ is given by

$$\sigma : \begin{array}{c} L \longrightarrow L \\ \alpha \longmapsto \zeta_n \alpha, \end{array}$$

where $\alpha = \sqrt[n]{d}$. Assume that complex conjugation induces an automorphism of $L/\mathbb{Q}$ which commutes with $\mathrm{Gal}(L/k)$. Then $\bar{\alpha} \in L$, and we have

$$\sigma(\bar{\alpha}\alpha) = \overline{\sigma(\alpha)}\sigma(\alpha) = \bar{\alpha}\alpha,$$

since $\overline{\zeta}_n \zeta_n = 1$. Thus $\overline{\alpha}\alpha \in k^\times$. Now we have

$$(\overline{\alpha}\alpha)^n = \overline{\alpha^n}\alpha^n = \overline{d}d,$$

so $\overline{d}d \in k^{\times n}$. Conversely, assume that $\overline{d}d \in k^{\times n}$. Then we have

$$(\overline{\alpha}\alpha)^n = \overline{d}d = c^n \text{ for some } c \in k^\times,$$

and therefore $\overline{\alpha} = \dfrac{c\zeta'}{\alpha}$ for some $\zeta' \in \mu_n$. In particular, $\overline{\alpha} \in L$ and complex conjugation is therefore a $\mathbb{Q}$-automorphism of L. Moreover, we have

$$\sigma(\overline{\alpha}) = \frac{c\zeta'}{\zeta_n \alpha} = \overline{\zeta}_n \frac{c\zeta'}{\alpha} = \overline{\zeta_n \alpha},$$

that is $\sigma(\overline{\alpha}) = \overline{\sigma(\alpha)}$. Hence complex conjugation commutes with σ and hence with $\mathrm{Gal}(L/k)$. This completes the proof. $\qquad\square$

We now introduce some notation that we will keep until the end of this chapter.

If $k = \mathbb{Q}(i)$, we denote by $S_3, S_1, \overline{S}_1$ the following subsets of $\mathcal{O}_k$:

$$S_3 = \{p \equiv 3[4], p \text{ prime number}\}$$

$$S_1 = \{\pi = a + bi \mid 0 < a < b, p_\pi = a^2 + b^2, p_\pi \text{ prime number}\}$$

$$\overline{S}_1 = \{\overline{\pi} \mid \pi \in S_1\}.$$

Then any prime element of $\mathcal{O}_k$ is associate either to $1 - i$ or to exactly one element of S_3, S_1 or $\overline{S}_1$.

If $k = \mathbb{Q}(j)$, we denote by $T_2, T_1, \overline{T}_1$ the following subsets of $\mathcal{O}_k$:

$$T_2 = \{p \equiv 2[3], p > 2 \text{ prime number}\}$$

$$T_1 = \{\pi = a + bj \mid 0 < a < b, p_\pi = a^2 + b^2 - ab, p_\pi \equiv 1[3], p_\pi \text{ prime}\}$$

$$\overline{T}_1 = \{\overline{\pi} \mid \pi \in T_1\}.$$

Then any prime element of $\mathcal{O}_k$ is associate either to $1 - j$, 2 or to exactly one element of T_2, T_1 or $\overline{T}_1$.

We then have the following result.

LEMMA VII.7.3.. *Let $k = \mathbb{Q}(i)$ or $\mathbb{Q}(j)$. Let $n \geq 2$ be an integer, and let $d \in \mathcal{O}_k$ be a non-zero element which is not divisible by any non trivial n^{th}-power. Then $\overline{d}d \in k^{\times n}$ if and only if the following properties hold:*

(1) $n_{1-i}(d) = 0$ *if* $n \not\equiv 0[8]$ *and* $2n_{1-i}(d) = 0$ *or n if* $n \equiv 0[8]$;

(2) $2n_p(d) = 0$ *or n for all* $p \in S_3$;

(3) $n_\pi(d) + n_{\overline{\pi}}(d) = 0$ *or n for all* $\pi \in S_1$

if $k = \mathbb{Q}(i)$ and

(1) $n_{1-j}(d) = 0$ *if* $n \not\equiv 0[12]$ *and* $2n_{1-j}(d) = 0$ *or* n *if* $n \equiv 0[12]$;

(2) $2n_p(d) = 0$ *or* n *for all* $p \in T_2$ *or* $p = 2$;

(3) $n_\pi(d) + n_{\overline{\pi}}(d) = 0$ *or* n *for all* $\pi \in T_1$

if $k = \mathbb{Q}(j)$.

Proof. Notice first that, since $\mathcal{O}_k$ is integrally closed, then $d\overline{d} \in \mathcal{O}_k$ lies in k^n if and only if it lies in $\mathcal{O}_k^n$.

Assume that $k = \mathbb{Q}(i)$, and write

$$d = u(1-i)^{n_{1-i}(d)} \Big(\prod_{p \in S_3} p^{n_p(d)} \Big) \Big(\prod_{\pi \in S_1} \pi^{n_\pi(d)} \overline{\pi}^{n_{\overline{\pi}}(d)} \Big), u \in \mathcal{O}_k^\times,$$

each power begin less or equal to $n - 1$ by assumption on d. Since $u\overline{u} = 1$, and taking into account that $1 + i = (1-i)i$, we get

$$d\overline{d} = i^{n_{1-i}(d)}(1-i)^{2n_{1-i}(d)} \Big(\prod_{p \in S_3} p^{2n_p(d)} \Big) \Big(\prod_{\pi \in S_1} (\pi\overline{\pi})^{n_\pi(d)+n_{\overline{\pi}}(d)} \Big).$$

If $d\overline{d}$ is the nth-power of an element of $\mathcal{O}_k$, then $2n_p(d), n_\pi(d)+n_{\overline{\pi}}(d)$ and $2n_{1-i}(d)$ are multiples of n. Taking into account the conditions on d, we get (2) and (3), as well as $2n_{1-i}(d) = 0$ or n.

Assume that $2n_{1-i}(d) = n$, so that n is even. Since $d\overline{d}$ is a nth-power, we also have that $i^{n_{1-i}(d)}$ is a n^{theo}-power of an element of $\mathcal{O}_k$, which is necessarily a unit. Since i generates the group of units of $\mathcal{O}_k$, we get an equality of the form $i^{n_{1-i}(d)} = i^{\ell n}, \ell \in \mathbb{Z}$. Therefore, $\ell n \equiv n_{1-i}(d)[4]$, that is $(2\ell - 1)\frac{n}{2} \equiv 0[4]$, which is equivalent to $n \equiv 0[8]$.

Conversely, it is clear that if d satisfies conditions $(1)-(3)$, then $d\overline{d}$ is a n^{theo}-power. The case $k = \mathbb{Q}(j)$ may be dealt with in a similar way, and is left to the reader. This completes the proof. $\qquad\square$

We start with the case $n = 2$. In this case, the Golden code constructed in Example VII.6.5 satisfies

$$|a_2|^2 d_{L/\mathbb{Q}(i)} = 5.$$

LEMMA VII.7.4. *Assume that there exists a perfect code based on a division k-algebra $A = (a, L/k, \sigma)$ of degree 2 such that $|a_2|^2 d_{L/k} < 5$. Then one may assume that $a = i$ if $k = \mathbb{Q}(i)$ and $a = -j$ if $k = \mathbb{Q}(j)$.*

Proof. By assumption, the cubic lattice may be constructed as an ideal lattice on L/k. By Corollary C.2.11, we have $d_{L/k} \geq 4$. Hence, $|a_2|^2 \leq 1$. Since $a_2 \in \mathcal{O}_k$, it follows that $|a_2|^2 = 1$, which is equivalent to $a_2 \in \mathcal{O}_k^\times$ (since $k = \mathbb{Q}(i)$ or $\mathbb{Q}(j)$). Therefore, $a \in \mathcal{O}_k^\times$ as well. It follows that a is a 4-th root of 1 if $k = \mathbb{Q}(i)$ and a 6-th root of 1 if $k = \mathbb{Q}(j)$.

Assume that $k = \mathbb{Q}(i)$. If $a = \pm 1$, then a is a square in k, hence is a norm of L/k. In particular, A is split by Proposition VII.1.9 (2). By the same proposition, since -1 is square in k, we have

$$(-i, L/k, \sigma) \simeq (i, L/k, \sigma).$$

Thus, $(-i, L/k, \sigma)$ is a division k-algebra if and only if $(i, L/k, \sigma)$ is. Therefore, one may assume that $a = i$.

Assume that $k = \mathbb{Q}(j)$. If $a = 1$ or j^2, then reasoning as previously shows that A is split. Notice now that we have $j \cdot j^2 = j^3 = 1$, hence if $a = j$, A is split as well. Finally, we have $-j \cdot j^2 = -1$, thus we get

$$(-j, L/k, \sigma) \simeq (-1, L/k, \sigma).$$

Thus, $(-j, L/k, \sigma)$ is a division k-algebra if and only if $(-1, L/k, \sigma)$ is. We may then assume that $a = -j$, and this concludes the proof. $\qquad\square$

THEOREM VII.7.5. *The Golden code is an optimal perfect code.*

Proof. Assume that we may construct a perfect code based on a division k-algebra $A = (a, L/k, \sigma)$ such that $|a_2|^2 d_{L/k} < 5$. By the previous lemma, one may assume that $a = \pm i$ if $k = \mathbb{Q}(i)$ and that $a = -j$ if $k = \mathbb{Q}(j)$. By Corollary C.2.11, we also have $d_{L/k} \geq 4$, and thus $d_{L/k} = 4$. By Lemma C.2.6, we have

$$N_{k/\mathbb{Q}}(\mathcal{D}_{L/k}) = 16 = 2^4.$$

By Remark B.2.23, the prime divisors of $\mathcal{D}_{L/k}$ necessarily lie above 2. Write $L = k(\sqrt{d})$, where $d \in \mathcal{O}_k$ is non-zero and square-free. The previous observations and Proposition B.2.29 show that $d = u\pi^\varepsilon$, where $u \in \mathcal{O}_k^\times, \varepsilon = 0$ or 1, $\pi = 1 - i$ if $k = \mathbb{Q}(i)$, and $\pi = 2$ if $k = \mathbb{Q}(j)$. By Propositions VII.5.3 and VII.5.7, A cannot be a division k-algebra. This concludes the proof. $\qquad\square$

We now examine the case $n = 3$. The code presented in Example VII.6.6 satisfies $|a_2|^2 d_{L/k} = 49$.

LEMMA VII.7.6. *Assume that there exists a perfect code based on a division k-algebra $A = (a, L/k, \sigma)$ of degree 3 such that $|a_2|^4 d_{L/k} < 49$. Then one may assume that $k = \mathbb{Q}(j)$ and $a = j$.*

Proof. By assumption, the cubic lattice may be constructed as an ideal lattice on L/k. By Corollary C.2.11, we have $d_{L/k} \geq 27$. Hence, $|a_2|^4 \leq \frac{49}{27} < 2$. Since $a_2 \in \mathcal{O}_k$, we get that $|a_2|^2 = 1$. As previously, it follows that a is a 4-th root of 1 if $k = \mathbb{Q}(i)$ and a 6-th root of 1 if $k = \mathbb{Q}(j)$.

Assume that $k = \mathbb{Q}(i)$. Reasoning as in the proof of Theorem VII.6.9, we get that $4[A] = 0$ since $a^4 = 1$ in this case. But we also have $3[A] = 0$ by Theorem V.3.1 (9). We then get that $[A] = 0$, meaning that A is split, which is a contradiction. Hence $k = \mathbb{Q}(j)$, and therefore $a = \pm 1, \pm j, \pm j^2$. Notice that $1, -1 \in k^{\times 3} \subset N_{L/k}(L^\times)$. Similar arguments to those used in the proof of Lemma VII.7.4 show that we may assume that $a = j$ or j^2. Notice now that we have

$$(j, L/k, \sigma) \otimes_k (j, L/k, \sigma) \sim (j^2, L/k, \sigma)$$

by Proposition VII.1.9. Since $(j, L/k, \sigma)$ and $(j^2, L/k, \sigma)$ have index 1 or 3, Theorem V.3.1 (11) and the fact that Brauer-equivalent central simple algebras have the same index imply that $(j, L/k, \sigma)$ is a division k-algebra if and only if $(j^2, L/k, \sigma)$ is. This concludes the proof. $\qquad\square$

THEOREM VII.7.7. *The code presented in Example VII.6.6 is an optimal perfect code.*

Proof.   Assume that there exists a perfect code based on a division k-algebra $A = (a, L/k, \sigma)$ such that $|a_2|^4 d_{L/k} < 49$. By the previous lemma, one may assume that $k = \mathbb{Q}(j)$ and $a = j$. Write $L = k(\sqrt[3]{d})$, where $d \in \mathcal{O}_k$ is a non-zero element which is not divisible by any third power of an element of $\mathcal{O}_k$. Since A is a division k-algebra, by Proposition VII.5.8, there exists a prime element π of $\mathcal{O}_k$ such that $n_\pi(d)$ is prime to 3 and satisfying one of the following conditions:

(1) $\pi = p$, where p is a prime number satisfying $p \equiv 2$ or $5 [9]$;
(2) $\pi = a + bj$, where $p_\pi = a^2 + b^2 - ab$ is a prime number satisfying $p_\pi \equiv 4$ or $7 [9]$.

A prime number p congruent to 2 or 5 modulo 9 lies in T_2 or is equal to 2. By Lemma VII.7.3, we have $n_p(d) = 0$ (since $n = 3$). Hence, π necessarily satisfies the second condition. In particular, π is associate to an element of T_1. Without any loss of generality, we may assume that $\pi \in T_1$. By Lemma VII.7.3, $n_{\bar\pi}(d)$ is also prime to 3. By Proposition B.2.29, π and $\bar\pi$ then totally ramify in L/k, and we have

$$\pi \mathcal{O}_L = \mathfrak{P}^3 \text{ and } \bar\pi \mathcal{O}_L = \bar{\mathfrak{P}}^3.$$

Notice that we have $N_{L/\mathbb{Q}}(\mathfrak{P}) = p_\pi$. Since π and $\bar\pi$ tamely ramify (since they do not lie above 3), Theorem B.2.31 shows that $\mathfrak{P}^2 \bar{\mathfrak{P}}^2 \mid \mathcal{D}_{L/k}$. Taking the square of absolute norms on both sides, we get $p_\pi^2 \mid d_{L/k}$ using Lemma C.2.6. In particular, we have $d_{L/k} \geq p_\pi^2 \geq 7^2$, and we get a contradiction. $\square$

Let us deal with the case $n = 4$. In this case, the code constructed in Example VII.6.7 satisfies $|a_2|^6 d_{L/k} = 1125$. The next lemma may be proved as previously.

LEMMA VII.7.8. *Assume that there exists a perfect code based on a division k-algebra $A = (a, L/k, \sigma)$ of degree 4 such that $|a_2|^6 d_{L/k} < 1125$. Then one may assume that $k = \mathbb{Q}(i)$ and $a = i$.*

We will also need the following result.

LEMMA VII.7.9. *Let $k = \mathbb{Q}(i)$, and let $L = k(\sqrt[4]{d})$, where $d \in \mathcal{O}_k$ is an non-zero element which is not divisible by any fourth power of an element of $\mathcal{O}_k$. Assume that complex conjugation commutes with the elements of $\mathrm{Gal}(L/k)$. Then the odd part of $d_{L/k}$ is*

$$\Big(\prod_{\substack{p \in S_3 \\ p \mid d}} p \Big)^2 \Big(\prod_{\substack{\pi \in S_1 \\ n_\pi(d) = 1, 3}} p_\pi \Big)^3 \Big(\prod_{\substack{\pi \in S_1 \\ n_\pi(d) = 2}} p_\pi \Big)^2 .$$

Proof.   Let $p \in S_3$ dividing d. By Lemma VII.7.3, $n_p(d) = 2$, so p ramifies but does not totally ramify by Proposition B.2.29. Hence

$$(p) = \mathfrak{P}_0^2 \text{ or } \mathfrak{P}_1^2 \mathfrak{P}_2^2 \text{ in } \mathcal{O}_L,$$

where $\mathfrak{P}_1, \mathfrak{P}_2$ form an orbit under the action of $\mathrm{Gal}(L/k)$. Now since p is odd, p tamely ramifies, and thus $n_{\mathfrak{P}_i}(\mathfrak{d}_{L/k}) = 2 - 1 = 1$.

If $(p) = \mathfrak{P}_0^2$, we have $N_{L/\mathbb{Q}}(\mathfrak{P}_0) = p^4$. By Lemma C.2.6, $n_p(d_{L/k}) = 2$.

If $(p) = \mathfrak{P}_1^2 \mathfrak{P}_2^2$, we have $N_{L/\mathbb{Q}}(\mathfrak{P}_i) = p^2$. By Lemma C.2.6, we get $n_p(d_{L/k}) = 2$ in this case as well.

Assume now that $\pi \in S_1$ divides d with an odd valuation. Then $\overline{\pi}$ also divides d with an odd valuation by Lemma VII.7.3. In this case, π and $\overline{\pi}$ totally ramify by Proposition B.2.29. We then have $(\pi) = \mathfrak{P}^4$ and $(\overline{\pi}) = \overline{\mathfrak{P}}^4$. Thus $N_{L/\mathbb{Q}}(\mathfrak{P}) = N_{L/\mathbb{Q}}(\overline{\mathfrak{P}}) = p_\pi$. Once again π and $\overline{\pi}$ are tamely ramified, and reasoning as before shows that $n_{p_\pi}(d_{L/k}) = 3$. Finally, assume that $n_\pi(d) = 2$. By Proposition B.2.29, π and $\overline{\pi}$ ramify but do not totally ramify. We then have

$$(\pi) = \mathfrak{P}_0^2 \text{ and } (\overline{\pi}) = \overline{\mathfrak{P}}_0^2$$

or

$$(\pi) = \mathfrak{P}_1^2 \mathfrak{P}_2^2 \text{ and } (\overline{\pi}) = \overline{\mathfrak{P}}_1^2 \overline{\mathfrak{P}}_2^2.$$

In the first case, we have $N_{L/\mathbb{Q}}(\mathfrak{P}_0) = N_{L/\mathbb{Q}}(\overline{\mathfrak{P}}_0) = p_\pi^2$. In the second case we have $N_{L/\mathbb{Q}}(\mathfrak{P}_i) = N_{L/\mathbb{Q}}(\overline{\mathfrak{P}}_i) = p_\pi$. We now finish the proof as before. $\square$

THEOREM VII.7.10. *The perfect code constructed in Example VII.6.7 is optimal.*

Proof. Assume that we may build a perfect on $A = (i, L/K, \sigma)$ with $d_{L/k} < 1125$. By Corollary VII.5.5, there exists $\pi \in S_1$ dividing d with an odd valuation such that $p_\pi \equiv 5[8]$. In particular, $p_\pi \geq 5$. Then by Lemma VII.7.9, we have $p_\pi^3 \mid d_{L/k}$. Thus, we necessarily have $\pi = 1 + 2i$ and $125 \mid d_{L/K}$ (otherwise, we would have $p_\pi \geq 13$ and $d_{L/k} \geq 13^3 > 1125$). Now if $\pi' \in S_1, \pi' \neq 1 + 2i$ divides d, we would have $125 \cdot p_{\pi'}^2 \mid d_{L/k}$, and thus $p_{\pi'}^2 \leq 9$, which is a contradiction since $p_{\pi'} \geq 5$. Similarly, if $p \in S_3$ divides d, we have $125 \cdot p^2 \mid d_{L/k}$ and thus necessarily $p = 3$.

Hence the only possible prime divisors for d are $1 - i, 3$ and $1 \pm 2i$. Noticing that conjugate values of d generate the same field extension, we see that the remaining possibilities for d are

$$d = u \cdot 3^m (1 + 2i)(1 - 2i)^3, m = 0, 2, u = \pm 1, \pm i.$$

Using Theorem B.2.33 or PARI GP [53] to compute the relative discriminants of the corresponding extensions, we see that the only possibility to have $d_{L/k} < 1125$ is

$$d = (1 + 2i)(1 - 2i)^3.$$

In this case, $d_{L/k} = 125 < 4^4$. Hence, using Corollary C.2.11, we see that no complex ideal lattice on L/k will be isomorphic to the cubic lattice, and this completes the proof. $\square$

We finally examine the case $n = 6$. The code presented in Example VII.6.7 satisfies $|a_2|^{10} d_{L/k} = 2^6 \cdot 7^5$. The proof of the next lemma is left to the reader as an exercise.

LEMMA VII.7.11. *Assume that there exists a perfect code based on a division k-algebra $A = (a, L/k, \sigma)$ of degree 6 such that $|a_2|^6 d_{L/k} < 2^6 \cdot 7^5$. Then one may assume that $k = \mathbb{Q}(j)$ and $a = -j$.*

We now provide an estimation of $d_{L/k}$, when L/k is a Kummer extension of degree 6.

LEMMA VII.7.12. *Let $k = \mathbb{Q}(j)$, and let $L = k(\sqrt[6]{d})$, where $d \in \mathcal{O}_k$ is an non-zero element which is not divisible by any sixth power of an element of $\mathcal{O}_k$. Assume that* *complex conjugation commutes with the elements of* $\mathrm{Gal}(L/k)$. *Then the prime-to-6 part of* $d_{L/k}$ *is*

$$\big(\prod_{\substack{p \in T_2 \\ p|d}} p \big)^3 \big(\prod_{\substack{\pi \in T_1 \\ n_\pi(d) = 1,5}} p_\pi \big)^5 \big(\prod_{\substack{\pi \in T_1 \\ n_\pi(d) = 3}} p_\pi \big)^3 \big(\prod_{\substack{\pi \in T_1 \\ n_\pi(d) = 2,4}} p_\pi \big)^4.$$

Moreover, the following properties hold:

(1) *if* 2 *ramifies in* L/k, *then* $n_2(d_{L/k}) \geq 6$;
(2) *if* $1 - j$ *ramifies in* L/k, *then* $n_3(d_{L/k}) \geq 4$.

Proof.   Let $p \in T_2$ dividing d. By Lemma VII.7.3, $n_p(d) = 3$, so p ramifies but does not totally ramify by Proposition B.2.29. Moreover, p tamely ramifies since p is prime to 6. Write $d = d'p^3$ with $d' \in \mathcal{O}_k, p \nmid d'$. Then p totally ramifies in $k_2 = k(\sqrt{d}) = k(\sqrt{d'p})$ by the same proposition, so we may write

$$(p) = \mathfrak{P}_0^2,$$

for some ideal $\mathfrak{P}_0$ of $\mathcal{O}_{k_2}$. Since L/k_2 is a Galois extension of prime degree 3 , either $\mathfrak{P}_0$ is totally ramified, inert or totally split. Since $\mathfrak{P}_0$ cannot be totally ramified (otherwise p would be totally ramified), we finally have

$$(p)\mathcal{O}_L = \mathfrak{P}_0^2 \text{ or } \mathfrak{P}_1^2\mathfrak{P}_2^2\mathfrak{P}_3^2 \text{ in } \mathcal{O}_L,$$

where $\mathfrak{P}_1, \mathfrak{P}_2, \mathfrak{P}_3$ form an orbit under the action of $\mathrm{Gal}(L/k_2)$.

If $(p) = \mathfrak{P}_0^2$, we have $N_{L/\mathbb{Q}}(\mathfrak{P}_0) = p^6$, and since $n_{\mathfrak{P}_0}(\mathfrak{d}_{L/k}) = 1$ by Theorem B.2.31, we get $n_p(d_{L/k}) = 3$.

If $(p) = \mathfrak{P}_1^2\mathfrak{P}_2^2\mathfrak{P}_3^2$, we have $N_{L/\mathbb{Q}}(\mathfrak{P}_i) = p^2$, and since $n_{\mathfrak{P}_i}(\mathfrak{d}_{L/k}) = 1$ by Theorem B.2.31, we also get $n_p(d_{L/k}) = 3$ in this case.

Now, let $\pi \in T_1$ such that $n_\pi(d) = 1$ or 5. Then $n_{\overline{\pi}}(d) = 5$ or 1 respectively by Lemma VII.7.3. By Proposition B.2.29, π and $\overline{\pi}$ totally ramify, so we have

$$(\pi) = \mathfrak{P}^6 \text{ and } (\overline{\pi}) = \overline{\mathfrak{P}}^6 \text{ in } \mathcal{O}_L,$$

with $N_{L/\mathbb{Q}}(\mathfrak{P}) = p_\pi$. We then have $n_{\mathfrak{P}}(\mathfrak{d}_{L/k}) = 5 = n_{\overline{\mathfrak{P}}}(\mathfrak{d}_{L/k})$ and thus $n_{p_\pi}(d_{L/k}) = 5$.

Let $\pi \in T_1$ such that $n_\pi(d) = 3$. Then we also have $n_{\overline{\pi}}(d) = 3$. Reasoning as above, we see that we have

$$(\pi) = \mathfrak{P}_0^2 \text{ or } \mathfrak{P}_1^2\mathfrak{P}_2^2\mathfrak{P}_3^2 \text{ in } \mathcal{O}_L,$$

and similarly for $(\overline{\pi})$, so that $n_{p_\pi}(d_{L/k}) = 3$.

Finally, let $\pi \in T_1$ such that $n_\pi(d) = 2$ or 4, and set $k_3 = k(\sqrt[3]{d})$. In this case, we have

$$(\pi) = \mathfrak{P}_0^3 \text{ or } \mathfrak{P}_1^3\mathfrak{P}_2^3 \text{ in } \mathcal{O}_L,$$

where $\mathfrak{P}_1, \mathfrak{P}_2$ form an orbit under the action of $\mathrm{Gal}(L/k_3)$. One may check as before that in both cases, we have $n_{p_\pi}(d_{L/k}) = 4$.

We now examine the case of the wildly ramified primes. Let us start with $1 - j$. Since $1 - j \nmid d$ and $1 - j \nmid 2$ by Lemma VII.7.3, it does not ramify in k_2 by Proposition B.2.29. Then it necessarily totally ramifies in k_3, so we have

$$(1 - j) = \mathfrak{P}_0^3 \text{ in } \mathcal{O}_{k_3},$$

where $N_{k_3/\mathbb{Q}}(\mathfrak{P}_0) = 3$. Now since $1 - j$ wildly ramifies, $n_{\mathfrak{p}_\circ}(\mathfrak{d}_{k_3/k}) \geq 3$. Therefore, 3^3 divides $N_{k_3/\mathbb{Q}}(\mathfrak{d}_{k_3/k}) = d_{k_3/k}^2$. Since $d_{k_3/k}$ is an integer, we get that $3^2 \mid d_{k_3/k}$. By Corollary C.2.7, we get $3^4 \mid d_{L/k}$.

Assume now that 2 ramifies in L/k. Since $n_2(d) = 0$ or 3 by Lemma VII.7.3, it does not ramify in k_3, hence it totally ramifies in k_2. We then have

$$(2) = \mathfrak{P}_0^2 \text{ in } \mathcal{O}_{k_2},$$

so that $N_{k_2/\mathbb{Q}}(\mathfrak{P}_0) = 2^2$. Since 2 wildly ramifies, $n_{\mathfrak{p}_\circ}(\mathfrak{d}_{k_2/k}) \geq 2$, and we get as before that $2^4 \mid d_{k_2/k}^2$. Hence $2^2 \mid d_{k_2/k}$, and by Corollary C.2.7, we get $2^6 \mid d_{L/k}$. $\qquad\square$

THEOREM VII.7.13. *The perfect code constructed in Example* VII.6.8 *is optimal.*

Proof. Assume that we may construct a perfect code on the k-algebra $A = (-j, L/k, \sigma)$ with $d_{L/k} < 2^6 \cdot 7^5$. Write $L = k(\sqrt[6]{d})$, where $d \in \mathcal{O}_k$ is an non-zero element which is not divisible by any sixth power of an element of $\mathcal{O}_k$. By Corollary VII.5.10, there exists $\pi \in T_1$ dividing d with an odd valuation, such that $p_\pi \equiv 7[12]$. If $p_\pi > 7$, then $p_\pi \geq 19$. Notice now that by Lemma VII.7.3, $n_{1-j}(d) = 0$. Corollary VII.5.10 then implies that there exists $\pi' \in T_1$ dividing d with a valuation which is prime to 3, such that $p_{\pi'} \equiv 4, 7[9]$. We then have $p_{\pi'} \geq 7$ and thus Lemma VII.7.12 implies that $7^4 \mid d_{L/k}$. Hence $7^4 \cdot 19^3 \mid d_{L/k}$, which is a contradiction since $7^4 \cdot 19^3 > 2^6 \cdot 7^5$.

Therefore, $p_\pi = 7$. We then get $7^3 \mid d_{L/k}$. If $p_{\pi'} > 7$, we have $p_{\pi'} \geq 13$ and thus $7^3 \cdot 13^4 \mid d_{L/k}$, which is again a contradiction. Hence $p_{\pi'} = 7$.

Hence we have proved that $\pi = \pi' = 2 + 3j$. Moreover, since $2 + 3j$ divides $d_{L/k}$ with an odd valuation, which is also prime to 3, then $n_{2+3j}(d_{L/k}) = 1$ or 5, and thus $7^5 \mid d_{L/k}$ by Lemma VII.7.12.

Assume that some $p \in T_2$ ramifies in L/k. Since $p \equiv 2[3]$, we have $p \geq 5$, and by Lemma VII.7.12, we get $5^3 \mid d_{L/k}$. We then get a contradiction, since $5^3 \cdot 7^5 > 2^6 \cdot 7^5$. If $\pi'' \in T_1 \cup \overline{T}_1, \pi'' \nmid 7$ ramifies in L/k, we have $p_{\pi''} \geq 13$ since $p_{\pi''} \equiv 1[3]$. In this case, we obtain that $13^3 \mid d_{L/k}$, which again yields a contradiction. Notice that $1 - j$ and 2 do not ramify in L/k either, since otherwise we would have $3^4 \cdot 7^5 \mid d_{L/k}$ or $2^6 \cdot 7^5 \mid d_{L/k}$, which is a contradiction.

The computations above then show that $d_{L/k} = 7^5 < 6^6$, so we may not construct the cubic lattice as a complex ideal lattice on L/k by Corollary C.2.11. $\qquad\square$

EXERCISES

1. Let k be a finite field.

   (a) Show that for any field extension L/k of finite degree, the norm map $N_{L/k} : L^\times \longrightarrow k^\times$ is surjective.

   (b) Use the previous question to recover the fact that $\mathrm{Br}(k) = 0$.

2. Let p, q be two odd prime numbers. Find necessary and sufficient conditions on p and q for the quaternion algebra $Q = (p, q)_{\mathbb{Q}}$ to be a division algebra.

3. Let $A = (7, 13)_{\mathbb{Q}} \otimes_{\mathbb{Q}} (3, 5)_{\mathbb{Q}}$.

   (a) Justify that $\exp(A) \leq 2$, and deduce that A is not a division algebra.

   (b) Show that $\operatorname{ind}(A) = 2$.

   Hint: Show that $A \otimes_{\mathbb{Q}} \mathbb{Q}_p$ is not split for a suitable p.

4. Let p be a prime number satisfying $p \equiv 5[8]$, and let $L = \mathbb{Q}(i, \sqrt{p})$. Recall that we have $\mathcal{O}_L = \mathbb{Z}[i, \theta]$, where $\theta = \dfrac{1 + \sqrt{p}}{2}$.

   Write $p = u^2 + v^2$, where u is odd and v is even, and let I be the ideal of $\mathcal{O}_L$ generated by $\dfrac{p-1}{2} + \theta$ and $u + iv$. Recall that $\dfrac{p-1}{2} + \theta$ and $u + iv$ form a $\mathbb{Z}[i]$-basis of I.

   (a) Compute the representative matrix of $(\mathcal{O}_L, h_1)$ in the $\mathbb{Z}[i]$-basis $(1, \theta)$ and deduce that $d_{L/\mathbb{Q}(i)} = p$.

   (b) Compute the representative matrix of $(I, h_{\frac{1}{p}})$ in the $\mathbb{Z}[i]$-basis $(\dfrac{p-1}{2} + \theta, u + iv)$ and check that $(I, h_{\frac{1}{p}})$ is totally positive of determinant 1.

   (c) Show that $\dfrac{u + iv + \sqrt{p}}{2}$ and $\dfrac{u + iv - \sqrt{p}}{2}$ are elements of I, and that they form a $\mathbb{Z}[i]$-orthonormal basis of $(I, h_{\frac{1}{p}})$.

   (d) Write the elements of the corresponding code.

   (e) Class field theory shows that I is a principal ideal. Check that I is generated by $\alpha = 2 - i - \theta$ if $p = 13$, and by $\alpha = -2 + i - \theta$ if $p = 29$.

5. Compute $d_{L/k}$ in the following cases:

   (a) $k = \mathbb{Q}(i), L = k(\sqrt[4]{d}), d = u \cdot 3^m (1 + 2i)(1 - 2i)^3, m = 0, 2$ and $u = \pm 1, \pm i$;

   (b) $k = \mathbb{Q}(j), L = k(\zeta_7 + \zeta_7^{-1})$;

   (c) $k = \mathbb{Q}(j), L = k(\zeta_{28} + \zeta_{28}^{-1})$.

CHAPTER VIII

Central simple algebras of degree 4

The goal of this chapter is to use the results we have proved so far to determine completely the structure of central simple k-algebras of degree 4. We are going to restrict ourselves to the case $\mathrm{char}(k) \neq 2$ for simplicity, even if the following results are true in any characteristic. The reader will refer to [**2**] or [**25**] for a proof in the general case.

VIII.1. A theorem of Albert

Let A be a central simple k-algebra of degree 4 over a field k of characteristic different from 2. Since $\mathrm{ind}(A) \mid \deg(A)$, we get $\mathrm{ind}(A) = 1, 2$ or 4. Let us examine the first two cases.

If $\mathrm{ind}(A) = 1$, then $A \cong_k \mathrm{M}_4(k)$.

If $\mathrm{ind}(A) = 2$, then $A \cong_k \mathrm{M}_2(Q)$ for some division quaternion k-algebra in view of the following lemma.

LEMMA VIII.1.1. *Let k be a field of characteristic different from 2. Then every central simple k-algebra of degree 2 is isomorphic to a quaternion k-algebra.*

Proof. If A is split, then $A \cong_k (1,1)_k$ by Lemma II.1.1 (3). If A is not split, then $\mathrm{ind}(A) = 2$ necessarily since $\mathrm{ind}(A) \mid \deg(A)$ and therefore A is a division algebra. We know that in this case A has a maximal subfield L by Corollary IV.1.9. Since $\mathrm{char}(k) \neq 2$ and $[L : k] = 2$, L/k is a Galois extension of degree 2. Write $L = k(\sqrt{b}), b \in k^\times$. The non-trivial element of $\mathrm{Gal}(L/k)$ is then the map

$$\sigma \colon \begin{array}{c} L \longrightarrow L \\ x + y\sqrt{b} \longmapsto x - y\sqrt{b}. \end{array}$$

By Proposition VI.1.2, we have $A = 1_A L \oplus iL$ where $i \in A^\times$ satisfies

$$\lambda i = i \lambda^\sigma \text{ for all } \lambda \in L.$$

In particular, if we set $j = \sqrt{b}$, we get $j^2 = b$ and $ji = -ij$. Finally, since $\sigma^2 = \mathrm{Id}$, we have

$$\lambda i^2 = i^2 \lambda \text{ for all } \lambda \in L,$$

and therefore $i^2 \in Z(A) = k$. Since $i \in A^\times$, we then have $a = i^2 \in k^\times$. Therefore, D is generated over k by two elements i, j satisfying

$$i^2 = a, j^2 = b, ij = -ji$$

for some $a, b \in k^\times$, that is $A \cong_k (a,b)_k$. This concludes the proof. $\square$

It remains to examine the case of division algebras. We are going to prove the following result.

THEOREM VIII.1.2 (Albert). *Assume that* $\mathrm{char}(k) \neq 2$. *Then every central division k-algebra of degree* 4 *contains a biquadratic extension of k.*

Proof. Let D such a division k-algebra. We are going to prove successively the following facts:

(1) if D contains a quadratic extension of k, then it contains a biquadratic extension of k;

(2) every central division k-algebra of degree 4 contains a quadratic extension of k.

The theorem will easily follow from (1) and (2).

(1) Let us assume that D contains a quadratic extension K of k, and let $\sigma \in \mathrm{Gal}(K/k)$ be the non-trivial k-automorphism of K. By Skolem-Noether's Theorem, there exists $u \in D^\times$ such that $\mathrm{Int}(u)_{|K} = \sigma$. Note that $\sigma \neq \mathrm{Id}_K$, and thus $u \notin C_D(K)$. Since $\sigma^2 = \mathrm{Id}_K$, we have $\mathrm{Int}(u^2)_{|K} = \mathrm{Id}_K$, and therefore $v = u^2 \in C_D(K)$. Notice that $v \neq 0$ since $u \in D^\times$. Now we have

$$k(v) = k(u^2) \subset k(u).$$

Since $k(v) \subset C_D(K)$ but $u \notin C_D(K)$, $k(v)$ is a proper subfield of $k(u)$. Since $k(u)$ is a subfield of D, we have $[k(u) : k] \mid 4$ by Lemma IV.1.4 and therefore $[k(v) : k] = 1$ or 2 since $k(v)$ is a proper subfield of $k(u)$.

Assume first that $v \notin k$. Then $k(v)/k$ is a quadratic extension. Notice that $k(v) \cap K = k$ or K, since K/k is quadratic. If $k(v) \cap K = K$, then we get $K \subset k(v)$ and thus $K = k(v)$. In particular, we get $v \in K$, and therefore

$$v = u^2 = uu^2u^{-1} = uvu^{-1} = \sigma(v).$$

Therefore, $v \in k$ and we get a contradiction. Hence, $k(v) \cap K = k$. Moreover, $k(v)/k$ is quadratic and the elements of $k(v)$ and K commute since $v \in C_D(K)$, so the compositum $L = k(v)K$ is a biquadratic extension of k contained in D.

Assume now that $v \in k$. Write $K = k(\sqrt{d})$ and let $z = \sqrt{d}$. Then we have $u^2 = v \in k^\times, z^2 = d \in k^\times$, and $uzu^{-1} = \sigma(z) = -z$. Hence the subalgebra Q generated by u and z is isomorphic to $(v, d)_k$. By the Centralizer Theorem, we have

$$D \cong_k Q \otimes_k C_D(Q).$$

Since $C_D(Q)$ is a central simple k-algebra of degree 2, $C_D(Q)$ is isomorphic to a quaternion k-algebra $(a, b)_k$ by Lemma VIII.1.1. Notice that by Example II.1.6, v and a are not squares in k since D is a division k-algebra. Let i, j be the generators of $(v, d)_k$ and let be i', j' be the generators of $(a, b)_k$. Suppose that v and a belong to the same square classes. In this case, we would have

$$(v, d)_k \otimes (a, b)_k \cong_k (a, d)_k \otimes (a, b)_k \cong_k \mathrm{M}_2((a, db)_k)$$

by Lemma II.1.1 (2) and Proposition II.1.9, which is absurd since D is a division algebra. Hence v and a belong to different square classes, and therefore $i \otimes 1$ and $1 \otimes i'$ generate a subfield of D isomorphic to $k(\sqrt{v}, \sqrt{a})$.

(2) The proof presented here is due to Haile, and taken from [**17**].

We first proceed to show that there exists a k-linear subspace W of D of dimension 3 such that
$$\mathrm{Trd}_D(w) = \mathrm{Trd}_D(w^{-1}) = 0 \ \text{ for all } w \in W \setminus \{0\}.$$

Let L/k be a maximal subfield of D (which exists by Corollary IV.1.9). For all $d \in D$, the map
$$\theta_d : \begin{array}{c} L \longrightarrow k \\ x \longmapsto \mathrm{Trd}_D(xd) \end{array}$$
is k-linear. The map
$$\theta : \begin{array}{c} D \longrightarrow \mathrm{Hom}_k(L, k) \\ d \longmapsto \theta_d \end{array}$$
is then k-linear.

Since $\dim_k(D) = 16$ and $\dim_k(\mathrm{Hom}_k(L,k)) = [L : k] = 4$, θ has a non-trivial kernel. Thus, there exists $d \in D^\times$ such that
$$\mathrm{Trd}_D(xd) = 0 \ \text{ for all } x \in L.$$

Consider the k-linear map
$$f : \begin{array}{c} L \longrightarrow k \\ x \longmapsto \mathrm{Trd}_D(d^{-1}x). \end{array}$$

Then $\ker(f)$ is a k-linear subspace of dimension ≥ 3. Let V be a linear subspace of dimension 3 of $\ker(f)$ and set $W = d^{-1}V$. Then $\dim_k(W) = 3$. Let us show that W satisfies the desired property. For all $w = d^{-1}v \in W, w \neq 0$, we have
$$\mathrm{Trd}_D(w) = \mathrm{Trd}_D(d^{-1}v) = 0,$$
since $v \in V \subset \ker(f)$. Notice now that if $v \in V, \setminus \{0\}$, then $v \in L^\times$ and thus $v^{-1} \in L^\times$. Therefore, we have
$$\mathrm{Trd}_D(w^{-1}) = \mathrm{Trd}_D(v^{-1}d) = 0$$
by choice of d, and W has the required property.

We now prove that W contains an element generating a quadratic extension. Since $\dim_k(W) = 3$, one may choose an element $w \in W \setminus k$. Hence $[k(w) : k] \geq 2$. Since $k(w)$ is a subfield of D, $[k(w) : k] \mid 4$ by Lemma IV.1.4, and we get $[k(w) : k] = 2$ or $[k(w) : k] = 4$. In the first case, we are done, so let us assume that $[k(w) : k] = 4$, and let
$$\mu_{w,k} = X^4 + a_3 X^3 + a_2 X^2 + a_1 X + a_0$$
be the minimal polynomial of w over k. Now $k(w) = k(w^{-1})$, so $\mu_{w^{-1},k}$ has degree 4. Since $1 + a_3 w^{-1} + a_2 w^{-2} + a_1 w^{-3} + a_0 w^{-4} = 0$, we get
$$\mu_{w^{-1},k} = X^4 + a_1 a_0^{-1} X^3 + a_2 a_0^{-1} X^2 + a_3 a_0^{-1} X + a_0^{-1}.$$

By Lemma IV.2.4, we have $\mathrm{Prd}_D(w) = \mu_{w,k}$ and $\mathrm{Prd}_D(w^{-1}) = \mu_{w^{-1},k}$. It follows that $a_3 = \mathrm{Trd}_D(w)$ and $a_1 a_0^{-1} = \mathrm{Trd}_D(w^{-1})$. By choice of W, we then get $a_1 = a_3 = 0$, that is
$$\mu_{w,k} = X^4 + a_2 X^2 + a_0.$$
It implies that $\mu_{w^2,k} = X^2 + a_2 X + a_0$, and $k(w^2)/k$ is a quadratic subextension of k contained in D. This concludes the proof. $\qquad\square$

REMARK VIII.1.3. If k has no biquadratic extension, a central simple k-algebra of degree 4 is either split or isomorphic to $\mathrm{M}_2((-1,-1)_k)$.

Indeed, Theorem VIII.1.2 shows that there is no central division k-algebra of degree 4 in this case. In particular, a central simple k-algebra of degree 4 is either split or isomorphic to $\mathrm{M}_2(Q)$ where Q is a division quaternion k-algebra. Now write $Q = (a,b)_k$. Since Q is a division algebra, then a and b are not squares by Example II.1.6, and since k as no biquadratic extensions, a and b belong to the same square class. Hence $Q \cong_k (a,a)_k$ by Lemma II.1.1 (2). If -1 is a square in k, then $Q \cong_k (a,-a)_k$ by Example II.1.6, which is not a division algebra. Hence -1 is not a square, so a and -1 belong to the same square class, meaning that $Q \cong_k (-1,-1)_k$. $\qquad\square$

In view of the previous remark, we will assume that k has at least one biquadratic extension until the end of this chapter (char$(k) \neq 2$). In this case, we have the following result.

COROLLARY VIII.1.4. *Every central simple k-algebra of degree 4 contains a biquadratic extension of k.*

Proof. If A is a division algebra, this is nothing but Theorem VIII.1.2. If $A \cong_k \mathrm{M}_4(k)$ is split, let L/k be a biquadratic extension of k. Write $L = k(\alpha)$, let P be the minimal polynomial of α over k and let M be the companion matrix of P. Then the image of M under the isomorphism $\mathrm{M}_4(k) \cong_k A$ is an element of A generating a subalgebra isomorphic to L. Assume now that

$$A \cong_k \mathrm{M}_2((a,b)_k) \cong_k \mathrm{M}_2(k) \otimes_k (a,b)_k,$$

where $(a,b)_k$ is a division k-algebra. By Example II.1.6, a is not a square in k. We claim that there exists an element $a' \in k^\times \setminus k^{\times 2}$ which does not belong to the square class of a. Otherwise, $k^\times$ would only have one non-trivial square-class (namely the class of a), contradicting the fact that k has a biquadratic extension. Let $M \in \mathrm{M}_2(k)$ be the companion matrix of $X^2 - a'$. Then $M \otimes 1$ and $1 \otimes i$ generate a subfield of $\mathrm{M}_2((a,b)_k)$ isomorphic to $k(\sqrt{a},\sqrt{a'})$, and then A contains a biquadratic extension of k. $\qquad\square$

VIII.2. Structure of central simple algebras of degree 4

It follows from Corollary VIII.1.4 and Proposition VI.2.1 that every central simple k-algebra of degree 4 is a crossed product over a biquadratic extension of k. Our next goal is to give a nicer description of these algebras.

LEMMA VIII.2.1. *Let L/k be a biquadratic extension, with Galois group $G = \{\mathrm{Id}, \sigma, \tau, \sigma\tau\}$, and let $A = (\xi, L/k, G)$ be a crossed product over L/k, with L-basis $e_{\mathrm{Id}} = 1_A, e_\sigma, e_\tau, e_{\sigma\tau}$. We will identify L and $1_A L$. Then the elements*

$$e = e_\sigma, f = e_\tau, a = \xi_{\sigma,\sigma}, b = \xi_{\tau,\tau}, u = \frac{\xi_{\tau,\sigma}}{\xi_{\sigma,\tau}}$$

satisfy the following properties:

(1) $A = L \oplus eL \oplus fL \oplus efL$;
(2) $e^2 = a, f^2 = b, fe = efu, \lambda e = e\lambda^\sigma, \lambda f = f\lambda^\tau$, *for all $\lambda \in L$;*

(3) $a^\sigma = a, b^\tau = b, uu^\sigma = \dfrac{a}{a^\tau}, uu^\tau = \dfrac{b^\sigma}{b}$.

Proof. Notice that we have

$$ef = e_\sigma e_\tau = e_{\sigma\tau}\xi_{\sigma,\tau},$$

and

$$fe = e_\tau e_\sigma = e_{\tau\sigma}\xi_{\tau,\sigma}.$$

Since $\tau\sigma = \sigma\tau$, we have

$$fe = e_{\sigma\tau}\xi_{\sigma,\tau}u = efu.$$

Now

$$e^2 = e_\sigma e_\sigma = e_{\mathrm{Id}}\xi_{\sigma,\sigma} = 1_A a = a,$$

and also

$$f^2 = e_\tau e_\tau = e_{\mathrm{Id}}\xi_{\tau,\tau} = 1_A b = b.$$

Moreover,

$$\lambda e = \lambda e_\sigma = e_\sigma \lambda^\sigma = e\lambda^\sigma,$$

and similarly

$$\lambda f = \lambda e_\tau = e_\tau \lambda^\tau = f\lambda^\tau,$$

hence (2). Since $1_A, e_\sigma, e_\tau, e_{\sigma\tau}$ is an L-basis of A, so is $1_A, e_\sigma, e_\tau, e_{\sigma\tau}\xi_{\sigma,\tau}$. Hence $1_A, e, f, ef$ is an L-basis of A, which proves (1). We now prove (3). Recall that for $\sigma_1, \sigma_2, \sigma_3 \in G$, we have

$$\xi_{\sigma_1,\mathrm{Id}} = \xi_{\mathrm{Id},\sigma_2} = 1$$

and

$$\xi_{\sigma_1,\sigma_2\sigma_3}\xi_{\sigma_2,\sigma_3} = \xi_{\sigma_1\sigma_2,\sigma_3}\xi_{\sigma_1,\sigma_2}^{\sigma_3}.$$

For $\sigma_1 = \sigma_2 = \sigma_3 = \rho \in G$, we get $\xi_{\rho,\rho} = \xi_{\rho,\rho}^\rho$, and then we have

$$a = a^\sigma \text{ and } b = b^\tau.$$

For $\sigma_1 = \sigma_2 = \tau, \sigma_3 = \sigma$, we get

$$\xi_{\tau,\tau}^\sigma = \xi_{\tau,\tau\sigma}\xi_{\tau,\sigma}.$$

For $\sigma_1 = \sigma_3 = \tau, \sigma_2 = \sigma$, we get

$$\xi_{\tau,\sigma\tau}\xi_{\sigma,\tau} = \xi_{\tau\sigma,\tau}\xi_{\tau,\sigma}^\tau,$$

and therefore

$$\xi_{\tau,\sigma\tau} = \frac{\xi_{\tau\sigma,\tau}\xi_{\tau,\sigma}^\tau}{\xi_{\sigma,\tau}}.$$

Since $\tau\sigma = \sigma\tau$, plugging into the previous equation gives

$$\xi_{\tau,\tau}^\sigma = \frac{\xi_{\sigma\tau,\tau}\xi_{\tau,\sigma}^\tau\xi_{\tau,\sigma}}{\xi_{\sigma,\tau}}.$$

For $\sigma_1 = \sigma, \sigma_2 = \sigma_3 = \tau$, we get

$$\xi_{\tau,\tau} = \xi_{\sigma\tau,\tau}\xi_{\sigma,\tau}^\tau.$$

Dividing these two equations, we get

$$\frac{\xi_{\tau,\tau}^\sigma}{\xi_{\tau,\tau}} = \frac{\xi_{\tau,\sigma}^\tau\xi_{\tau,\sigma}}{\xi_{\sigma,\tau}^\tau\xi_{\sigma,\tau}},$$

that is

$$\frac{b^\sigma}{b} = uu^\tau.$$

Exchanging the roles of σ and τ, we get

$$\frac{\xi_{\sigma,\sigma}^\tau}{\xi_{\sigma,\sigma}} = \frac{\xi_{\sigma,\tau}^\sigma \xi_{\sigma,\tau}}{\xi_{\tau,\sigma}^\sigma \xi_{\tau,\sigma}}.$$

We then have

$$\frac{\xi_{\sigma,\sigma}}{\xi_{\sigma,\sigma}^\tau} = \frac{\xi_{\tau,\sigma}^\sigma \xi_{\tau,\sigma}}{\xi_{\sigma,\tau}^\sigma \xi_{\sigma,\tau}},$$

that is

$$\frac{a}{a^\tau} = uu^\sigma.$$

$\square$

Conversely, given 3 elements $a, b, u \in L^\times$ satisfying (3), we are going to construct a central simple k-algebra of degree 4 satisfying (1) and (2).

Let L/k be a biquadratic extension, with Galois group $G = \langle \sigma, \tau \rangle$, let $a, b, u \in L^\times$ satisfying

$$a^\sigma = a, b^\tau = b, uu^\sigma = \frac{a}{a^\tau}, uu^\tau = \frac{b^\sigma}{b},$$

and let $\xi^{a,b,u} : G \times G \longrightarrow L^\times$ defined by

$$\xi_{\mathrm{Id},\mathrm{Id}}^{a,b,u} = 1, \xi_{\mathrm{Id},\sigma}^{a,b,u} = 1, \xi_{\mathrm{Id},\tau}^{a,b,u} = 1, \xi_{\mathrm{Id},\sigma\tau}^{a,b,u} = 1,$$

$$\xi_{\sigma,\mathrm{Id}}^{a,b,u} = 1, \xi_{\sigma,\sigma}^{a,b,u} = a, \xi_{\sigma,\tau}^{a,b,u} = 1, \xi_{\sigma,\sigma\tau}^{a,b,u} = a^\tau,$$

$$\xi_{\tau,\mathrm{Id}}^{a,b,u} = 1, \xi_{\tau,\sigma}^{a,b,u} = u, \xi_{\tau,\tau}^{a,b,u} = b, \xi_{\tau,\sigma\tau}^{a,b,u} = \frac{b^\sigma}{u},$$

$$\xi_{\sigma\tau,\mathrm{Id}}^{a,b,u} = 1, \xi_{\sigma\tau,\sigma}^{a,b,u} = \frac{a}{u^\sigma}, \xi_{\sigma\tau,\tau}^{a,b,u} = b, \xi_{\sigma\tau,\sigma\tau}^{a,b,u} = abu^\tau.$$

LEMMA VIII.2.2. *The map $\xi^{a,b,u}$ is a 2-cocycle.*

Proof. Lengthy case by case verifications. $\square$

The two previous lemmas show that, given $a, b, u \in L^\times$ satisfying

$$a^\sigma = a, b^\tau = b, uu^\sigma = \frac{a}{a^\tau}, uu^\tau = \frac{b^\sigma}{b},$$

there exists a central simple k-algebra $(a, b, u, L/K)$ containing L as k-subalgebra, generated by two elements e and f satisfying

$$(a, b, u, L/k) = L \oplus eL \oplus fL \oplus efL$$

and subject to the relations

$$\lambda e = e\lambda^\sigma, \lambda f = f\lambda^\tau, e^2 = a, f^2 = b, fe = efu.$$

In fact, this k-algebra is just $(\xi^{a,b,u}, L/k, G)$.

Notice that we may also construct such an algebra directly, by taking the k-subalgebra of $\mathrm{M}_4(L)$ generated by the matrices

$$\begin{pmatrix} \lambda & 0 & 0 & 0 \\ 0 & \lambda^\sigma & 0 & 0 \\ 0 & 0 & \lambda^\tau & 0 \\ 0 & 0 & 0 & \lambda^{\sigma\tau} \end{pmatrix}, \lambda \in L,$$

$$e = \begin{pmatrix} 0 & 1 & 0 & 0 \\ a & 0 & 0 & 0 \\ 0 & 0 & 0 & u^{\sigma\tau} \\ 0 & 0 & a^{\tau}(u^{\sigma\tau})^{-1} & 0 \end{pmatrix} \text{ and } f = \begin{pmatrix} 0 & 0 & 1 & 0 \\ 0 & 0 & 0 & 1 \\ b & 0 & 0 & 0 \\ 0 & b^{\sigma} & 0 & 0 \end{pmatrix}.$$

Therefore, Albert's Theorem may be reformulated as follows.

THEOREM VIII.2.3. *Every central simple k-algebra A of degree 4 over k is isomorphic to some k-algebra $(a, b, u, L/k)$, where L/k is a biquadratic subfield of A.*

Elements a, b, u satisfying the required conditions may seem to be hard to find. The next lemma shows that everything boils down to finding elements of norm 1 in L.

LEMMA VIII.2.4. *Let L/k be a biquadratic extension, with Galois group $G = \{\mathrm{Id}, \sigma, \tau, \sigma\tau\}$. For $u \in L$, the following conditions are equivalent:*

(1) $N_{L/k}(u) = 1$;

(2) *there exists $a \in L^{\times}$ such that*

$$a^{\sigma} = a, \ uu^{\sigma} = \frac{a}{a^{\tau}};$$

(3) *there exists $b \in L^{\times}$ such that*

$$b^{\tau} = b, \ uu^{\tau} = \frac{b^{\sigma}}{b}.$$

Moreover, if $a', b' \in L^{\times}$ are two other elements satisfying (2) and (3) respectively, then there exist $\lambda, \mu \in k^{\times}$ such that

$$a' = \lambda a \text{ and } b' = \mu b.$$

Proof. If $N_{L/k}(u) = 1$, then $uu^{\sigma}u^{\tau}u^{\sigma\tau} = 1$, so that

$$N_{k(\sqrt{d})/k}(uu^{\sigma}) = 1, N_{k(\sqrt{d'})/k}(uu^{\tau}) = 1,$$

and thus we get both (2) and (3) by Hilbert 90. Now, if (2) holds, then

$$N_{L/k}(u) = uu^{\sigma}u^{\tau}u^{\sigma\tau} = \frac{a}{a^{\tau}}\left(\frac{a}{a^{\tau}}\right)^{\tau} = 1,$$

and similarly (3) implies (1).

Now let u be given, and consider $a, a' \in L^{\times}$ such that

$$uu^{\sigma} = \frac{a}{a^{\tau}} = \frac{a'}{a'^{\tau}},$$

so that $a'a^{\tau} = aa'^{\tau}$. In other words, $a'a^{-1} = (a'a^{-1})^{\tau}$. Since we also have

$$(a'a^{-1})^{\sigma} = (a')^{\sigma}(a^{\sigma})^{-1} = a'a^{-1},$$

we have $a'a^{-1} \in k$. Since $a, a' \in L^{\times}$, we have in fact $a'a^{-1} \in k^{\times}$. Similar arguments show that $b'b^{-1} \in k^{\times}$; this concludes the proof. $\quad\square$

This lemma and the description of the relations of the generators of $(a, b, u, L/k)$ show that the parameter $u \in L$ satisfies $N_{L/k}(u) = 1$.

LEMMA VIII.2.5. *Let L/k be a biquadratic extension, with Galois group $G = \{\mathrm{Id}, \sigma, \tau, \sigma\tau\}$. Write $L^{\langle\sigma\rangle} = k(\sqrt{d})$ and $L^{\langle\tau\rangle} = k(\sqrt{d'})$.*

Let $u \in L$ such that $N_{L/k}(u) = 1$, and let $a, b \in L^{\times}$ satisfying

$$a^{\sigma} = a, \, b^{\tau} = b, \, uu^{\sigma} = \frac{a}{a^{\tau}}, \, uu^{\tau} = \frac{b^{\sigma}}{b}.$$

Then we have

$$a = \begin{cases} \lambda\sqrt{d} & if \quad uu^{\sigma} = -1 \\ \lambda(1 + uu^{\sigma}) & if \quad uu^{\sigma} \neq -1 \end{cases}$$

and

$$b = \begin{cases} \mu\sqrt{d'} & if \quad uu^{\tau} = -1 \\ \dfrac{\mu}{1 + uu^{\tau}} & if \quad uu^{\tau} \neq -1 \end{cases},$$

for some $\lambda, \mu \in k^{\times}$.

Proof. If $uu^{\sigma} = -1$, this is obvious. Now, assume that $uu^{\sigma} \neq -1$ and set $a_0 = 1 + uu^{\sigma}$. We have that $uu^{\sigma} + N_{L/k}(u) = uu^{\sigma}(1 + uu^{\sigma})^{\tau}$, so that

$$uu^{\sigma} = \frac{uu^{\sigma} + N_{L/k}(u)}{(1 + uu^{\sigma})^{\tau}} = \frac{uu^{\sigma} + 1}{(1 + uu^{\sigma})^{\tau}} = \frac{a_0}{a_0^{\tau}}.$$

Now use the last part of the previous lemma to conclude that $a = \lambda a_0$ for some $\lambda \in k^{\times}$. The other half of the lemma can be proven in a similar way. $\quad\square$

We end this section by describing completely the relative Brauer group $\mathrm{Br}(L/k)$ for a biquadratic extension.

PROPOSITION VIII.2.6. *Let L/k be a biquadratic extension, with Galois group $G = \{\mathrm{Id}, \sigma, \tau, \sigma\tau\}$. Let $\mathbb{T}(k)$ be the multiplicative group*

$$\mathbb{T}(k) = \Big\{(a, b, u) \in L^{\times 3} \mid a^{\sigma} = a, b^{\tau} = b, uu^{\sigma} = \frac{a}{a^{\tau}}, uu^{\tau} = \frac{b^{\sigma}}{b}\Big\}.$$

The map

$$\varphi\colon \begin{array}{c} \mathbb{T}(k) \longrightarrow \mathrm{Br}(L/k) \\ (a, b, u) \longmapsto [(a, b, u, L/k)] \end{array}$$

is a surjective group morphism with kernel

$$\ker(\varphi) = \Big\{\big(w_1 w_1^{\sigma}, w_2 w_2^{\tau}, \frac{w_1}{w_1^{\tau}} \cdot \frac{w_2^{\sigma}}{w_2}\big) \mid w_1, w_2 \in L^{\times}\Big\}.$$

Proof. Let $[A] \in \mathrm{Br}(L/k)$. Then A is split by L, and by Proposition IV.1.12, there exists a central simple k-algebra A' Brauer equivalent to A which contains a maximal subfield L' isomorphic to L (notice that A' has degree 4 over k). Hence A' is isomorphic to a crossed product over L'/k by Proposition VI.2.1, and is therefore isomorphic to a crossed product over L/k by Remark VI.1.9. The surjectivity of φ is then a consequence of Theorem VIII.2.3. To prove that φ is a group morphism, it is enough to notice that we have

$$\xi^{a,b,u}\xi^{a',b',u'} = \xi^{aa',bb',uu'},$$

and to use Proposition VI.2.3. It remains to compute $\ker(\varphi)$. By Theorem VI.2.7 (or Lemma VI.2.5), $(a, b, u, L/k)$ is split if and only if $\xi^{a,b,u}$ is cohomologous to the trivial 2-cocycle. Assume first that we have

$$\xi^{a,b,u}_{\rho,\rho'} = z_{\rho'} z_\rho^{\rho'} z_{\rho\rho'}^{-1} \text{ for all } \rho, \rho' \in G$$

for some map $z : G \longrightarrow L^\times$ satisfying $z_{\mathrm{Id}} = 1$.

Taking $(\rho, \rho') = (\sigma, \sigma), (\tau, \tau), (\sigma, \tau)$ and (τ, σ) yields

$$a = z_\sigma z_\sigma^\sigma, b = z_\tau z_\tau^\tau, 1 = z_\tau z_\sigma^\tau z_{\sigma\tau}^{-1} \text{ and } u = z_\sigma z_\tau^\sigma z_{\sigma\tau}^{-1}.$$

We then have $z_{\sigma\tau} = z_\tau z_\sigma^\tau$, and plugging into the last equation gives

$$u = \frac{z_\sigma}{z_\sigma^\tau} \cdot \frac{z_\tau^\sigma}{z_\tau}.$$

Therefore, we have

$$\ker(\varphi) \subset \left\{ \left(w_1 w_1^\sigma, w_2 w_2^\tau, \frac{w_1}{w_1^\tau} \cdot \frac{w_2^\sigma}{w_2} \right) \mid w_1, w_2 \in L^\times \right\}.$$

Conversely, assume that we have

$$a = w_1 w_1^\sigma, b = w_2 w_2^\tau \text{ and } u = \frac{w_1}{w_1^\tau} \cdot \frac{w_2^\sigma}{w_2},$$

for some $w_1, w_2 \in L^\times$, and set

$$z_{\mathrm{Id}} = 1, z_\sigma = w_1, z_\tau = w_2 \text{ and } z_{\sigma\tau} = w_2 w_1^\tau.$$

One can check that we have

$$(\xi^{a,b,u})_{\rho,\rho'} = z_{\rho'} z_\rho^{\rho'} z_{\rho\rho'}^{-1} \text{ for all } \rho, \rho' \in G,$$

and therefore $\xi^{a,b,u}$ is cohomologous to the trivial 2-cocycle. This concludes the proof.   □

REMARK VIII.2.7. The result above can be obtained in an slightly easier way by computing $H^2(G, L^\times)$ using homological algebra techniques. In fact, we have a nice description of $H^2(\mathrm{Gal}(L/k), L^\times)$ when L/k is an abelian extension. See for example [54] for more details.   □

To end this section, we would like to give a necessary and sufficient condition on $(a, b, u, L/k)$ to be a division algebra, when k is a number field. We start with a lemma.

LEMMA VIII.2.8. Let $A = (a, b, u, L/k)$, and let $G = \{\mathrm{Id}, \sigma, \tau, \sigma\tau\}$ be the Galois group of L/k. Write $L^{\langle\sigma\rangle} = k(\sqrt{d}), L^{\langle\tau\rangle} = k(\sqrt{d'})$, and let

$$n_a = N_{L^{\langle\sigma\rangle}/k}(a), n_b = N_{L^{\langle\tau\rangle}/k}(b).$$

Then we have

$$2[A] = [(n_a, d')_k] = [(n_b, d)_k].$$

Proof. Elementary Galois theory shows that we have $L = k(\sqrt{d}, \sqrt{d'})$, and that σ and τ satisfy

$$\sigma(\sqrt{d}) = \sqrt{d}, \sigma(\sqrt{d'}) = -\sqrt{d'}$$

and

$$\tau(\sqrt{d}) = -\sqrt{d}, \tau(\sqrt{d'}) = \sqrt{d'}.$$

In particular, the restriction of τ to $L^{\langle\sigma\rangle}$ is the unique non-trivial k-automorphism ι of $L^{\langle\sigma\rangle}/k$, and therefore $n_a = a\iota(a) = a\tau(a)$.

We are now going to use the Inflation theorem to prove the lemma. Let $F = L^{\langle\tau\rangle}$, and set $H = \mathrm{Gal}(F/k) = \langle\sigma_{|F}\rangle$. By Example VI.1.3, we have

$$(n_a, d)_k \cong_k (\xi, F/k, \langle\iota\rangle),$$

where $\xi \in Z^2(G, F^\times)$ is the 2-cocycle defined by

$$\xi_{\mathrm{Id,Id}} = \xi_{\mathrm{Id},\iota} = \xi_{\iota,\mathrm{Id}} = 1, \xi_{\iota,\iota} = n_a = a\tau(a).$$

Therefore, for all $\rho, \rho' \in G$, we have

$$\mathrm{Inf}_F^L(\xi)_{\rho,\rho'} = \begin{cases} a\tau(a) & \text{if } \rho_{|F} = \rho'_{|F} = \iota \\ 1 & \text{otherwise} . \end{cases}$$

By Theorem VI.2.7, we have

$$2[A] = 2[(\xi^{a,b,u}, L/k, G)] = [((\xi^{a,b,u})^2, L/k, G)].$$

By the Inflation theorem, we also have

$$[(n_a, d)_k] = [(\xi, F/k, \langle\iota\rangle)] = [(\mathrm{Inf}_F^L(\xi), L/k, G)].$$

Hence, by Theorem VI.2.7 again, we need to prove that ξ and $(\xi^{a,b,u})^2$ are cohomologous, or in other words, that there exists a map

$$z : G \longrightarrow L^\times$$

such that $z_{\mathrm{Id}} = 1$ and

$$(\xi^{a,b,u})_{\rho,\rho'}^2 = \mathrm{Inf}_F^L(\xi)_{\rho,\rho'} z_{\rho'}^\rho z_\rho^{-1} z_{\rho\rho'}^{-1} \text{ for all } \rho, \rho' \in G.$$

Set $z_{\mathrm{Id}} = 1, z_\sigma = u, z_\tau = b$ and $z_{\sigma\tau} = bu^\tau$.

If $\rho = \mathrm{Id}$ or $\rho' = \mathrm{Id}$, the equality above is trivial. We then have 9 remaining cases to consider. We will not check all of them, since they follow from straightforward computations and the relations between a, b and u. Let us give an example. Recall that we have $\sigma_{|F} = \sigma\tau_{|F} = \iota$ and $\tau_{|F} = \mathrm{Id}_F$. Let us explain the case $(\rho, \rho') = (\tau, \sigma\tau)$ in details. By definition, we have

$$\mathrm{Inf}_F^L(\xi)_{\tau,\sigma\tau} = \xi_{\mathrm{Id},\iota} = 1,$$

and therefore we have to check the equality

$$(\xi_{a,b,u})_{\tau,\sigma\tau}^2 = z_{\sigma\tau} z_\tau^{\sigma\tau} z_\sigma^{-1}.$$

But we have

$$z_{\sigma\tau} z_\tau^{\sigma\tau} z_\sigma^{-1} = \frac{bu^\tau \cdot b^{\sigma\tau}}{u} = \frac{bb^\sigma u^\tau}{u} = \frac{bb^\sigma uu^\tau}{u^2},$$

and thus

$$z_{\sigma\tau} z_\tau^{\sigma\tau} z_\sigma^{-1} = \frac{bb^\sigma}{u^2} \cdot \frac{b^\sigma}{b} = \left(\frac{b^\sigma}{u}\right)^2 = (\xi^{a,b,u})_{\tau,\sigma\tau}^2.$$

The remaining cases may be checked in a similar way, and are left to the reader. This proves the equality

$$2[A] = [(n_a, d')_k].$$

The equality $2[A] = [(n_b, d)_k]$ is obtained by switching the roles of σ and τ, or by similar arguments. $\qquad \square$

COROLLARY VIII.2.9. *Let $A = (a, b, u, L/k)$, where k is a number field. Then we have the following properties:*

(1) *if $uu^\sigma = -1$, then A is a division algebra if and only if the quaternion k-algebra $(-d, d')_k$ is not split;*

(2) *if $uu^\sigma \neq -1$, then A is a division algebra if and only if the quaternion k-algebra $(2 + \mathrm{Tr}_{L\langle\sigma\rangle/k}(uu^\sigma), d')_k$ is not split.*

Proof. Since k is a number field, $\exp(A) = \mathrm{ind}(A)$, so A will be a division algebra if and only if $2[A] \neq 0$. By Lemma VIII.2.8, we have $2[A] = (n_a, d')$.

If $uu^\sigma = -1$, then we have $n_a = -\lambda^2 d$ for some $\lambda \in k^\times$ by Lemma VIII.2.5. Hence, we get

$$2[A] = [(-\lambda^2 d, d')_k] = [(-d, d')_k].$$

Now, assume that $uu^\sigma \neq -1$, so that $a = \lambda(1 + uu^\sigma)$ for some $\lambda \in k^\times$ by the same lemma. We then have

$$
\begin{aligned}
n_a &= \lambda^2 (uu^\sigma + 1)(uu^\sigma + 1)^\tau \\
&= \lambda^2 (1 + \mathrm{Tr}_{L\langle\sigma\rangle/k}(u\sigma(u)) + N_{L/K}(u)) \\
&= \lambda^2 (2 + \mathrm{Tr}_{L\langle\sigma\rangle/k}(u\sigma(u))).
\end{aligned}
$$

We then have

$$2[A] = [(-2 + \mathrm{Tr}_{L\langle\sigma\rangle/k}(u\sigma(u)), d')_k],$$

and this concludes the proof. $\qquad \square$

EXAMPLES VIII.2.10. Let $k = \mathbb{Q}(i)$.

(1) Assume that $L = k(\sqrt{3}, \sqrt{5})$, and let $\sigma, \tau \in G$ be the automorphisms of L/k defined by

$$\sigma(\sqrt{3}) = \sqrt{3}, \sigma(\sqrt{5}) = -\sqrt{5},$$

and

$$\sigma(\sqrt{3}) = -\sqrt{3}, \sigma(\sqrt{5}) = \sqrt{5}.$$

The elements $u = i, a = \sqrt{3}, b = \sqrt{5}$ of L satisfy all the required relations (this follows for example from Lemma VIII.2.5, or from direct computations). By Corollary VIII.2.9, we have $2[A] = (-3, 5)_k$. Since $1 + 2i$ totally ramifies and does not lie above 2, $2[A]$ is not split by Lemma VII.4.6, since -3 is not a square modulo 5. Hence A is a division k-algebra.

(2) Assume that $L = k(\sqrt{2}, \sqrt{5})$, and let $\sigma, \tau \in G$ be the automorphisms of L/k defined by

$$\sigma(\sqrt{2}) = \sqrt{2}, \sigma(\sqrt{5}) = -\sqrt{5},$$

and

$$\sigma(\sqrt{2}) = -\sqrt{2}, \sigma(\sqrt{5}) = \sqrt{5}.$$

It follows from direct computations that the elements

$$u = i, a = \zeta_8, b = \frac{1 + 2i}{\sqrt{5}}$$

of L satisfy all the required relations.   By Corollary VIII.2.9, we have $2[A] = (-2, 5)_k$. Since $1 + 2i$ totally ramifies and does not lie above 2, $2[A]$ is not split by Lemma VII.4.6, since -2 is not a square modulo 5. Hence A is a division algebra.

$\square$

VIII.3. Albert's Theorem

We are now going to prove a theorem of Albert, which says that a central simple k-algebra of degree 4 has exponent at most 2 if and only if it is isomorphic to a tensor product of two quaternion k-algebras. In fact, we are going to prove a slightly more precise result. Recall that we denote by $\mathbb{T}(k)$ the multiplicative group

$$\mathbb{T}(k) = \big\{(a, b, u) \in L^{\times 3} \mid a^\sigma = a, b^\tau = b, uu^\sigma = \frac{a}{a^\tau}, uu^\tau = \frac{b^\sigma}{b}\big\}.$$

We then have the following theorem.

THEOREM VIII.3.1. *Let L/k be a biquadratic extension, with Galois group $G = \langle \sigma, \tau \rangle$. Let $L^{\langle \sigma \rangle} = k(\sqrt{d})$ and $L^{\langle \tau \rangle} = k(\sqrt{d'})$, and let $(a, b, u) \in \mathbb{T}(k)$. Then $A = (a, b, u, L/k)$ has exponent at most 2 if and only if we have*

$$u = \frac{w_1}{w_1^\tau} \cdot \frac{w_2^\sigma}{w_2},$$

for some $w_1, w_2 \in L^\times$. In this case, we have $a = \lambda w_1 w_1^\sigma, b = \mu w_2 w_2^\sigma$ for some $\lambda, \mu \in k^\times$, and A is isomorphic to

$$(\lambda, \mu d')_k \otimes_k (\mu, \lambda d)_k.$$

Proof.   Assume that $\exp(A) = 1$ or 2, that is $2[A] = 0$. By Lemma VIII.2.8, the quaternion algebra (n_a, d') splits. By Corollary II.1.5, n_a is a norm of the extension $L^{\langle \tau \rangle} = k(\sqrt{d'})/k$. Since the unique non-trivial k-automorphism of this field extension is the restriction of σ, we get $n_a = x_1 x_1^\sigma$ for some $x_1 \in (L^{\langle \tau \rangle})^\times$, that is

$$aa^\tau = x_1 x_1^\sigma.$$

Hence we have

$$uu^\sigma = \frac{a}{a^\tau} = \frac{a^2}{x_1 x_1^\sigma} = \frac{a}{x_1} \cdot \frac{a^\sigma}{x_1^\sigma},$$

since $a = a^\sigma$. Since $L/L^{\langle \sigma \rangle}$ is cyclic with Galois group generated by σ, we get

$$N_{L/L^{\langle \sigma \rangle}} \big(\frac{ux_1}{a}\big) = 1.$$

Therefore, Hilbert 90 gives

$$\frac{ux_1}{a} = \frac{x_2^\sigma}{x_2},$$

for some $x_2 \in L^\times$. Hence

$$u = \frac{a}{x_1} \cdot \frac{x_2^\sigma}{x_2}.$$

Now notice that $L/L^{\langle \sigma\tau \rangle}$ is cyclic with Galois group generated by $\sigma\tau$. Since $a = a^\sigma$ and $x_1 = x_1^\tau$, the equality $aa^\tau = x_1 x_1^\sigma$ implies that we have

$$N_{L/L^{\langle \sigma\tau \rangle}} \big(\frac{a}{x_1}\big) = 1.$$

By Hilbert 90, we have
$$\frac{a}{x_1} = \frac{x_3}{x_3^{\sigma\tau}},$$
for some $x_3 \in L^\times$. Putting things together, we get
$$u = \frac{x_3}{x_3^{\sigma\tau}} \cdot \frac{x_2^\sigma}{x_2} = \frac{x_3^\sigma}{(x_3^\sigma)^\tau} \cdot \frac{x_2^\sigma x_3}{x_2 x_3^\sigma}.$$

Now, set $w_1 = x_3^\sigma$ and $w_2 = x_2 x_3^\sigma$. We then have $w_2^\sigma = x_2^\sigma x_3$ since $\sigma^2 = \mathrm{Id}_L$, and therefore
$$u = \frac{w_1}{w_1^\tau} \cdot \frac{w_2^\sigma}{w_2}.$$

It is easy to check that we have
$$uu^\sigma = \frac{w_1 w_1^\sigma}{(w_1 w_1^\sigma)^\tau} \quad \text{and} \quad uu^\tau = \frac{(w_2 w_2^\tau)^\sigma}{w_2 w_2^\tau}.$$

The last part of Lemma VIII.2.4 then implies the existence of $\lambda, \mu \in k^\times$ satisfying $a = \lambda w_1 w_1^\sigma$ and $b = \mu w_2 w_2^\tau$.

Conversely, assume that we have
$$a = \lambda w_1 w_1^\sigma, b = \mu w_2 w_2^\sigma, u = \frac{w_1}{w_1^\tau} \cdot \frac{w_2^\sigma}{w_2},$$
for some $\lambda, \mu \in k^\times, w_1, w_2 \in L^\times$. By Proposition VIII.2.6, we then have
$$[(a, b, u, L/k)] = [(\lambda, \mu, 1), L/k)].$$

Let e, f be the two generators of $(\lambda, \mu, 1, L/k)$. By definition, we have
$$e^2 = \lambda, f^2 = \mu, fe = ef, xe = ex^\sigma \text{ and } xf = fx^\tau$$
for all $x \in L$. Set $i = e$ and $j = \sqrt{d'}$, and let Q be the k-subalgebra generated by i and j. We have
$$i^2 = e^2 = \lambda, j^2 = \sqrt{d'}\sqrt{d'} = d'$$
and
$$ji = \sqrt{d'}e = -e\sqrt{d'} = -ij.$$

Hence $Q \cong_k (\lambda, d')_k$. Since Q is therefore a central simple k-algebra, the Centralizer theorem implies that $C_A(Q)$ is a central simple k-algebra of degree 2, and that we have
$$A \cong_k Q \otimes_k C_A(Q).$$

It remains to determine $C_A(Q)$. We set $i' = f$ and $j' = \sqrt{d}$. We have
$$i'i = fe = ef = ii', i'j = f\sqrt{d'} = \sqrt{d'}f = ji',$$
so $i' \in C_A(Q)$. Moreover, we have
$$j'i = \sqrt{d}e = e\sqrt{d} = ij',$$
and
$$j'j = \sqrt{d}\sqrt{d'} = \sqrt{d'}\sqrt{d} = jj',$$
hence $j' \in C_A(Q)$. Thus $C_A(Q)$ contains the subalgebra Q' generated by i' and j'. We have
$$i'^2 = f^2 = \mu, j'^2 = d,$$
and
$$j'i' = \sqrt{d}f = -f\sqrt{d} = -i'j'.$$

Therefore $Q' \cong_k (\mu, d)_k$.

Since $Q' \subset C_A(Q)$ and $\deg(Q') = \deg(C_A(Q)) = 2$, we get $C_A(Q) = Q'$. Therefore, we obtain

$$A \cong_k Q \otimes_k Q' \cong_k (\lambda, d')_k \otimes_k (\mu, d)_k.$$

Now we have

$$2[A] = 2([Q] + [Q']) = 2[Q] + 2[Q'] = 0,$$

since a quaternion k-algebra has index 1 or 2, and therefore exponent 1 or 2 by Theorem V.3.5. This concludes the proof.   □

COROLLARY VIII.3.2 (Albert). *Let k be an arbitrary field of characteristic different from 2. Then every central simple k-algebra of degree 4 and exponent at most 2 is isomorphic to a tensor product of two quaternion k-algebras.*

Proof. If k has no biquadratic extensions, then by Remark VIII.1.3, we have either

$$A \cong_k \mathrm{M}_4(k) \cong_k \mathrm{M}_2(k) \otimes_k \mathrm{M}_2(k) \cong_k (1,1)_k \otimes_k (1,1)_k$$

or

$$A \cong_k \mathrm{M}_2((-1,-1)_k) \cong_k \mathrm{M}_2(k) \otimes_k (-1,-1)_k \cong_k (1,1)_k \otimes_k (-1,-1)_k.$$

Hence the result is true in this case. If k has at least one biquadratic extension, then $A \cong_k (a, b, u, L/k)$ by Theorem VIII.2.3. Now use the previous theorem to conclude.   □

VIII.4. Codes over biquadratic crossed products

We are going to apply the results of the previous section to construct codes with good performances over crossed-product algebras of degree 4.

Consider a Galois extension L/k of degree 4. Its Galois group is either cyclic of order 4 or a product of two cyclic groups of order 2. Since we already dealt with the cyclic case in Section VII.6, we focus on the latter, and consider the case where L/k is a biquadratic extension, with Galois group $G = \mathrm{Gal}(L/k) = \{1, \sigma, \tau, \sigma\tau\}$.

If we write $L^{\langle \sigma \rangle} = k(\sqrt{d})$ and $L^{\langle \tau \rangle} = k(\sqrt{d'})$, then we have

$$L = k(\sqrt{d}, \sqrt{d'}).$$

and σ, τ are then defined by

$$\sigma(\sqrt{d}) = \sqrt{d}, \sigma(\sqrt{d'}) = -\sqrt{d'}$$
$$\tau(\sqrt{d}) = -\sqrt{d}, \tau(\sqrt{d'}) = \sqrt{d'}.$$

By the results of Section VIII.2, a crossed product algebra A over L/k has the form $(a, b, u, L/k, \sigma)$, for some $a, b, u \in L^\times$, where

$$a^\sigma = a, b^\tau = b, uu^\sigma = \frac{a}{a^\tau}, uu^\tau = \frac{b^\sigma}{b}.$$

Notice that these relations imply that $(abu^\tau)^{\sigma\tau} = abu^\tau$.

We will simply denote by $\xi : G \times G \longrightarrow L^\times$ the corresponding cocycle, instead of $\xi^{a,b,u}$. Recall from Section VIII.2 that we have

$$\xi_{\mathrm{Id},\mathrm{Id}} = 1, \xi_{\mathrm{Id},\sigma} = 1, \xi_{\mathrm{Id},\tau} = 1, \xi_{\mathrm{Id},\sigma\tau} = 1,$$
$$\xi_{\sigma,\mathrm{Id}} = 1, \xi_{\sigma,\sigma} = a, \xi_{\sigma,\tau} = 1, \xi_{\sigma,\sigma\tau} = a^{\tau},$$
$$\xi_{\tau,\mathrm{Id}} = 1, \xi_{\tau,\sigma} = u, \xi_{\tau,\tau} = b, \xi_{\tau,\sigma\tau} = bu^{\tau},$$
$$\xi_{\sigma\tau,\mathrm{Id}} = 1, \xi_{\sigma\tau,\sigma} = a^{\tau}u, \xi_{\sigma\tau,\tau} = b, \xi_{\sigma\tau,\sigma\tau} = abu^{\tau}.$$

In order for the encoding to fulfill the shaping constraint, we need all the cocycle values to have modulus 1 by Proposition VI.3.5. This is equivalent to

$$|a|^2 = |\tau(a)|^2 = |b|^2 = |\sigma(b)|^2 = |u|^2 = |\sigma(u)|^2 = |\tau(u)|^2 = 1.$$

When complex conjugation commutes with the elements of G, this may be rewritten as

$$|a|^2 = |b|^2 = |u|^2 = 1.$$

We now compute the left multiplication matrix M_x.

PROPOSITION VIII.4.1. *The multiplication matrix M_x of the element $x = x_1 + ex_\sigma + fx_\tau + efx_{\sigma\tau}$ in the k-algebra $(a, b, u, L/k)$ is given by*

$$\begin{pmatrix} x_1 & ax_\sigma^\sigma & bx_\tau^\tau & abu^\tau x_{\sigma\tau}^{\sigma\tau} \\ x_\sigma & x_1^\sigma & bx_{\sigma\tau}^\tau & bu^\tau x_\tau^{\sigma\tau} \\ x_\tau & a^\tau u x_{\sigma\tau}^\sigma & x_1^\tau & a^\tau x_\sigma^{\sigma\tau} \\ x_{\sigma\tau} & ux_\tau^\sigma & x_\sigma^\tau & x_1^{\sigma\tau} \end{pmatrix}.$$

Proof. This follows from Lemma VI.3.1. □

Let us consider the case of a biquadratic crossed-product algebra containing $L = \mathbb{Q}(i)(\sqrt{3}, \sqrt{5})/\mathbb{Q}(i)$ as a maximal subfield. Notice that complex conjugation commutes with σ and τ, and therefore, condition (1) reduces to

$$|a|^2 = |b|^2 = |u|^2 = 1.$$

Example VIII.2.10 (1) shows that if $u = i$, A is a division algebra and that moreover, we can take $a = \sqrt{3}$ and $b = \sqrt{5}$. Hence we have $|u|^2 = 1$, but $|a|^2 \neq 1$ and $|b|^2 \neq 1$. Thus we need to change the values of the parameters a and b. Lemma VIII.2.4 shows that the possible choices of a have the form $\lambda\sqrt{3}, \lambda \in \mathbb{Q}(i)^\times$. We then need to find $\lambda \in \mathbb{Q}(i)^\times$ such that $3|\lambda|^2 = 1$. This easily implies that 3 is a sum of 2 squares in $\mathbb{Q}$, which is known to be impossible.

Hence, there is no code on a crossed-product based on $L/\mathbb{Q}(i)$ with $u = i$ and satisfying the shaping constraint.

Assume now that $L = \mathbb{Q}(i)(\sqrt{2}, \sqrt{5})$. As before, complex conjugation commutes with σ and τ, so we again need

$$|a|^2 = 1, \ |b|^2 = 1, \ |u|^2 = 1.$$

Example VIII.2.10 (2) shows if $u = i$, A is a division algebra, and that we can take $a = \zeta_8$ and $b = \dfrac{1 + 2i}{\sqrt{5}}$.

We are thus left with making sure that we may construct the cubic lattice as an hermitian ideal lattice on L. Set

$$\theta = \frac{1 + \sqrt{5}}{2}, \ \alpha = 1 + i - i\theta.$$

Let $I = \alpha \mathcal{O}_L$, and $\lambda = \dfrac{1}{10}$. Then the hermitian ideal lattice (I, h_λ) is isomorphic to the cubic lattice, since the $\mathcal{O}_k$-basis

$$\omega_1 = \alpha, \ \omega_\sigma = \alpha\theta, \ \omega_\tau = \alpha\zeta_8, \ \omega_{\sigma\tau} = \alpha\theta\zeta_8$$

is orthonormal, as we may check easily.

This finally yields a code $\mathcal{C}_{A,\lambda,I}$ on the division algebra A, satisfying the shaping constraint. The reader will find a generalization of this construction in Exercise 4.

We would like now to give an estimation of the minimum determinant of the code, so we need to compute Δ_ξ in our case. Clearly we have

$$\Delta_\xi^{(\mathrm{Id})} = \Delta_\xi^{(\sigma)} = 1.$$

Moreover, we have

$$\mathcal{E}_\xi^{(\tau)} = \mathcal{E}_\xi^{(\sigma\tau)} = \{c \in \mathbb{Z}[i] \mid cb \in \mathcal{O}_L\}.$$

We claim that $\Delta_\xi^{(\tau)} = 5$. Notice that $1 - 2i \in \mathcal{E}_\xi^{(\tau)}$ since $(1 - 2i)b = \sqrt{5}$. Hence $\Delta_\xi^{(\tau)}$ divides $|1 - 2i|^2 = 5$. Since $b \notin \mathcal{O}_L$ (its minimal polynomial over $\mathbb{Q}$ is $X^4 - \frac{6}{5}X^2 + 1 \notin \mathbb{Z}[X]$), we have $\Delta_\xi^{(\tau)} \neq 1$ and therefore $\Delta_\xi^{(\tau)} = 5$. We also have $\Delta_\xi^{(\sigma\tau)} = 5$, and thus $\Delta_\xi = 25$.

In particular, Proposition VI.3.6 shows that the minimum determinant of any code $\mathcal{C} \subset \mathcal{A}, \lambda, \mathcal{I}$ satisfies

$$\delta_{min}(\mathcal{C}) \geq \frac{1}{10000}.$$

REMARK VIII.4.2. Notice that in [**8**], the better bound $\dfrac{1}{2500}$ was announced. This bound is not correct, since it was obtained by writing $b = \dfrac{\sqrt{1 + 2i}}{\sqrt{1 - 2i}}$, and taking the denominator outside the multiplication matrix. However, the conclusion that the determinant of the remaining matrix was an element of $\mathbb{Z}[i]$ was not correct, since $\sqrt{1 + 2i} \notin \mathcal{O}_L$. $\qquad\square$

The reader may wonder why we bother considering codes on biquadratic crossed-products, since every central simple k-algebra is isomorphic to a cyclic algebra. The point is that the existence of a suitable code based on a division algebra depends on the presentation of the algebra by generators and relations, as the following example shows.

EXAMPLE VIII.4.3. Let $k = \mathbb{Q}(i)$, $L = k(\sqrt{2}, \sqrt{5})$ and consider the k-algebra $A = (\zeta_8, \frac{1+2i}{\sqrt{5}}, i, L/k)$. As we have seen previously, one may construct a complex ideal lattice on L/k which is isomorphic to the cubic lattice, and therefore a code base on A satisfying the shaping condition. Now, we have an isomorphism of k-algebras

$$A \simeq (i, k(\sqrt[4]{d})/k, \rho),$$

where $d = (1 + 2i)(1 - 2i)^3$ and ρ is defined by

$$\rho(\sqrt[4]{d}) = -i\sqrt[4]{d}.$$

Indeed, the generators e and f of A satisfy

$$e^4 = \zeta_8^2 = i, f^4 = \frac{(1+2i)^2}{5} = \frac{1+2i}{1-2i} \text{ and } fe = efi.$$

Set $L' = k((1 - 2i)f) \simeq_k k(\sqrt[4]{d})$. The equality $fe = efi$ then may be rewritten as

$$\sqrt[4]{d}e = e\rho^{-1}(\sqrt[4]{d}) = e(\sqrt[4]{d})^\rho.$$

It easily follows that we have

$$\lambda'e = e\lambda'^\rho \text{ for all } \lambda' \in L',$$

and we get the desired isomorphism.

However, there is no perfect code built on this cyclic k-algebra, because the cubic lattice is not isomorphic to a complex hermitian lattice on L'/k, as already shown at the end of the proof of Theorem VII.7.10. $\qquad\square$

To end this chapter, we are going to prove that upper bound for the probability error obtained for this code is optimal. More precisely, we have the following result.

THEOREM VIII.4.4. *Let $k = \mathbb{Q}(i)$. If $\mathcal{C} \subset \mathcal{C}_{A,\lambda,I}$ is an energy-preserving code built on a division k-algebra $A = (a, b, u, L/k, \sigma, \tau)$, then we have*

$$d_{L/k}\Delta_\xi \geq 10000.$$

The proof detailed here is (up to minor modifications) the one presented in [**10**] (see also [**50**]).

We will assume in the sequel that $k = \mathbb{Q}(i)$. We start with the study of the ramification of biquadratic extensions of k.

The following lemma is a direct consequence of Lemma VII.7.2.

LEMMA VIII.4.5. *Let $d \in \mathcal{O}_k$ be a non-zero squarefree element of $\mathcal{O}_k$, and let $F = k(\sqrt{d})$. Then complex conjugation induces a $\mathbb{Q}$-automorphism of L which commutes with $\mathrm{Gal}(F/k)$ if and only if all the following conditions are satisfied:*

(1) $v_{1-i}(d) = 0$.
(2) $v_p(d) = 0$ *or* 1, *for all* $p \in S_3$.
(3) $v_\pi(d) = v_{\bar{\pi}}(d) = 1$ *for all* $\pi \in S_1$ *dividing* d.

Let $L = k(\sqrt{d}, \sqrt{d'})$ be a biquadratic extension, whose Galois group commutes with complex conjugation. Then d and d' have the form m or mi, where m is a squarefree odd integer (apply twice the previous lemma). Moreover, we have $4mi = 2m(1 + i)^2$, so we may in fact assume that d and d' are squarefree integers. Since -1 is a square, we may also assume that d and d' are positive.

We will then assume from now on that $L = k(\sqrt{d}, \sqrt{d'})$, where d and d' are square-free positive integers.

Notice for later use that if π is an irreducible element lying above the prime number p, and if π divides d and d', then $p \mid d$ and $p \mid d'$ (This is clear if π is a prime number, and if $p \equiv 1[4]$, it follows from the fact that $\bar{\pi}$ also divides d and d').

PROPOSITION VIII.4.6. *Let $L = k(\sqrt{d}, \sqrt{d'})$, where d and d' are squarefree positive integers. Then the following properties hold:*

(1) *the odd part of $d_{L/k}$ is $\prod_{p} p^2$, where p runs through the odd prime numbers that divide d or d'.*

(2) *the element $1 - i$ ramifies in L if and only if d or d' is even. In this case, $2^4 \mid d_{L/k}$.*

Proof. Let p be an odd prime integer and let π be an irreducible element lying above p. Assume that p does not divide d and d'. Then $\pi \nmid d$ and $\pi \nmid d'$ (this comes from Lemma VIII.4.5) , and thus π does not ramify in $k(\sqrt{d})$ and $k(\sqrt{d'})$. Therefore, π does not ramify in L. Assume now that $p \mid d$ for example. Replacing d' by $\dfrac{dd'}{p^2}$ if necessary, one may assume that $p \nmid d'$, so that π does not ramify in $M' = k(\sqrt{d'})$. Since π divides d, it totally ramifies in $M = k(\sqrt{d})$. Write $(\pi) = \mathfrak{p}^2$. Since π does not ramify in M', $\mathfrak{p}$ does not ramify in L/M. Hence π ramifies but does not totally ramify in L.

We then either have $(\pi) = \mathfrak{P}_0^2$ or $(\pi) = \mathfrak{P}_1^2 \mathfrak{P}_2^2$ in $\mathcal{O}_L$. Reasoning as in the proof of Lemma VII.7.12, we may show that $v_p(d_{L/k}) = 2$.

We now study the ramification of $1 - i$ in $M = k(\sqrt{d})$.

Assume first that d is odd. If $d \equiv 1 [4]$ (resp. $d \equiv 3 [4]$), then $x = 1$ (resp. $x = i$) is a solution of the equation $x^2 \equiv d \mod 4\mathcal{O}_k$. Hence $1 - i$ does not ramify in $k(\sqrt{d})$ by [**31**, Theorem 6.8.6].

If now d is even, then the equation $x^2 \equiv d \mod 4\mathcal{O}_k$ has no solution. Assume to the contrary that $x \in \mathcal{O}_k$ is a solution. Since $d = 2m$, m odd, we have $d \equiv 2 [4]$, so $x^2 \equiv 2 \mod 4\mathcal{O}_k$. Writing $x = a + bi, a, b \in \mathbb{Z}$ and comparing real parts show that $a^2 - b^2 \equiv 2 [4]$. But $a^2 - b^2$ is always congruent to 0 or ± 1 modulo 4, hence we have a contradiction. Thus $1 - i$ totally ramifies in M in this case.

It follows as before that $1 - i$ ramifies in L if and only if d or d' is even. It remains to prove that $2^4 \mid d_{L/k}$ in this case. Assume for example that d is even, so that $1 - i$ totally ramifies in M. By [**22**, Theorem 1], a $\mathbb{Z}$-basis of $\mathcal{O}_M$ is

$$1, i, \sqrt{d}, \frac{1-i}{2}\sqrt{d}.$$

Now let $\mathfrak{P}$ be the unique prime ideal of $\mathcal{O}_M$ lying above $1 - i$, and consider the third ramification group of $\mathfrak{P}$

$$G_3(\mathfrak{P}) = \{\rho \in \mathrm{Gal}(M/k) \mid \rho(\alpha) \equiv \alpha \mod \mathfrak{P}^4 \text{ for all } \alpha \in \mathcal{O}_M\}$$
$$= \{\rho \in \mathrm{Gal}(M/k) \mid \rho(\alpha) \equiv \alpha \mod 2\mathcal{O}_M \text{ for all } \alpha \in \mathcal{O}_M\}$$

Now any $\alpha \in \mathcal{O}_M$ has the form $\alpha = \alpha_1 + \alpha_2 i + \alpha_3 \sqrt{d} + \alpha_4 \dfrac{1-i}{2}\sqrt{d}$, where $\alpha_i \in \mathbb{Z}$. If ι is the unique non-trivial automorphism of M/k, we have

$$\iota(\alpha) - \alpha = -2(\alpha_3 \sqrt{d} + \alpha_4 \frac{1-i}{2}\sqrt{d}) \in 2\mathcal{O}_M.$$

Therefore G_3 is non-trivial. Theorem B.2.33 then implies that

$$v_{\mathfrak{P}}(\mathfrak{d}_{k(\sqrt{d})/k}) = \sum_{i \geq 0}(|G_i| - 1) \geq 4,$$

since the ramification groups form a decreasing sequence. This then gives as usual $2^4|d_{M/k}^2$, which by Proposition C.2.7 implies that $2^4|d_{L/k}$. $\square$

EXAMPLE VIII.4.7. Let $L = k(\sqrt{2}, \sqrt{5})$. In this case, the only prime ideals which ramify in L are those generated by the prime elements $1 - i, 1 + 2i$ and $1 - 2i$. Set $M = k(\sqrt{2})$ and $M' = k(\sqrt{5})$. Notice that 2 remains inert in $\mathbb{Q}(\sqrt{5})/\mathbb{Q}$ and then totally ramifies $M/\mathbb{Q}(\sqrt{5})$. It follows easily that $1 - i$ is inert in M/k. In particular, $\mathcal{D}_{L/M}$ is not divisible by any prime ideal lying above $1 - i$. Proposition C.2.7 then implies that $v_2(d_{L/k}) = v_2(d_{M/k}^2)$.

As pointed in the proof of the previous proposition, $1, i, \sqrt{2}, \frac{1-i}{2}\sqrt{2}$ is a $\mathbb{Z}$-basis of $\mathcal{O}_M$. One may then check that $\iota(\zeta_8) - \zeta_8 \notin (1 - i)^5\mathcal{O}_M$, where ι is the unique non-trivial automorphism of M/k. Thus the fourth ramification group of M/k is trivial. Hence $v_{\mathfrak{P}}(\mathfrak{d}_{k(\sqrt{d})/k}) = 4$ by Theorem B.2.31, and it follows that $v_2(d_{L/k}) = v_2(d_{M/k}^2) = 4$. We then get $d_{L/k} = 2^45^2 = 400$. $\square$

The following lemma shows that the existence of a code satisfying the shaping condition, built on a G-crossed product does not depend on the choice of the two generators of G.

LEMMA VIII.4.8. Let L/k be a biquadratic extension, and let σ, τ be two generators of the Galois group of L/k. Then we have

$$(a, b, u, L/k, \sigma, \tau) \simeq (a, abu^\tau, u, L/k, \sigma, \sigma\tau) \simeq (abu^\tau, b, u^\tau, L/k, \sigma\tau, \tau).$$

Proof.   If e, f are the generators of the first k-algebra, the isomorphisms with the second and the third one are obtained by taking e, ef and ef, f as new sets of generators. $\square$

In particular, if any of these three k-algebras is division, so are the other two. Moreover, if complex conjugation commutes with the elements of G, we have

$$|a|^2 = |b|^2 = |u|^2 = 1 \iff |a|^2 = |abu^\tau|^2 = |u|^2 = 1$$
$$\iff |abu^\tau|^2 = |b|^2 = |u^\tau|^2 = 1.$$

It follows that if one may build a code satisfying the shaping condition on a G-crossed product k-algebra for a particular choice of generators of G, one may also build a suitable code for another choice of generators.

From now on, if $L = k(\sqrt{d}, \sqrt{d'})$, we set

$$\sigma(\sqrt{d}) = \sqrt{d}, \sigma(\sqrt{d'}) = -\sqrt{d'}$$
$$\tau(\sqrt{d}) = -\sqrt{d}, \tau(\sqrt{d'}) = \sqrt{d'}$$

PROPOSITION VIII.4.9. Let $A = (a, b, u, L/k, \sigma, \tau)$. Assume that A is a division k-algebra. Then d or d' is divisible by an irreducible element π lying above a prime $p \equiv 1[4]$.

Proof.   Let M be any quadratic subfield of L. Since M is a quadratic k-subalgebra of A, A_M is not a division k-algebra (if A is not division, this is clear, and if A is division, it follows from Proposition V.3.2). In particular, A_M has index at most 2 and $2[A]_M = 0 \in \text{Br}(M)$. Hence $2[A]$ is split by any quadratic subfield of L.

It follows that any field extension k'/k in which at least one of the elements d, d' or dd' is a square splits $2[A]$. Indeed, in this case, k' contains at least one quadratic subfield M of L, and since M splits $2[A]$, so does k'.

Assume that d and d' are only divisible by prime elements $p \equiv 3[4]$ and eventually by 2, and let us prove that $2[A] = 0$ in this case, showing that A is not a division k-algebra.

Let $\pi \neq 1 - i$ be an irreducible element of $\mathcal{O}_k$. We are going to prove that $2[A]$ splits over k_π.

If π is lying above the prime number p and π divides d and d', then p divides d and d'. Thus replacing d by $\dfrac{dd'}{p^2}$ if necessary, we may assume that $\pi \nmid d$. Assume first that $\pi \nmid d$ and $\pi \nmid d'$. If d or d' is a square modulo $\pi \mathcal{O}_k$, since π does not lie above 2, applying Hensel's lemma shows that d or d' is a square in k_π. If d and d' are not squares modulo $\pi \mathcal{O}_k$, they both represent the unique non-trivial square class of the finite field $\mathcal{O}_k/\pi \mathcal{O}_k$, hence dd' is a (non-zero) square modulo $\pi \mathcal{O}_k$. Once again, we may use Hensel's lemma to conclude.

Assume now that $\pi \nmid d$ and $\pi \mid d'$. We are going to show that d is a square in k_π. Since $\pi \mid d'$, then by assumption $\pi = p$, where p is a prime number which is congruent to 3 modulo 4.

If $d \in \mathbb{Z}$ is a square modulo $p\mathbb{Z}$, then d is a square modulo $p\mathcal{O}_k$. If d is not a square modulo $p\mathbb{Z}$, then d represents the unique non-trivial square-class modulo $p\mathbb{Z}$, which is the class of -1, since $p \equiv 3[4]$. Hence $-d$ is a square modulo $p\mathbb{Z}$, hence a square modulo $p\mathcal{O}_k$. Then $d = i^2(-d)$ is a square modulo $p\mathcal{O}_k$. As before, Hensel's lemma implies that d is a square in the corresponding completion of k in both cases.

Therefore, $2[A]$ splits over k_π for all $\pi \neq 1 - i$. By the Brauer-Hasse-Noether's theorem, we get $2[A] = 0$.   □

LEMMA VIII.4.10. *Let* $F = k(\sqrt{\Delta})$, *where* Δ *is a square free positive integer. Let* $x \in \mathcal{O}_F$ *such that* $|x|^2 = 1$. *Then* x *is a root of 1.*

More precisely:

(1) *If* $\Delta \neq 2$ *or 3, then* x *is a 4th root of 1.*
(2) *If* $\Delta = 2$, x *is an 8th root of 1.*
(3) *If* $\Delta = 3$, x *is a 4th root of 1 or a 6th root of 1.*

Proof. Since F is stable by conjugation, we have $\bar{x} \in \mathcal{O}_F$. Hence $x \in \mathcal{O}_F^\times$. Since $F/\mathbb{Q}$ is totally imaginary, Dirichlet's unit theorem shows that $x = \zeta \varepsilon_F^r, r \in \mathbb{Z}$, where $\varepsilon_F \in \mathcal{O}_F^\times$ is a fundamental unit and $\zeta \in L$ is a root of 1. Since $|\varepsilon_F| > 1$ and $|x| = 1$, we get $r = 0$, so x is a root of 1. Write $x = e^{\frac{2ik\pi}{\ell}}, gcd(k, \ell) = 1$. Then $\mathbb{Q}(x) = \mathbb{Q}(\zeta_\ell) \subset F$. Since $[F : \mathbb{Q}] = 4$, it implies that $\varphi(\ell) \leq 4$, so we get $\ell = 1, 2, 3, 4, 5, 6$ or 8. If $\ell = 5$, we get $\mathbb{Q}(\zeta_5) = F$, which is impossible as the Galois group of $\mathbb{Q}(\zeta_5)/\mathbb{Q}$ is cyclic, while the Galois group of $F/\mathbb{Q}$ is the Klein group. This implies that x is a m-th root of 1, with $m = 6$ or 8 in any case.

If $\ell = 1, 2$ or 4, we get that x is in fact a 4th root of 1.

If $\ell = 3$ or 6, x is in both cases a 6th root of 1. Moreover, we get $\mathbb{Q}(j) = \mathbb{Q}(i\sqrt{3}) \subset F$, so $\mathbb{Q}(i\sqrt{3})$ is one of the three quadratic subfields of F. The only possibility is that $\mathbb{Q}(i\sqrt{3}) = \mathbb{Q}(i\sqrt{\Delta})$, and since Δ is positive and squarefree, we get $\Delta = 3$. Finally, if $\ell = 8$, we get $\mathbb{Q}(\zeta_8) = \mathbb{Q}(i, \sqrt{2}) = F$. Comparing quadratic subfields shows that $\Delta = 2$. This concludes the proof. $\qquad\square$

LEMMA VIII.4.11. *Assume that $A = (a, b, u, L/k, \sigma, \tau)$ is a division k-algebra. Then the elements a, b and abu^τ do not lie in k. Moreover, if one may build a code on A satisfying the shaping condition, then at most one of these elements lies in $\mathcal{O}_L$.*

Proof. If $a \in k$, the elements e and $\sqrt{d'}$ generate a k-subalgebra of A which is isomorphic to the quaternion k-algebra $A_1 = (a, d')$. Since A has degree 4, the centralizer A_2 of A_1 in A has degree 2. Now as A_2 is central simple, the centralizer theorem shows that $A \simeq A_1 \otimes_k A_2$. Since A_1 and A_2 have degree 2, we get $2[A] = 2[A_1] + 2[A_2] = 0 \in \mathrm{Br}(k)$. Hence A is not a division k-algebra. If $b \in k$ or $abu^\tau \in k$, similar arguments show that A is not a division k-algebra in these two cases (consider the elements f and $\sqrt{d}$ for the first case, and the elements ef and $\sqrt{dd'}$ for the second one).

Recall now that $a \in k(\sqrt{d})$, $b \in k(\sqrt{d'})$ and $abu^\tau \in k(\sqrt{dd'})$. Hence a, b and abu^τ all lie in a different quadratic subfield of L. Assume that one may build a code on the division k-algebra A satisfying the shaping condition, so a, b and u (hence abu^τ) have modulus 1.

Assume that at least two of the elements above lie in $\mathcal{O}_L$. Then they are units of the ring of integers of the quadratic subfield of L they belong to. If one of them is a 4th root of 1, it lies in k, and therefore A is not a division k-algebra by the previous point, which is a contradiction. Since they lie in a different quadratic subfield of F, Lemma VIII.4.10 implies that $L = k(\sqrt{2}, \sqrt{3})$. However, $d_{L/k} < 256$ in this case, contradicting the existence of a code built on A by Corollary C.2.11. This completes the proof. $\qquad\square$

PROPOSITION VIII.4.12. *Assume that there exists an energy-preserving code on the division k-algebra $A = (a, b, u, L/k, \sigma, \tau)$ with $d_{L/k}\Delta_\xi < 10000$. Then we have $256 \leq d_{L/k} < 2500$, and L is one of the three following extensions:*

$$k(\sqrt{2}, \sqrt{5}), k(\sqrt{5}, \sqrt{7}), k(\sqrt{3}, \sqrt{13}),$$

whose relative discriminants are respectively equal to $400, 1125$ and 1521.

Proof. By the previous lemma, at least two elements among a, b and abu^τ do not lie in $\mathcal{O}_L$. Assume first that $a \notin \mathcal{O}_L$. By examining the multiplication matrix given in Lemma VIII.4.1, we deduce that the ideals $\mathcal{E}_\xi^{(\sigma)}$ and $\mathcal{E}_\xi^{(\sigma\tau)}$ are proper ideals of $\mathcal{O}_k$. Hence $\Delta_\xi^{(\sigma)} \geq 2$ and $\Delta_\xi^{(\sigma\tau)} \geq 2$. If $a \in \mathcal{O}_L$, then $b \notin \mathcal{O}_L$ and $abu^\tau \notin \mathcal{O}_L$ and we get $\Delta_\xi^{(\tau)} \geq 2$ and $\Delta_\xi^{(\sigma\tau)} \geq 2$ in a similar way. In both cases, we then obtain $\Delta_\xi \geq 4$, and thus $d_{L/k} < 2500$. The lower bound follows from Corollary C.2.11.

Let us prove the second part of the proposition. Replacing d' by $\dfrac{dd'}{4}$ if necessary, one may assume that d' is odd. Assume first that d is even. Then $2^4 \mid d_{L/k}$ by Proposition VIII.4.6. Since A is a division k-algebra, d or d' is divisible by a prime

$p \equiv 1[4]$, and thus $p^2 \mid d_{L/k}$ by the same proposition. If d or d' were divisible by an odd prime number $\ell \neq p$, we would have in the same way $\ell^2 \mid d_{L/k}$ and thus

$$d_{L/k} \geq 2^4 p^2 \ell^2 \geq 2^4 5^2 3^2 > 2500,$$

hence a contradiction. Thus p is the only odd prime divisor of d and d'. It follows easily that $L = k(\sqrt{2}, \sqrt{p})$. The upper bound on $d_{L/k}$ immediately implies that $p = 5$. Hence $L = k(\sqrt{2}, \sqrt{5})$ and $d_{L/k} = 400$ by Example VIII.4.7.

Assume now that d is odd. Let $p \equiv 1[4]$ be a prime number dividing d or d'. We may assume without loss of generality that $p \mid d$ and $p \nmid d'$. Since d' is an odd positive integer, it has another odd prime divisor ℓ. Assume that d or d' is divisible by a prime number $q \neq p, \ell$. Since $q \neq \ell$, one of them is necessarily ≥ 5. Since $p \geq 5$, we get

$$d_{L/k} \geq p^2 q^2 \ell^2 \geq 5^4 3^2 > 2500,$$

which is a contradiction. Thus d and d' are only divisible by p and ℓ, so $L = k(\sqrt{p}, \sqrt{\ell})$ and $d_{L/k} = p^2 \ell^2$. Since $\ell \geq 3$, the upper bound on $d_{L/k}$ shows that $p = 5$ or 13. If $p = 5$, we get that $\ell = 3$ or 7. The first possibility has to be discarded since $3^2 5^2 < 256$. Hence $L = k(\sqrt{5}, \sqrt{7})$ and $d_{L/k} = 1125$. If $p = 13$, then necessarily $\ell = 3$. In this case, $L = k(\sqrt{3}, \sqrt{13})$ and $d_{L/k} = 1521$. $\square$

LEMMA VIII.4.13. *Let F/k be a quadratic extension such that complex conjugation is a $\mathbb{Q}$-automorphism of F which commutes with $\mathrm{Gal}(F/k)$. Assume that there is only one prime ideal of $\mathcal{O}_F$ lying above 2. Let $x \in F^\times, x \notin \mathcal{O}_F$ satisfying $|x|^2 = 1$ and let $\delta \in \mathcal{O}_k$ such that $\delta x \in \mathcal{O}_F$. Then $|\delta|^2 \geq 5$.*

Proof. Write $F_0 = F \cap \mathbb{R} = \mathbb{Q}(\sqrt{\Delta}), \Delta > 0$. Since $x \notin \mathcal{O}_F$, δ is not a unit of $\mathcal{O}_k$, and thus $|\delta|^2 \neq 1$. Moreover, the equation $|\delta|^2 = 3$ has no solution in $\mathcal{O}_k$, so we need to prove that $|\delta|^2 \neq 2, 4$.

Assume to the contrary that $|\delta|^2 = 2$ or 4, and set $y = \delta x \in \mathcal{O}_F$. By assumption, we have $|y|^2 = |\delta|^2$. This may be rewritten as $N_{F/\mathbb{Q}(\sqrt{\Delta})}(y) = N_{F/\mathbb{Q}(\sqrt{\Delta})}(\delta) = 2$ or 4. In particular, $N_{F/\mathbb{Q}}(y)$ and $N_{F/\mathbb{Q}}(\delta)$ are equal to the same power of 2. Hence the prime ideals of $\mathcal{O}_F$ dividing $\delta \mathcal{O}_F$ and $y \mathcal{O}_F$ all lie above 2. The assumption then implies that $\delta \mathcal{O}_F$ and $y \mathcal{O}_F$ are powers of the same prime ideal, and since they have same absolute norms, we get that $y \mathcal{O}_F = \delta \mathcal{O}_F$. It follows that there exists $v \in \mathcal{O}_F^\times$ such that $y = \delta v$, that is $\delta x = \delta v$. Thus $x = v \in \mathcal{O}_F$, which is a contradiction. $\square$

We are finally ready to prove Theorem VIII.4.4. Assume that there exists a code on the division k-algebra $(a, b, u, L/k, \sigma, \tau)$ with $d_{L/k} \Delta_\xi < 10000$. By Proposition VIII.4.12, we have, up to a change of generators

$$L = k(\sqrt{5}, \sqrt{10}), k(\sqrt{5}, \sqrt{7}), \text{ or } k(\sqrt{13}, \sqrt{39}).$$

In each case, $L = k(\sqrt{p}, \sqrt{\Delta})$, where p is a prime number satisfying $p \equiv 5[8]$, and $\Delta \geq 7$. Notice for later use that there is only one prime ideal lying above 2 in $\mathcal{O}_{k(\sqrt{p})}$. Indeed, 2 is inert in $\mathbb{Q}(\sqrt{p})/\mathbb{Q}$ by assumption on p, and then totally ramifies in $k(\sqrt{p})/\mathbb{Q}(\sqrt{p})$.

If $\rho \in G$ has order 2, we will denote by z_ρ the element among a, b and abu^τ which belongs to $L^{\langle \rho \rangle}$.

Let $\rho, \rho' \in G$ such that $L^{\langle \rho \rangle} = k(\sqrt{p})$ and $L^{\langle \rho' \rangle} = k(\sqrt{\Delta})$. Assume z_ρ is a unit. Then $z_\rho \in k$ by Lemma VIII.4.10, and thus A is not a division k-algebra by Lemma VIII.4.11. Hence z_ρ is not a unit and by Lemma VIII.4.13, we get that $\Delta_\xi^{(\rho)} \geq 5$.

If $\Delta = 10$, then $1 - i$ totally ramifies in $L^{\langle \rho' \rangle}/k$, so 2 totally ramifies in $L^{\langle \rho' \rangle}/\mathbb{Q}$, and the same reasoning shows that $\Delta_\xi^{(\rho')} \geq 5$. If $\Delta = 7$ or 39, one may show as above that $z_{\rho'}$ is not a unit, and then $\Delta_\xi^{(\rho')} \geq 2$. In all cases, we then get that

$$d_{L/k}\Delta_\xi \geq d_{L/k}\Delta_\xi^{(\rho)}\Delta_\xi^{(\rho')} \geq 10000,$$

hence a contradiction. This concludes the proof.

REMARK VIII.4.14. Similar arguments show that the bound 10000 is also optimal if $k = \mathbb{Q}(j)$. $\qquad\Box$

EXERCISES

1. Prove that the map $\xi^{a,b,u}$ defined in Section 2 is indeed a 2-cocycle.

2. Let k be a field of characteristic 2. Let $a \in k^\times, b \in k$. Recall that $(a, b]_2$ is the k-algebra generated by two elements e, f subject to the relations

$$e^2 = a, f^2 + f = b, ef = fe + e.$$

This is a central simple k-algebra of degree 2.

Prove that every central simple k-algebra of degree 2 is isomorphic to some k-algebra $(a, b]_2$, and describe all the central simple algebras of degree 4 which are not division algebras.

3. Let k be a field of characteristic 2. Prove that every central division k-algebra of degree 4 contains a separable biquadratic extension of k (that is, the compositum of two linearly disjoint separable quadratic extensions of k).

4. Let $k = \mathbb{Q}(i)$, and let $L = k(\sqrt{d}, \sqrt{d'})$ be a biquadratic extension. Assume that $d = m^2 + n^2$ and $d' = u^2 + v^2$, where $u, v, m, n \in \mathbb{Z}$.

   (a) Show that we may define the k-algebra $A = (a, b, u, L/k, \sigma, \tau)$, where

   $$a = \frac{m + in}{\sqrt{d}}, \quad b = \frac{u + iv}{\sqrt{d'}} \quad \text{and } u = i,$$

   and that A is a division k-algebra if and only if $(-d, d')_k$ is a division k-algebra.

   Assume now that $d = 2$ and $d' = p$ is a prime number satisfying $p \equiv 1[4]$, so we may write $p = u^2 + v^2, u, v, \in \mathbb{Z}$, where u is odd and v is even.

   (b) Check that the k-algebra $A = (a, b, u, L/k, \sigma, \tau)$ is a division k-algebra in this case.

   (c) Use Proposition B.2.28 to show that $\mathcal{O}_L = \mathbb{Z}[\zeta_8, \theta]$, where $\theta = \dfrac{1 + \sqrt{p}}{2}$.

   Set $e_1 = \dfrac{u + iv + \sqrt{p}}{2}$ and $e_2 = \dfrac{u + iv - \sqrt{p}}{2}$.

(d) Using the results of Chapter VII Exercise 4, show that the $\mathbb{Z}[i]$-module J spanned by $e_1, e_2, \zeta_8 e_1$ and $\zeta_8 e_2$ is an ideal of $\mathcal{O}_L$, and that the complex ideal lattice $(J, h_{\frac{1}{2p}})$ is isomorphic to the cubic lattice.

(e) Write the elements of the corresponding code.

CHAPTER IX

Central simple algebras with unitary involutions

In this chapter, k is a field, and A is a central simple k-algebra. To simplify the exposition, we will assume that $\operatorname{char}(k) \neq 2$.

IX.1. Basic concepts.

DEFINITION IX.1.1. An **involution** on A is a ring anti-automorphism of A of order at most 2.

In other words, an involution is a map $\sigma : A \longrightarrow A$ satisfying for all $x, y \in A$:

(1) $\sigma(x + y) = \sigma(x) + \sigma(y)$;

(2) $\sigma(1) = 1$;

(3) $\sigma(xy) = \sigma(y)\sigma(x)$;

(4) $\sigma(\sigma(x)) = x$.

For example, the transposition is an involution on $\mathrm{M}_n(k)$. Notice that Id_A is never an involution unless A is commutative, which implies that $A = k$. Therefore, if $A \neq k$, an involution on A has order 2.

It is easy to check that for every $\lambda \in k$, we have $\sigma(\lambda) \in k$. Hence $\sigma_{|_k}$ is an automorphism of order at most 2 of k.

We set
$$k_0 = \{\lambda \in k \,|\, \sigma(\lambda) = \lambda\}.$$

We say that σ is an **involution of the first kind** if $\sigma_{|_k} = \operatorname{Id}_k$, that is if $k = k_0$, and an **involution of the second kind (or unitary)** otherwise. In the latter case, k/k_0 is a quadratic field extension, and $\sigma_{|_k}$ is the unique non-trivial k_0-automorphism of k/k_0. Conversely, if k/k_0 is a quadratic field extension, we will say that a unitary involution σ on a central simple k-algebra A is a k/k_0-**involution** if $\sigma_{|_k}$ is the unique non-trivial k_0-automorphism of k.

An element $x \in A$ is called **symmetric** if $\sigma(x) = x$, and **skew-symmetric** if $\sigma(x) = -x$. We denote by $\operatorname{Sym}(A, \sigma)$ the set of symmetric elements of A, and by $\operatorname{Skew}(A, \sigma)$ the set of skew-symmetric elements of A. Both have a natural structure of a k_0-vector space. We also set
$$\operatorname{Sym}(A, \sigma)^{\times} = \operatorname{Sym}(A, \sigma) \cap A^{\times} \text{ and } \operatorname{Skew}(A, \sigma)^{\times} = \operatorname{Skew}(A, \sigma) \cap A^{\times}.$$

We say that two central simple k-algebras with involutions (A, σ) and (A', σ') are **isomorphic** if there exists an isomorphism of k-algebras $f : A \xrightarrow{\sim} A'$ such that
$$\sigma' \circ f = f \circ \sigma.$$

In this case, one may verify that σ and σ' are involutions of the same kind. Moreover, f then induces isomorphisms of k_0-vector spaces

$$\mathrm{Sym}(A, \sigma) \simeq \mathrm{Sym}(A', \sigma') \text{ and } \mathrm{Skew}(A, \sigma) \simeq \mathrm{Skew}(A', \sigma').$$

PROPOSITION IX.1.2. *Let σ, σ' be two involutions on A having the same restriction to k.*

(1) *If σ, σ' are involutions of the first kind, there exists $u \in A^\times$ such that $\sigma' = \mathrm{Int}(u) \circ \sigma$ and $\sigma(u) = \pm u$;*

(2) *if σ, σ' are two k/k_0-involutions of the second kind, there exists $u \in A^\times$ such that $\sigma' = \mathrm{Int}(u) \circ \sigma$ and $\sigma(u) = u$.*

In both cases, u is uniquely determined up to multiplication by an element of $k_0^\times$.

Proof.   Since σ and σ' have the same restriction to k, $\sigma' \circ \sigma^{-1}$ is a k-algebra automorphism of A. By Skolem-Noether's Theorem, there exists $v \in A^\times$ such that $\sigma' \circ \sigma^{-1} = \mathrm{Int}(v)$, that is

$$\sigma' = \mathrm{Int}(v) \circ \sigma.$$

Easy computations show that we have $\sigma \circ \mathrm{Int}(v) = \mathrm{Int}(\sigma^{-1}(v)) \circ \sigma$. Thus, we get

$$\sigma'^2 = \mathrm{Int}(v) \circ \mathrm{Int}(\sigma^{-1}(v)) \circ \sigma^2 = \mathrm{Int}(v\sigma^{-1}(v)) \circ \sigma^2.$$

Since σ and σ' have order 2, we get $\mathrm{Id}_A = \mathrm{Int}(v\sigma^{-1}(v))$, so $v\sigma^{-1}(v)$ lies in the center of A. Hence there exists $\lambda \in k^\times$ such that

$$v = \lambda\sigma(v).$$

We then have

$$v = \lambda\sigma(\lambda\sigma(v)) = \lambda v\sigma(\lambda) = \lambda\sigma(\lambda)v,$$

since $\sigma(\lambda)$ lies in the center of A.

Since $v \in A^\times$, we get $\lambda\sigma(\lambda) = 1$. If σ is an involution of the first kind, we get $\lambda^2 = 1$, that is $\lambda = \pm 1$. It follows that $\sigma(v) = \pm v$. We then set $u = v$.

Assume now that σ is an involution of the second kind. In this case, the equality above may be rewritten as $N_{k/k_0}(\lambda) = 1$, since $\sigma_{|k}$ is the non-trivial k_0-automorphism of k/k_0. By Hilbert 90, there exists $z \in k$ such that $\lambda = \dfrac{\sigma_{|k}(z)}{z} = \dfrac{\sigma(z)}{z}$. Set $u = vz$. We then have

$$\sigma(u) = \sigma(vz) = \sigma(z)\sigma(v) = z\lambda\sigma(v) = zv = vz = u.$$

Moreover, we have $\mathrm{Int}(u) = \mathrm{Int}(vz) = \mathrm{Int}(v)$, since $z \in k$. Therefore, $\sigma' = \mathrm{Int}(u) \circ \sigma$ with $u \in A^\times$ satisfying $\sigma(u) = u$.

Let us prove the last part of the proposition. Assume that we have

$$\sigma' = \mathrm{Int}(u_1) \circ \sigma = \mathrm{Int}(u_2) \circ \sigma,$$

where u_1, u_2 are either symmetric or skew-symmetric if σ and σ' are involutions of the first kind, and where u_1, u_2 are both symmetric if σ and σ' are involutions of the second kind. Then we have

$$\mathrm{Int}(u_2 u_1^{-1}) \circ \sigma = \sigma.$$

It implies that $\text{Int}(u_2 u_1^{-1}) = \text{Id}_A$. Thus, as before, there exists $\lambda \in k^\times$ such that $u_2 = \lambda u_1$. If σ and σ' are involutions of the first kind, we have $k = k_0$ and we are done.

If σ and σ' are involutions of the second kind, since u_1, u_2 are both symmetric in this case, applying σ to the previous equality yields

$$u_2 = u_1 \sigma(\lambda) = \sigma(\lambda) u_1.$$

We deduce that $\sigma(\lambda) = \lambda$, that is $\lambda \in k_0^\times$. This concludes the proof. $\square$

EXAMPLE IX.1.3. Let k/k_0 be a quadratic field extension, and let

$$\overline{} : \begin{array}{c} k \longrightarrow k \\ \lambda \longmapsto \overline{\lambda} \end{array}$$

be its non-trivial k_0-automorphism. If $n \geq 1$, the map

$$M_n(k) \longrightarrow M_n(k)$$
$$M = (a_{ij}) \longmapsto M^* = (\overline{a}_{ji})$$

is a unitary involution on $M_n(k)$. The previous proposition then shows that every k/k_0-involution on $M_n(k)$ has the form

$$\sigma_H = \text{Int}(H) \circ {}^*,$$

where $H \in \text{GL}_n(k)$ satisfies $H^* = H$. $\square$

Given a central simple k-algebra A, it is natural to ask whether A carries a unitary involution or not. The goal of the next section is to give a full answer to this question.

IX.2. The corestriction algebra.

We now fix once for all a quadratic field extension k/k_0. If $\lambda \in k$, we will denote by $\overline{\lambda}$ its image under the unique non-trivial k_0-automorphism of k. We will assume in the sequel that all the unitary involutions considered are k/k_0-involutions.

To investigate the existence of unitary involutions, we need first to introduce the corestriction of a central simple k-algebra A. This algebra, denoted by $\text{Cor}_{k/k_0}(A)$, is a central simple k_0-algebra which measures the obstruction to the existence of a k/k_0-involution on A. This construction may be viewed as a substitute for the norm map for Brauer classes. Indeed, in the same way the norm map N_{k/k_0} induces a group morphism

$$N_{k/k_0} : k^\times \longrightarrow k_0^\times,$$

we will see that the corestriction map induces a group morphism

$$\text{Cor}_{k/k_0} : \text{Br}(k) \longrightarrow \text{Br}(k_0).$$

We start with preliminary results on semilinear maps.

DEFINITION IX.2.1. Let V, W be two k-vector spaces. We say that a map $f : V \longrightarrow W$ is **semilinear** if for all $v_1, v_2, v \in V$ and all $\lambda \in k$, we have

$$f(v_1 + v_2) = f(v_1) + f(v_2), f(\lambda v) = \overline{\lambda} f(v).$$

EXAMPLE IX.2.2. The map

$$k^n \longrightarrow k^n$$

$$\begin{pmatrix} \lambda_1 \\ \vdots \\ \lambda_n \end{pmatrix} \longmapsto \begin{pmatrix} \overline{\lambda}_1 \\ \vdots \\ \overline{\lambda}_n \end{pmatrix}$$

is semilinear. $\qquad\qquad\qquad\qquad\qquad\qquad\qquad\qquad\qquad\qquad\qquad$ □

Let V be a k-vector space. For any $v \in V$, we associate a new symbol $\overline{v}$, and we define a set $\overline{V}$ by

$$\overline{V} = \{\overline{v} \mid v \in V\}.$$

We then have an obvious bijection of sets

$$\iota_V : \begin{array}{c} V \xrightarrow{\sim} \overline{V} \\ v \longmapsto \overline{v}. \end{array}$$

REMARK IX.2.3. From a set-theoretical point of view, $\overline{V}$ is just a copy of V, but we would like to distinguish the elements of $\overline{V}$ from the elements of V for the same reasons we have introduced a copy of a given algebra to construct the associated opposite algebra. $\qquad\qquad\qquad\qquad\qquad\qquad\qquad\qquad\qquad\qquad$ □

Our next goal is to define a structure of a k-vector space on $\overline{V}$ such that the map above is semilinear. It is easy to check that the operations

$$\begin{array}{ccc} \overline{V} \times \overline{V} \longrightarrow \overline{V} & & k \times \overline{V} \longrightarrow \overline{V} \\ (\overline{v}, \overline{w}) \longmapsto \overline{v + w} & \text{and} & (\lambda, \overline{v}) \longmapsto \overline{\overline{\lambda}v} \end{array}$$

satisfy the required conditions. Roughly speaking, $\overline{V}$ is the vector space obtained from V by twisting the external law on V by the non-trivial k_0-automorphism of k.

REMARK IX.2.4. Notice that by definition of the k-vector space structure on $\overline{V}$, we have

$$\overline{\lambda v} = \overline{\lambda}\,\overline{v} \text{ and } \overline{v_1 + v_2} = \overline{v_1} + \overline{v_2} \text{ for all } \lambda \in k, v, v_1, v_2 \in V,$$

and that the map

$$\iota_V : \begin{array}{c} V \longrightarrow \overline{V} \\ v \longmapsto \overline{v} \end{array}$$

is bijective and semilinear. $\qquad\qquad\qquad\qquad\qquad\qquad\qquad\qquad\qquad\qquad\qquad\qquad$ □

We continue with some easy lemmas.

LEMMA IX.2.5. *Let V be a k-vector space, and let $s : V \longrightarrow V$ be a semilinear map satisfying $s^2 = \mathrm{Id}_V$. Then the set*

$$V^{\langle s \rangle} = \{v \in V \mid s(v) = v\}$$

is a k_0-vector space. Moreover, if V is finite-dimensional, so is $V^{\langle s \rangle}$ and we have

$$\dim_{k_0}(V^{\langle s \rangle}) = \dim_k(V).$$

Proof. It follows from the definition that s is k_0-linear, hence the first part. Write $k = k_0(\sqrt{d})$, and assume that V is finite-dimensional over k, hence over k_0. By the Rank Theorem, we have

$$2\dim_k(V) = \dim_{k_0}(V) = \dim_{k_0}(V^{\langle s \rangle}) + \dim_{k_0}(\operatorname{Im}(\operatorname{Id}_V - s)).$$

Let us prove now that

$$V^{\langle s \rangle} = \sqrt{d}\operatorname{Im}(\operatorname{Id}_V - s).$$

For, we have

$$s(\sqrt{d}(v - s(v))) = -\sqrt{d}(s(v) - s^2(v)) = \sqrt{d}(v - s(v)),$$

hence the inclusion $' \supset '$. Conversely, if $s(v) = v$, we may write

$$v = \sqrt{d}\left(\frac{v}{2\sqrt{d}} - s\left(\frac{v}{2\sqrt{d}}\right)\right),$$

and we get the other inclusion. Thus we have

$$\dim_{k_0}(V^{\langle s \rangle}) = \dim_{k_0}(\operatorname{Im}(\operatorname{Id}_V - s)).$$

The desired equality follows. $\square$

LEMMA IX.2.6. *Let V, W be two k-vector spaces. For every semilinear map $f : V \longrightarrow W$, the map*

$$\overline{f}: \begin{array}{c} \overline{V} \longrightarrow W \\ \overline{v} \longmapsto f(v) \end{array}$$

is a k-linear map.

Proof. Since V and $\overline{V}$ have the same additive group structure, we just have to check that $\overline{s}$ preserves scalar multiplication. For all $v \in V, \lambda \in k$, we have

$$\overline{f}(\lambda\overline{v}) = \overline{f}(\overline{\overline{\lambda}v}) = f(\overline{\lambda}v) = \overline{\overline{\lambda}}f(v) = \lambda f(v),$$

and we are done. $\square$

We now prove a twisted version of the universal property of the tensor product of two vector spaces.

LEMMA IX.2.7. *Let V_1, V_2, W be k-vector spaces, and let*

$$\varphi : V_1 \times V_2 \longrightarrow W$$

be a biadditive map satisfying

$$\varphi(\lambda_1 v_1, \lambda_2 v_2) = \overline{\lambda}_1\overline{\lambda}_2\varphi(v_1, v_2),$$

for all $v_1 \in V_1, v_2 \in V_2, \lambda_1, \lambda_2 \in k$. Then there exists a unique semilinear map $\rho : V_1 \otimes_k V_2 \longrightarrow W$ satisfying

$$\rho(v_1 \otimes v_2) = \varphi(v_1, v_2) \text{ for all } v_1 \in V_1, v_2 \in V_2.$$

Proof. Notice that the map

$$b: \begin{array}{c} V_1 \times V_2 \longrightarrow \overline{W} \\ (v_1, v_2) \longmapsto \overline{\varphi(v_1, v_2)} \end{array}$$

is k-bilinear. Indeed, it is clearly biadditive. Moreover, we have

$$
\begin{aligned}
b(\lambda_1 v_1, \lambda_2 v_2) &= \overline{\varphi(\lambda_1 v_1, \lambda_2 v_2)} \\
&= \overline{\overline{\lambda_1}\lambda_2 \varphi(v_1, v_2)} \\
&= \overline{\overline{\lambda_1}\lambda_2}\,\overline{\varphi(v_1, v_2)} \\
&= \lambda_1 \overline{\lambda_2}\,\overline{\varphi(v_1, v_2)} \\
&= \lambda_1 \lambda_2 b(v_1, v_2),
\end{aligned}
$$

for all $v_1 \in V_1, v_2 \in V_2, \lambda_1, \lambda_2 \in k$. Hence, there exists a unique k-linear map $f : V_1 \otimes_k V_2 \longrightarrow \overline{W}$ such that

$$
f(v_1 \otimes v_2) = b(v_1, v_2) = \overline{\varphi(v_1, v_2)}
$$

for all $v_1, v_2 \in V$. Straightforward computations show that the map $\rho = \iota_W^{-1} \circ f$ satisfies the required conditions (where ι_W is defined is Remark IX.2.4). The uniqueness of ρ follows from the fact $V_1 \otimes_k V_2$ is generated by elementary tensors as an additive group. $\qquad\square$

EXAMPLE IX.2.8. Let V be a k-vector space. Then the map

$$
\varphi_V : \begin{array}{c} \overline{V} \times V \longrightarrow \overline{V} \otimes_k V \\ (\overline{v}_1, v_2) \longmapsto \overline{v}_2 \otimes v_1 \end{array}
$$

satisfies the assumptions of Lemma IX.2.7. Indeed, it is clearly biadditive. Moreover, for all $\lambda_1, \lambda_2 \in k$, we have

$$
\begin{aligned}
\varphi_V(\lambda_1 \overline{v}_1, \lambda_2 v_2) &= \varphi_V(\overline{\overline{\lambda_1} v_1}, \lambda_2 v_2) \\
&= (\overline{\lambda_2 v_2}) \otimes (\overline{\lambda_1} v_1) \\
&= (\overline{\lambda_2}\,\overline{v}_2) \otimes (\overline{\lambda_1} v_1) \\
&= \overline{\lambda_1}\lambda_2 (\overline{v}_2 \otimes v_1) \\
&= \overline{\lambda_1}\lambda_2 \varphi_V(\overline{v}_1, v_2).
\end{aligned}
$$

Thus, there exists a unique semilinear map $s_V : \overline{V} \otimes_k V \longrightarrow \overline{V} \otimes_k V$ satisfying

$$
s_V(\overline{v}_1 \otimes v_2) = \overline{v}_2 \otimes v_1 \text{ for all } v_1, v_2 \in V.
$$

$\qquad\square$

Let A be a k-algebra. Then the operations

$$
\begin{array}{cc} \overline{A} \times \overline{A} \longrightarrow \overline{A} & \overline{A} \times \overline{A} \longrightarrow \overline{A} \\ (\overline{a}, \overline{b}) \longmapsto \overline{a + b} & , \quad (\overline{a}, \overline{b}) \longmapsto \overline{ab} \end{array}
$$

and

$$
k \times \overline{A} \longrightarrow \overline{A}
$$

$$
(\lambda, \overline{a}) \longmapsto \overline{\overline{\lambda} a}
$$

endow $\overline{A}$ with the structure of a k-algebra on $\overline{A}$ such that the semilinear bijective map $\iota_A : A \longrightarrow \overline{A}$ is a ring morphism. In particular, ι_A induces a bijection between the set of ideals of A and the set of ideals of $\overline{A}$. Moreover, ι_A induces a ring isomorphism $Z(A) \simeq Z(\overline{A})$. It follows that $\overline{A}$ is a central simple k-algebra whenever A is.

Finally, straightforward computations show that the semilinear map $s_A : \overline{A} \otimes_k A \longrightarrow \overline{A} \otimes_k A$ is a semilinear ring morphism.

LEMMA IX.2.9. *Let V, W be two k-vector spaces.*

(1) *For every k-linear map $f : V \longrightarrow W$, the map*

$$\theta(f) : \begin{array}{c} \overline{V} \longrightarrow \overline{W} \\ \overline{v} \longmapsto \overline{f(v)} \end{array}$$

is k-linear. Moreover, if f is an isomorphism of k-vector spaces, so is $\theta(f)$;

(2) *the map*

$$u : \begin{array}{c} \overline{\mathrm{End}_k(V)} \longrightarrow \mathrm{End}_k(\overline{V}) \\ \overline{f} \longmapsto \theta(f) \end{array}$$

is an isomorphism of k-algebras.

Proof. Since f is k-linear, the map

$$\begin{array}{c} V \longrightarrow \overline{W} \\ v \longmapsto \overline{f(v)} \end{array}$$

is semilinear. By Lemma IX.2.6 it induces a k-linear map

$$\begin{array}{c} \overline{V} \longrightarrow \overline{W} \\ \overline{v} \longmapsto \overline{f(v)}, \end{array}$$

which is nothing but $\theta(f)$. Clearly, if f is bijective, so is $\theta(f)$; this proves (1).

Now it is easy to check using the definitions that for all $f, g \in \overline{\mathrm{End}_k(V)}$ and $\lambda \in k$, we have

$$\theta(f + g) = \theta(f) + \theta(g), \theta(f \circ g) = \theta(f) \circ \theta(g) \text{ and } \theta(\lambda f) = \overline{\lambda}\theta(f).$$

In particular, the map $\theta : \mathrm{End}_k(V) \longrightarrow \mathrm{End}_k(\overline{V})$ is a k_0-algebra morphism which is semilinear, and therefore, the induced map

$$u : \begin{array}{c} \overline{\mathrm{End}_k(V)} \longrightarrow \mathrm{End}_k(\overline{V}) \\ \overline{f} \longmapsto \theta(f) \end{array}$$

is a k-algebra morphism between central simple k-algebras of same degrees, hence an isomorphism. $\qquad\square$

DEFINITION IX.2.10. The k_0-vector space $\mathrm{Cor}_{k/k_0}(V)$ defined by

$$\mathrm{Cor}_{k/k_0}(V) = \{z \in \overline{V} \otimes_k V \mid s_V(z) = z\}$$

is called the **corestriction** of V.

Similarly, the k_0-algebra $\mathrm{Cor}_{k/k_0}(A)$ defined by

$$\mathrm{Cor}_{k/k_0}(A) = \{z \in \overline{A} \otimes_k A \mid s_A(z) = z\}$$

is called the **corestriction** of A.

REMARK IX.2.11. Let V be a finite dimensional k-vector space. It follows from Lemma IX.2.5 that we have

$$\dim_{k_0}(\mathrm{Cor}_{k/k_0}(V)) = \dim_k(V)^2.$$

Similarly, if A is a central simple k-algebra, we have

$$\dim_{k_0}(\mathrm{Cor}_{k/k_0}(A)) = \dim_k(A)^2.$$

$\square$

We now establish some useful properties of the corestriction.

PROPOSITION IX.2.12. *Let A, B be two central simple k-algebras, and let V be a finite dimensional k-vector space. Then the following properties hold:*

(1) $\mathrm{Cor}_{k/k_0}(A) \otimes_{k_0} k \cong_k \overline{A} \otimes_k A$;
(2) *if $A \cong_k B$, then $\mathrm{Cor}_{k/k_0}(A) \cong_{k_0} \mathrm{Cor}_{k/k_0}(B)$;*
(3) $\mathrm{Cor}_{k/k_0}(A \otimes_k B) \cong_{k_0} \mathrm{Cor}_{k/k_0}(A) \otimes_{k_0} \mathrm{Cor}_{k/k_0}(B)$;
(4) $\mathrm{Cor}_{k/k_0}(\mathrm{End}_k(V)) \cong_{k_0} \mathrm{End}_{k_0}(\mathrm{Cor}_{k/k_0}(V))$.

In particular, the corestriction algebra $\mathrm{Cor}_{k/k_0}(A)$ is a central simple k_0-algebra of degree $\deg(A)^2$, and whose Brauer class only depends on the Brauer class of A.

Proof. The inclusions $\mathrm{Cor}_{k/k_0}(A) \subset \overline{A} \otimes_k A$ and $k \subset \overline{A} \otimes_k A$ are two k_0-algebra morphisms with commuting images, and therefore induce a k_0-algebra morphism

$$\phi : \mathrm{Cor}_{k/k_0}(A) \otimes_{k_0} k \longrightarrow \overline{A} \otimes_k A$$

which satisfies

$$\phi(z \otimes \lambda) = z\lambda \text{ for all } \lambda \in k, z \in \mathrm{Cor}_{k/k_0}(A).$$

Straightforward computations show that ϕ is in fact k-linear. Moreover, ϕ is bijective. Indeed, write $k = k_0(\sqrt{d})$. For $a \in \overline{A} \otimes_k A$, we have

$$\frac{a + s_A(a)}{2}, \frac{a - s_A(a)}{2\sqrt{d}} \in \mathrm{Cor}_{k/k_0}(A).$$

One may verify that the map

$$\phi' : \begin{array}{c} \overline{A} \otimes_k A \longrightarrow \mathrm{Cor}_{k/k_0}(A) \otimes_{k_0} k \\[2mm] a \longmapsto \left(\dfrac{a + s_A(a)}{2}\right) \otimes 1 + \left(\dfrac{a - s_A(a)}{2\sqrt{d}}\right) \otimes \sqrt{d} \end{array}$$

is a k-algebra morphism, and that ϕ and ϕ' are mutually inverse. This proves (1). Notice also that, since A and $\overline{A}$ are central simple k-algebras, so is $\overline{A} \otimes_k A$ by Corollary III.1.8 (1). It follows that $\mathrm{Cor}_{k/k_0}(A) \otimes_{k_0} k$ is a central simple k_0-algebra, and thus $\mathrm{Cor}_{k/k_0}(A)$ is a central simple k_0-algebra by Corollary III.1.8 (2). Moreover, since the degree is preserved under scalar extension, we have

$$\deg(\mathrm{Cor}_{k/k_0}(A)) = \deg(\overline{A} \otimes_k A) = \deg(A)^2.$$

This proves the first half of the last statement.

We now prove (2). Let $f : A \xrightarrow{\sim} B$ be a k-algebra isomorphism. Then the map

$$\theta(f) : \begin{array}{c} \overline{A} \longrightarrow \overline{B} \\[2mm] \overline{a} \longmapsto \overline{f(a)} \end{array}$$

is also a k-algebra isomorphism. Indeed, we already know by Lemma IX.2.9 that $\theta(f)$ is an isomorphism of k-vector spaces, and one may see that $\theta(f)$ is in fact a k-algebra morphism. Therefore, we have a k-algebra isomorphism

$$\rho = \theta(f) \otimes f : \overline{A} \otimes_k A \xrightarrow{\sim} \overline{B} \otimes_k B.$$

Now we have

$$\rho \circ s_A = s_B \circ \rho,$$

as we may check easily on elementary tensors. It readily follows that the isomorphism above restricts to an isomorphism of k_0-algebras

$$\mathrm{Cor}_{k/k_0}(A) \cong_{k_0} \mathrm{Cor}_{k/k_0}(B).$$

We now prove (3). Applying Lemma IX.2.7 to the map

$$A \times B \longrightarrow \overline{A} \otimes_k \overline{B}$$

$$(a, b) \longmapsto \overline{a} \otimes \overline{b}$$

shows the existence of a unique semilinear map

$$\varphi : A \otimes_k B \longrightarrow \overline{A} \otimes_k \overline{B}$$

satisfying

$$\varphi(a \otimes b) = \overline{a} \otimes \overline{b} \text{ for all } a \in A, b \in B.$$

By Lemma IX.2.6, it therefore induces a k-linear map

$$\psi : \overline{A \otimes_k B} \longrightarrow \overline{A} \otimes_k \overline{B}$$

satisfying

$$\psi(\overline{a \otimes b}) = \overline{a} \otimes \overline{b} \text{ for all } a \in A, b \in B.$$

One may check that ψ is a k-algebra morphism. Since $A \otimes_k B$ is central simple, so is $\overline{A \otimes_k B}$. Moreover, $\overline{A \otimes_k B}$ and $\overline{A} \otimes_k \overline{B}$ have the same degree, so ψ is in fact an isomorphism of k-algebras. By associativity and commutativity of the tensor product, we have an isomorphism of k-algebras

$$(\overline{A} \otimes_k \overline{B}) \otimes_k (A \otimes_k B) \cong_k (\overline{A} \otimes_k A) \otimes_k (\overline{B} \otimes_k B),$$

and therefore we get an isomorphism of k-algebras

$$f : (\overline{A} \otimes_k \overline{B}) \otimes_k (A \otimes_k B) \xrightarrow{\sim} (\overline{A} \otimes_k A) \otimes_k (\overline{B} \otimes_k B)$$

satisfying

$$f((\overline{a_1 \otimes b_1}) \otimes (a_2 \otimes b_2)) = (\overline{a}_1 \otimes a_2) \otimes (\overline{b}_1 \otimes b_2) \text{ for all } a_1, a_2 \in A, b_1, b_2 \in B.$$

In particular, for all $a_1, a_2 \in A, b_1, b_2 \in B$, we have

$$\begin{aligned}
f(s_{A \otimes_k B}((\overline{a_1 \otimes b_1}) \otimes (a_2 \otimes b_2))) &= f((\overline{a_2 \otimes b_2}) \otimes (a_1 \otimes b_1)) \\
&= (\overline{a}_2 \otimes a_1) \otimes (\overline{b}_2 \otimes b_1) \\
&= (s_A \otimes s_B)((\overline{a_1} \otimes a_2) \otimes (\overline{b}_1 \otimes b_2)).
\end{aligned}$$

Therefore,

$$f(s_{A \otimes_k B}((\overline{a_1 \otimes b_1}) \otimes (a_2 \otimes b_2))) = (s_A \otimes s_B)(f((\overline{a_1 \otimes b_1}) \otimes (a_2 \otimes b_2))),$$

and it follows that we have

$$f \circ s_{A \otimes_k B} = (s_A \otimes s_B) \circ f.$$

This easily implies that f restricts to a k_0-algebra isomorphism

$$\mathrm{Cor}_{k/k_0}(A \otimes_k B) \cong_{k_0} \{z \in (\overline{A} \otimes_k A) \otimes_k (\overline{B} \otimes_k B) \mid (s_A \otimes s_B)(z) = z\}.$$

But this last subalgebra is isomorphic to $\mathrm{Cor}_{k/k_0}(A) \otimes_{k_0} \mathrm{Cor}_{k/k_0}(B)$ (this is left to the reader as an exercise).

We finally prove (4). By Lemma IX.2.9, we have an isomorphism of k-algebras

$$u: \begin{array}{c} \overline{\mathrm{End}_k(V)} \xrightarrow{\sim} \mathrm{End}_k(\overline{V}) \\ \overline{f} \longmapsto \theta(f), \end{array}$$

where

$$\theta(f)(\overline{v}) = \overline{f(v)} \text{ for all } v \in V.$$

We then get a morphism of k-algebras

$$\varphi : \overline{\mathrm{End}_k(V)} \otimes_k \mathrm{End}_k(V) \xrightarrow{\sim} \mathrm{End}_k(\overline{V} \otimes_k V)$$

satisfying

$$\varphi(\overline{f} \otimes g) = \theta(f) \otimes g \text{ for all } f, g \in \mathrm{End}_k(V),$$

which is in fact an isomorphism since two central simple k-algebras $\overline{\mathrm{End}_k(V)} \otimes_k \mathrm{End}_k(V)$ and $\mathrm{End}_k(\overline{V} \otimes_k V)$ have same degree. It is easy to see that we have

$$\varphi \circ s_{\mathrm{End}_k(V)} = \mathrm{Int} s_V \circ \varphi.$$

In particular, φ restricts to an injective morphism of k_0-algebras

$$\mathrm{Cor}_{k/k_0}(\mathrm{End}_k(V)) \longrightarrow \mathrm{End}_{k_0}(\mathrm{Cor}_{k/k_0}(V)).$$

Now we have

$$\deg(\mathrm{Cor}_{k/k_0}(\mathrm{End}_k(V))) = \dim_k(V)^2 = \deg(\mathrm{End}_{k_0}(\mathrm{Cor}_{k/k_0}(V))),$$

so the morphism above is an isomorphism.

The fact that $\mathrm{Cor}_{k/k_0}(A)$ only depends on the Brauer equivalence class of A readily follows from (2), (3) and (4). $\square$

IX.3. Existence of unitary involutions.

We are now ready to investigate the existence of unitary involutions.

LEMMA IX.3.1. *Assume that B is a central simple k-algebra carrying a unitary involution τ. Then there exists a unique structure of a right $\overline{B} \otimes_k B$-module $*_\tau$ on B satisfying*

$$x *_\tau (\overline{b}_1 \otimes b_2) = \tau(b_1) x b_2 \text{ for all } x, b_1, b_2 \in B.$$

Moreover, $*_\tau$ induces a right $\mathrm{Cor}_{k/k_0}(B)$-module structure on $\mathrm{Sym}(B, \tau)$ of dimension $\deg(B)^2$ over k_0. In particular, $\mathrm{Cor}_{k/k_0}(B)$ is split.

Proof. The map

$$\overline{B} \times B \longrightarrow \mathrm{End}_k(B)$$

$$(\overline{b}_1, b_2) \longmapsto \ell_{\tau(b_1)} \circ r_{b_2}$$

is k-bilinear, where $\ell_{\tau(b_1)}$ is the left multiplication by $\tau(b_1)$, and r_{b_2} is the right multiplication by b_2. Hence, it induces a k-linear map $\varphi : \overline{B} \otimes_k B \longrightarrow \mathrm{End}_k(B)$ satisfying

$$\varphi(\overline{b}_1 \otimes b_2)(x) = \tau(b_1) x b_2 \text{ for all } x, b_1, b_2 \in B.$$

Moreover, for all $z, z' \in \overline{B} \otimes_k B$, we have

$$\varphi(zz') = \varphi(z')\varphi(z).$$

Indeed, a distributivity argument allows us to reduce to the case where z and z' are elementary tensors, for which the equality becomes clear. It follows easily that the external product

$$B \times (\overline{B} \otimes_k B) \longrightarrow B$$
$$(x, z) \longmapsto x *_\tau z = \varphi(z)(x)$$

endows B with the structure of a right $\overline{B} \otimes_k B$-module structure. By definition, $*_\tau$ satisfies the required condition. The uniqueness part follows from the fact that elementary tensors span $\overline{B} \otimes_k B$.

Let us notice now that we have

$$\tau(x *_\tau z) = \tau(x) *_\tau s_B(z) \text{ for all } x \in B, z \in \overline{B} \otimes_k B,$$

as we may check easily by reducing to the case where z is an elementary tensor. In particular, if $z \in \mathrm{Cor}_{k/k_0}(B)$ and $x \in \mathrm{Sym}(B, \tau)$, then $x *_\tau z \in \mathrm{Sym}(B, \tau)$, and $*_\tau$ induces a right $\mathrm{Cor}_{k/k_0}(B)$-module structure on $\mathrm{Sym}(B, \tau)$. Now Lemma IX.2.6 implies that we have

$$\dim_{k_0}(\mathrm{Sym}(B, \tau)) = \deg(B)^2 = \deg(\mathrm{Cor}_{k/k_0}(B)),$$

since τ is in particular a semilinear map satisfying $\tau^2 = \mathrm{Id}_B$. By Corollary V.2.5, $\mathrm{Cor}_{k/k_0}(B)$ is split. $\square$

Our next goal is to establish a correspondence between unitary involutions on B and right ideals of $\mathrm{Cor}_{k/k_0}(B)$ satisfying certain properties. Let us introduce some notation first. Recall from Proposition IX.2.12 (1) that we have an isomorphism of k-algebras

$$\phi : \mathrm{Cor}_{k/k_0}(B) \otimes_{k_0} k \xrightarrow{\sim} \overline{B} \otimes_k B$$

satisfying

$$\phi(z \otimes \lambda) = z\lambda \text{ for all } z \in \mathrm{Cor}_{k/k_0}(B), \lambda \in k.$$

If I is a right ideal of $\mathrm{Cor}_{k/k_0}(B)$, we denote by I_k the image of $I \otimes_{k_0} k$ under ϕ. This is a right ideal of $\overline{B} \otimes_k B$ satisfying $\dim_k(I_k) = \dim_{k_0}(I)$. Moreover, I_k is preserved under the map s_B, and we have

$$\phi(I \otimes_{k_0} 1) = \{z \in I_k \mid s_B(z) = z\}.$$

Indeed, if $z \in I_k$, then $z = \sum_{i=1}^{r} z_i \lambda_i$ for some $z_i \in I, \lambda_i \in k$ by definition of I_k. We then have

$$s_B(z) = \sum_{i=1}^{r} s_B(z_i)\overline{\lambda}_i = \sum_{i=1}^{r} z_i \overline{\lambda}_i \in I_k,$$

since each z_i lies in $\mathrm{Cor}_{k/k_0}(B)$ and is therefore fixed under s_B. Moreover, the inclusion $\phi(I \otimes_{k_0} 1) \subset \{z \in I_k \mid s_B(z) = z\}$ is clear, and since these k_0-vector spaces have the same dimension (as we may check using Lemma IX.2.5), we get the desired equality.

In particular, if I and J are two ideals of $\mathrm{Cor}_{k/k_0}(B)$ satisfying $I_k = J_k$, then we have $I \otimes_{k_0} 1 = J \otimes_{k_0} 1$, and thus $I = J$.

LEMMA IX.3.2. *Let B be a central simple k-algebra of degree n carrying a unitary involution τ. Then the set*

$$I_\tau = \{z \in \mathrm{Cor}_{k/k_0}(B) \mid 1 *_\tau z = 0\}$$

is a right ideal of $\mathrm{Cor}_{k/k_0}(B)$ of dimension $n^4 - n^2$ over k_0 satisfying

$$\overline{B} \otimes_k B = (I_\tau)_k \oplus (1 \otimes_k B).$$

Moreover, we have

$$(I_\tau)_k = \{z \in \overline{B} \otimes_k B \mid 1 *_\tau z = 0\}.$$

Proof. Since the map

$$f_\tau \colon \begin{array}{c} \mathrm{Cor}_{k/k_0}(B) \longrightarrow \mathrm{Sym}(B, \tau) \\ z \longmapsto 1 *_\tau z \end{array}$$

is $\mathrm{Cor}_{k/k_0}(B)$-linear, its kernel I_τ is a right ideal of $\mathrm{Cor}_{k/k_0}(B)$. Moreover, f_τ is surjective. Indeed, if $u \in \mathrm{Sym}(B, \tau)$, then $\dfrac{\overline{1} \otimes u + \overline{u} \otimes 1}{2}$ lies in $\mathrm{Cor}_{k/k_0}(B)$, and we have

$$f_\tau\left(\frac{\overline{1} \otimes u + \overline{u} \otimes 1}{2}\right) = \frac{\tau(1)u + \tau(u)1}{2} = u.$$

In particular, $\dim_{k_0}(I_\tau) = n^4 - n^2$, where $n = \deg(B)$. Consequently, we have

$$\dim_k((I_\tau)_k) = \dim_{k_0}(I_\tau) = n^2 - n.$$

Therefore, to show that $\overline{B} \otimes_k B = (I_\tau)_k \oplus (1 \otimes_k B)$, it is enough to check that we have

$$(I_\tau)_k \cap (1 \otimes_k B) = \{0\}.$$

If $1 \otimes b \in (I_\tau)_k$, then we have $1 \otimes b = \displaystyle\sum_{i=1}^r z_i \lambda_i$ for some $z_i \in I_\tau, \lambda_i \in k$. We then get

$$1 *_\tau (1 \otimes b) = \sum_{i=1}^r 1 *_\tau z_i \lambda_i = \sum_{i=1}^r (1 *_\tau z_i) *_\tau \lambda_i,$$

the last equality coming from the definition of a module. Since $z_i \in I_\tau$, we have $1 *_\tau z_i = 0$ and thus $1 *_\tau (1 \otimes b) = 0$. Now, by definition of $*_\tau$, we have $b = 1 *_\tau (1 \otimes b) = 0$.

Let us prove the last part of the lemma. Notice first that the map

$$g_\tau \colon \begin{array}{c} \overline{B} \otimes_k B \longrightarrow B \\ z \longmapsto 1 *_\tau z \end{array}$$

is $\overline{B} \otimes_k B$-linear and surjective, since we have

$$g_\tau(1 \otimes b) = b.$$

Hence, $\ker(g_\tau)$ is a right ideal of dimension $n^4 - n^2$ over k. Now if

$$z = \sum_{i=1}^r z_i \lambda_i \in (I_\tau)_k,$$

we get as before $1 *_\tau z = 0$, and thus we have the inclusion $(I_\tau)_k \subset \ker(g_\tau)$. Therefore, we get equality since these two ideals have the same dimension over k. This concludes the proof. $\qquad\square$

This lemma allows us to associate to a unitary involution on B a right ideal I_τ of $\mathrm{Cor}_{k/k_0}(B)$ satisfying $\overline{B} \otimes_k B = (I_\tau)_k \oplus (1 \otimes_k B)$. Conversely, we have the following lemma:

LEMMA IX.3.3. *Let B be a central simple k-algebra, and let I be a right ideal of* $\mathrm{Cor}_{k/k_0}(B)$ *satisfying*

$$\overline{B} \otimes_k B = I_k \oplus (1 \otimes_k B).$$

Then the map $\tau_I : B \longrightarrow B$ uniquely determined by the conditions

$$\overline{x} \otimes 1 - 1 \otimes \tau_I(x) \in I_k \text{ for all } x \in B$$

is a unitary involution on B.

Proof. First, it is clear from the definition that τ_I is semilinear. Now if $x, y \in B$, we have

$$\overline{xy} \otimes 1 - 1 \otimes \tau_I(y)\tau_I(x) = (\overline{x} \otimes 1 - 1 \otimes \tau_I(x))(\overline{y} \otimes 1) + (\overline{y} \otimes 1 - 1 \otimes \tau_I(y))(1 \otimes \tau_I(x)).$$

Thus $\overline{xy} \otimes 1 - 1 \otimes \tau_I(y)\tau_I(x) \in I_k$ since I_k is a right ideal of $\overline{B} \otimes_k B$, and therefore we have

$$\tau_I(xy) = \tau_I(y)\tau_I(x) \text{ for all } x, y \in B.$$

It remains to show that $\tau_I^2 = \mathrm{Id}_B$. Since I_k is preserved by s_B, we have

$$s_B(\overline{x} \otimes 1 - 1 \otimes \tau_I(x)) = 1 \otimes x - \overline{\tau_I(x)} \otimes 1 \in I_k \text{ for all } x \in B.$$

It follows that $\overline{\tau_I(x)} \otimes 1 - 1 \otimes x \in I_k$ and that $\tau_I(\tau_I(x)) = x$ for all $x \in B$. This proves that τ_I is a unitary involution on B. $\qquad\square$

We can now state the main result of this section.

THEOREM IX.3.4. *Let B be a central simple k-algebra. Then the maps*

$$\tau \longmapsto I_\tau$$
$$\tau_I \longleftarrow\!\shortmid I$$

establish a one-to-one correspondence between the set of unitary involutions on B and the set of right ideals I of $\mathrm{Cor}_{k/k_0}(B)$ *satisfying*

$$\overline{B} \otimes_k B = I_k \oplus (1 \otimes_k B).$$

Proof. In view of the two previous lemmas, it remains to show that the two constructions are mutually inverse. Let τ be a unitary involution on B. First of all, notice that we have

$$1 *_\tau (\overline{x} \otimes 1 - 1 \otimes \tau(x)) = \tau(x)1 - 1\tau(x) = 0 \text{ for all } x \in B,$$

and therefore $\overline{x} \otimes 1 - 1 \otimes \tau(x) \in (I_\tau)_k$ for all $x \in B$, by the last part of Lemma IX.3.2. Thus we get $\tau_{I_\tau} = \tau$. Conversely, assume that I is a right ideal of $\mathrm{Cor}_{k/k_0}(B)$ satisfying

$$\overline{B} \otimes_k B = I_k \oplus (1 \otimes_k B).$$

In particular, we have $\dim_{k_0}(I) = n^4 - n^2$. To show that $I_{\tau_I} = I$, it is enough to check that $(I_{\tau_I})_k = I_k$. Since these ideals have the same dimension over k, it is enough to prove the inclusion $(I_{\tau_I})_k \subset I_k$. Since elements of $(I_{\tau_I})_k$ are linear combinations of elements of I_{τ_I} with coefficients in k, it is enough to prove that $I_{\tau_I} \subset I_k$.

Let $z \in I_{\tau_I}$, and write

$$z = \overline{x}_1 \otimes y_1 + \cdots + \overline{x}_m \otimes y_m.$$

Since $z \in I_{\tau_I}$, we have

$$1 *_{\tau_I} z = \tau_I(x_1)y_1 + \cdots + \tau_I(x_m)y_m = 0.$$

Therefore, we get

$$z = \sum_{i=1}^{m} (\overline{x}_i \otimes y_i - 1 \otimes \tau_I(x_i)y_i) = \sum_{i=1}^{m} (\overline{x}_i \otimes 1 - 1 \otimes \tau_I(x_i))(1 \otimes y_i).$$

Now by definition of τ_I, we have

$$\overline{x}_i \otimes 1 - 1 \otimes \tau_I(x_i) \in I_k \quad \text{for all } i = 1, \ldots, m,$$

and thus we have $z \in I_k$, since I_k is a right ideal of $\overline{B} \otimes_k B$. This concludes the proof. □

COROLLARY IX.3.5. *Let B be a central simple k-algebra. Then B carries a unitary involution if and only if $\mathrm{Cor}_{k/k_0}(B)$ is split.*

Proof. One implication is given by Lemma IX.3.1. To prove the converse, we first consider the case of division k-algebras.

Let D be a division k-algebra of degree d such that $\mathrm{Cor}_{k/k_0}(D)$ splits, and let $f \colon \mathrm{Cor}_{k/k_0}(D) \xrightarrow{\sim} \mathrm{M}_{d^2}(k_0)$ be an isomorphism of k_0-algebras. The subset $J \subset \mathrm{M}_{d^2}(k)$ consisting of matrices whose first row is zero is a right ideal of $\mathrm{M}_{d^2}(k_0)$ of dimension $d^4 - d^2$ over k_0. Therefore, $I = f^{-1}(J)$ is a right ideal of $\mathrm{Cor}_{k/k_0}(D)$ of dimension $d^4 - d^2$ over k_0. We claim that

$$I_k \cap 1 \otimes_k D = \{0\}.$$

Indeed, if this intersection were non-trivial, then I_k would contain an invertible element (since D is a division algebra), and we would have $I_k = \overline{D} \otimes_k D$ which would contradict the fact that we have

$$\dim_k(I_k) = \dim_{k_0}(I) = d^4 - d^2.$$

By dimension count, we get the decomposition

$$\overline{D} \otimes_k D = I_k \oplus (1 \otimes_k D).$$

The previous theorem then shows that D carries a unitary involution.

We now come back to the general case. Let B be a central simple k-algebra, and let $\varphi \colon B \xrightarrow{\sim} \mathrm{M}_r(D)$ be an isomorphism of k-algebras, where D is a division k-algebra. Assume that $\mathrm{Cor}_{k/k_0}(B)$ splits. By Proposition IX.2.12, $\mathrm{Cor}_{k/k_0}(D)$ also splits and D carries a unitary involution τ by the previous point. One may verify that the map

$$\tau' \colon \begin{array}{c} \mathrm{M}_r(D) \longrightarrow \mathrm{M}_r(D) \\ (m_{ij}) \longmapsto (\tau(m_{ji})) \end{array}$$

is a unitary involution on $\mathrm{M}_r(D)$. The map $\tau'' = \varphi^{-1} \circ \tau' \circ \varphi$ is then a unitary involution on B. This concludes the proof. □

IX.4. Unitary involutions on crossed products.

We now focus on the case of crossed product algebras. Let k/k_0 be a quadratic field extension. If $\lambda \in k$, recall that we denote by $\overline{\lambda}$ its image under the unique non-trivial k_0-automorphism of k. Let L/k be a Galois extension of group G, and let $B = (\xi, L/k, G)$ be a crossed product algebra.

In the sequel, we will assume the existence of a ring automorphism $\alpha : L \longrightarrow L$ satisfying the following conditions:

(1) $\alpha^2 = \mathrm{Id}_L$;

(2) $\alpha \circ \sigma = \sigma \circ \alpha$ for all $\sigma \in G$;

(3) $\alpha(\lambda) = \overline{\lambda}$ for all $\lambda \in k$.

Let $L_0 = L^{\langle \alpha \rangle}$. Elementary Galois theory then shows that L_0/k_0 and k/k_0 are linearly disjoint, that $L = kL_0$ and that we have a group isomorphism

$$G \xrightarrow{\sim} \mathrm{Gal}(L_0/k_0)$$

$$\sigma \longmapsto \sigma_{|L_0}.$$

We will keep the assumptions and notation above throughout this section. Our main goal is to compute the corestriction algebra of B. We first identify $\overline{B}$ to a crossed product algebra.

LEMMA IX.4.1. *Let $\xi \in Z^2(G, L^\times)$ be a 2-cocycle. Then the map $\alpha \circ \xi$ is a 2-cocycle, and there is a unique isomorphism of k-algebras*

$$\varphi : \overline{B} \xrightarrow{\sim} (\alpha \circ \xi, L/k, G)$$

satisfying

$$\varphi(\overline{f}_\sigma) = e_\sigma, \text{ for all } \sigma \in G \text{ and } \varphi_{|_L} = \alpha,$$

where $(\alpha \circ \xi, L/k, G) = \bigoplus_{\sigma \in G} e_\sigma L$ and $B = \bigoplus_{\sigma \in G} f_\sigma L$.

Proof. Since ξ is a 2-cocycle, we have

$$\xi_{\sigma, \tau\rho} \xi_{\tau, \rho} = \xi_{\sigma\tau, \rho} \xi_{\sigma, \tau}^\rho \text{ for all } \sigma, \tau, \rho \in G.$$

Now applying α to this equality, and using the fact that α commutes with the elements of G, it follows easily that $\alpha \circ \xi$ is also a 2-cocycle.

Let $\varphi : \overline{B} \xrightarrow{\sim} (\alpha \circ \xi, L/k, G)$ be an isomorphism of k-algebras satisfying

$$\varphi(\overline{f}_\sigma) = e_\sigma \text{ and } \varphi(\overline{\lambda}) = \alpha(\lambda) \text{ for all } \sigma \in G, \lambda \in L.$$

Then we have

$$
\begin{aligned}
\varphi(\overline{\sum_{\sigma \in G} f_\sigma \lambda_\sigma}) &= \varphi(\sum_{\sigma \in G} \overline{f}_\sigma \overline{\lambda}_\sigma) \\
&= \sum_{\sigma \in G} \varphi(\overline{f}_\sigma) \varphi(\overline{\lambda}_\sigma) \\
&= \sum_{\sigma \in G} e_\sigma \alpha(\lambda_\sigma).
\end{aligned}
$$

Hence, if such a φ exists, it is unique. Let us prove the existence of φ.

The map

$$B \longrightarrow (\alpha \circ \xi, L/k, G)$$
$$\rho \colon \sum_{\sigma \in G} f_\sigma \lambda_\sigma \longmapsto \sum_{\sigma \in G} e_\sigma \alpha(\lambda_\sigma)$$

is clearly a semilinear map of k-vector spaces. Let us prove that ρ is a k_0-algebra morphism. In order to prove that ρ preserves products, a distributivity argument shows that it is enough to check the equality

$$\rho(f_\sigma \lambda \cdot f_\tau \mu) = \rho(f_\sigma \lambda)\rho(f_\tau \mu) \text{ for all } \sigma, \tau \in G, \lambda, \mu \in L.$$

We have

$$\rho(f_\sigma \lambda \cdot f_\tau \mu) = \rho(f_\sigma f_\tau \lambda^\tau \mu) = \rho(f_{\sigma\tau} \xi_{\sigma,\tau} \lambda^\tau \mu) = e_{\sigma\tau} \alpha(\xi_{\sigma,\tau}) \alpha(\lambda^\tau) \alpha(\mu).$$

Moreover, we have

$$\begin{aligned} \rho(f_\sigma \lambda)\rho(f_\tau \mu) &= e_\sigma \alpha(\lambda) e_\tau \alpha(\mu) \\ &= e_\sigma e_\tau \alpha(\lambda)^\tau \alpha(\mu) \\ &= e_{\sigma\tau} \alpha(\xi_{\sigma,\tau}) \alpha(\lambda)^\tau \alpha(\mu). \end{aligned}$$

Since α commutes with τ, we have the desired equality. Hence ρ is a k_0-algebra morphism which is semilinear, and therefore induces a k-algebra morphism

$$\overline{B} \longrightarrow (\alpha \circ \xi, L/k, G)$$
$$\varphi \colon \overline{\sum_{\sigma \in G} f_\sigma \lambda_\sigma} \longmapsto \sum_{\sigma \in G} e_\sigma \alpha(\lambda_\sigma).$$

Since $\overline{B}$ and $(\alpha \circ \xi, L/k, G)$ have the same degree, φ is an isomorphism. □

By the lemma above, if ξ is a 2-cocycle, so is $\alpha \circ \xi$, and therefore $(\alpha \circ \xi)\xi \in Z^2(G, L^\times)$ is a 2-cocycle. Moreover, for all $\sigma, \tau \in G$, we have $\alpha(\xi_{\sigma,\tau})\xi_{\sigma,\tau} \in L_0^\times$. Hence the map

$$G \times G \longrightarrow L_0^\times$$
$$(\sigma_{L_0}, \tau_{L_0}) \longmapsto \alpha(\xi_{\sigma,\tau})\xi_{\sigma,\tau}$$

is a 2-cocycle, that we will still denote by $(\alpha \circ \xi)\xi$ by abuse of notation. We then have the following result:

PROPOSITION IX.4.2. *Let* $\xi \in Z^2(G, L^\times)$, *and let* $B = (\xi, L/k, G)$. *Then we have*

$$\mathrm{Cor}_{k/k_0}(B) \sim_{k_0} ((\alpha \circ \xi)\xi, L_0/k_0, G).$$

Proof. We set

$$\begin{aligned} A &= (\alpha \circ \xi, L/k, G) &&= \bigoplus_{\sigma \in G} e_\sigma L \\ B &= (\xi, L/k, G) &&= \bigoplus_{\sigma \in G} f_\sigma L \\ C &= ((\alpha \circ \xi)\xi, L/k, G) &&= \bigoplus_{\sigma \in G} g_\sigma L \\ C' &= ((\alpha \circ \xi)\xi, L_0/k_0, G) && . \end{aligned}$$

Notice that all these algebras have degree n, where

$$n = [L : k] = [L_0 : k_0].$$

To prove the proposition, we have to find a $\mathrm{Cor}_{k/k_0}(B) - C'$-bimodule of dimension n^3 over k_0, by Proposition V.2.3.

From Lemma IX.4.1, we get an isomorphism of k-algebras

$$\chi : \overline{B} \otimes_k B \xrightarrow{\sim} A \otimes_k B$$

satisfying

$$\chi(\overline{f_\sigma \lambda} \otimes f_\tau \mu) = e_\sigma \alpha(\lambda) \otimes f_\tau \mu \text{ for all } \sigma, \tau \in G, \lambda, \mu \in G.$$

Recall from the proof of Proposition VI.2.3 that the abelian group

$$M = A \otimes_L B = \bigoplus_{\sigma, \tau \in G} e_\sigma L \otimes_L f_\tau L = \bigoplus_{\sigma, \tau \in G} (e_\sigma \otimes_L f_\tau) L$$

has a structure of $A \otimes_k B - C$-bimodule uniquely defined by the equalities

$$(a \otimes b) \bullet (a' \otimes b') = aa' \otimes bb' \text{ for all } a, a' \in A, b, b'' \in B,$$

and

$$(a' \otimes b') * (g_\sigma \gamma) = a' e_\sigma \otimes b' f_\sigma \gamma \text{ for all } a' \in A, b' \in B, \gamma \in L, \sigma \in G.$$

We then get a structure of $\overline{B} \otimes_k B - C$ bimodule on M by letting act $\overline{B} \otimes_k B$ on M through the isomorphism χ. We will denote by $\circ$ the left action of $\overline{B} \otimes_k B$ on M. We then have

$$z \circ x = \chi(z) \bullet x \text{ for all } z \in \overline{B} \otimes_k B, x \in M.$$

Consider the following morphisms of abelian groups

$$M \longrightarrow M$$

$$u: \sum_{\sigma, \tau \in G} e_\sigma \lambda_\sigma \otimes f_\tau \mu_\tau \longmapsto \sum_{\sigma, \tau \in G} e_\tau \alpha(\mu_\tau) \otimes f_\sigma \alpha(\lambda_\sigma)$$

and

$$C \longrightarrow C$$

$$\tilde{\alpha}: \sum_{\sigma \in G} g_\sigma \lambda_\sigma \longmapsto \sum_{\sigma \in G} g_\sigma \alpha(\lambda_\sigma).$$

Claim: we have

$$u(z \circ x) = s_A(z) \circ u(x) \text{ for all } z \in \overline{B} \otimes_k B, x \in M$$

and

$$u(x * z') = u(x) * \tilde{\alpha}(z') \text{ for all } x \in M, z' \in C.$$

Assume that the claim is proved. Let $M^{\langle u \rangle} = \{x \in M \mid u(x) = x\}$. It follows from the claim that the structure of a $\overline{B} \otimes_k B - C$-bimodule on M induces a structure of a $\mathrm{Cor}_{k/k_0}(B) - C^{\langle \tilde{\alpha} \rangle}$-bimodule on $M^{\langle u \rangle}$, where

$$C^{\langle \tilde{\alpha} \rangle} = \{z' \in C \mid \tilde{\alpha}(z') = z'\}.$$

By the proof of Lemma VI.2.9, scalar multiplication induces an isomorphism of k-algebras

$$\psi : C' \otimes_{k_0} k \xrightarrow{\sim} C.$$

It is easy to check that $C^{\langle\tilde{\alpha}\rangle} = \psi(C' \otimes_{k_0} 1)$. We then easily get a structure of a $\mathrm{Cor}_{k/k_0}(B) - C'$-bimodule on $M^{\langle u \rangle}$, where the structure of right C'-module is induced from the structure of C-module via the k_0-algebra morphism

$$C' \longrightarrow C$$
$$z' \longmapsto \psi(z' \otimes 1).$$

To conclude the argument, we have to check that $\dim_{k_0}(M^{\langle u \rangle}) = n^3$. But this comes from Lemma IX.2.5, since u is semilinear and satisfies $u^2 = \mathrm{Id}_M$.

It remains to prove the claim. Let us prove the first equality. A distributivity argument shows that it is enough to check this relation for $z = \overline{f_\sigma \lambda} \otimes f_\tau \mu$ and $x = e_\rho \lambda' \otimes f_\nu \mu'$. We have

$$
\begin{aligned}
z \circ x &= (e_\sigma \alpha(\lambda) \otimes f_\tau \mu) \bullet (e_\rho \lambda' \otimes f_\nu \mu') \\
&= e_\sigma \alpha(\lambda) e_\rho \lambda' \otimes f_\tau \mu f_\nu \mu' \\
&= e_{\sigma\rho} \alpha(\xi_{\sigma,\rho}) \alpha(\lambda)^\rho \lambda' \otimes f_{\tau\nu} \xi_{\tau,\nu} \mu^\nu \mu',
\end{aligned}
$$

and therefore

$$u(z \circ x) = e_{\tau\nu} \alpha(\xi_{\tau,\nu}) \alpha(\mu^\nu) \alpha(\mu') \otimes f_{\sigma\rho} \xi_{\sigma,\rho} \lambda^\rho \alpha(\lambda').$$

Notice that we used implicitly the fact that $\alpha^2 = \mathrm{Id}_L$ and that α commutes with the elements of G.

By definition of the structure of L-vector space on M, this may be rewritten as

$$u(z \circ x) = e_{\tau\nu} \otimes f_{\sigma\rho} \alpha(\xi_{\tau,\nu}) \alpha(\mu^\nu) \alpha(\mu') \xi_{\sigma,\rho} \lambda^\rho \alpha(\lambda').$$

We now compute $s_A(z) \circ u(x)$. We have

$$
\begin{aligned}
s_A(z) \circ u(x) &= (\overline{f_\tau \mu} \otimes f_\sigma \lambda) \circ (e_\nu \alpha(\mu') \otimes f_\rho \alpha(\lambda')) \\
&= (e_\tau \alpha(\mu) \otimes f_\sigma \lambda) \bullet (e_\nu \alpha(\mu') \otimes f_\rho \alpha(\lambda')) \\
&= e_\tau \alpha(\mu) e_\nu \alpha(\mu') \otimes f_\sigma \lambda f_\rho \alpha(\lambda') \\
&= e_{\tau\nu} \alpha(\xi_{\tau,\nu}) \alpha(\mu)^\nu \alpha(\mu') \otimes f_{\sigma\rho} \xi_{\sigma,\rho} \lambda^\rho \alpha(\lambda') \\
&= e_{\tau\nu} \otimes f_{\sigma\rho} \alpha(\xi_{\tau,\nu}) \alpha(\mu)^\nu \alpha(\mu') \xi_{\sigma,\rho} \lambda^\rho \alpha(\lambda').
\end{aligned}
$$

Taking into account that α commutes with ν, we get the desired equality.

Let us prove now the second equality. Once again, it is enough to check it when $x = e_\sigma \lambda \otimes f_\tau \mu$, $z' = g_\rho \gamma$. We have

$$x * z' = e_\sigma \lambda e_\rho \otimes f_\tau \mu f_\rho \gamma = e_{\sigma\rho} \alpha(\xi_{\sigma,\rho}) \lambda^\rho \otimes f_{\tau\rho} \xi_{\tau,\rho} \mu^\rho \gamma,$$

and therefore

$$u(x * z') = e_{\tau\rho} \alpha(\xi_{\tau,\rho}) \alpha(\mu^\rho) \alpha(\gamma) \otimes f_{\sigma\rho} \xi_{\sigma,\rho} \alpha(\lambda^\rho),$$

which may be rewritten as

$$u(x * z') = e_{\tau\rho} \otimes f_{\sigma\rho} \alpha(\xi_{\tau,\rho}) \alpha(\mu^\rho) \alpha(\gamma) \xi_{\sigma,\rho} \alpha(\lambda^\rho).$$

Now we also have

$$
\begin{aligned}
u(x) * \tilde{\alpha}(z') &= (e_\tau \alpha(\mu) \otimes f_\sigma \alpha(\lambda)) * g_\rho \alpha(\gamma) \\
&= e_\tau \alpha(\mu) e_\rho \otimes f_\sigma \alpha(\lambda) f_\rho \alpha(\gamma) \\
&= e_{\tau\rho} \alpha(\xi_{\tau,\rho}) \alpha(\mu)^\rho \otimes f_{\sigma\rho} \xi_{\sigma,\rho} \alpha(\lambda)^\rho \alpha(\gamma) \\
&= e_{\tau\rho} \otimes f_{\sigma\rho} \alpha(\xi_{\tau,\rho}) \alpha(\mu)^\rho \xi_{\sigma,\rho} \alpha(\lambda)^\rho \alpha(\gamma).
\end{aligned}
$$

Since α commutes with ρ, we get the desired equality. This concludes the proof.
$\square$

In particular, if the 2-cocycle $\xi \in Z^2(G, L^\times)$ satisfies $(\alpha \circ \xi)\xi = 1$, it follows from Corollary IX.3.5 and the last statement of Proposition VI.2.3 that $B = (\xi, L/k, G)$ carries a unitary involution. The next lemma gives a concrete example of such an involution.

LEMMA IX.4.3. *Let $\xi \in Z^2(G, L^\times)$ be a 2-cocycle satisfying $(\alpha \circ \xi)\xi = 1$, and let $B = (\xi, L/k, G)$ be the corresponding crossed-product algebra. Then there is a unique unitary involution τ on B satisfying*

$$\tau(f_\sigma) = f_\sigma^{-1} \text{ for all } \sigma \in G \text{ and } \tau_{|L} = \alpha.$$

Moreover, if M_b is the matrix of left multiplication by b in the L-basis $(f_\sigma)_{\sigma \in G}$, then we have

$$M_{\tau(b)} = M_b^\sharp \text{ for all } b \in B,$$

where $\sharp$ is the unitary involution on $\mathrm{M}_n(L)$ defined by

$$\mathrm{M}_n(L) \longrightarrow \mathrm{M}_n(L)$$

$$M = (m_{\sigma\rho})_{\sigma,\rho \in G} \longmapsto M^\sharp = (\alpha(m_{\rho\sigma})_{\sigma,\rho \in G}).$$

Proof. Assume that an involution τ satisfying the properties of the lemma exists. Using the fact that τ is an anti-automorphism, we get that

$$\tau\left(\sum_{\sigma \in G} f_\sigma \lambda_\sigma\right) = \sum_{\sigma \in G} \alpha(\lambda_\sigma) f_\sigma^{-1} \text{ for all } \lambda_\sigma \in L, \sigma \in G.$$

This proves the uniqueness of τ. We now have to prove that the map τ defined by the formula above is indeed a unitary involution on B. Clearly, τ is semilinear. We now check that we have

$$\tau(xy) = \tau(y)\tau(x) \text{ for all } x, y \in B.$$

The usual distributivity argument shows that it is enough to prove it for $x = f_\sigma \lambda, y = f_\rho \mu, \sigma, \rho \in G, \lambda, \mu \in L$. We have

$$\tau(f_\sigma \lambda f_\rho \mu) = \tau(f_{\sigma\tau} \xi_{\sigma,\rho} \lambda^\rho \mu) = \alpha(\xi_{\sigma,\rho}) \alpha(\lambda^\rho) \alpha(\mu) f_{\sigma\rho}^{-1}.$$

On the other hand, we have

$$\tau(f_\rho \mu)\tau(f_\sigma \lambda) = \alpha(\mu) f_\rho^{-1} \alpha(\lambda) f_\sigma^{-1}.$$

From the relation $\lambda f_\sigma = f_\sigma \lambda^\sigma$, we get

$$f_\sigma^{-1} \lambda = \lambda^\sigma f_\sigma^{-1}.$$

Therefore, we get

$$\begin{aligned}
\tau(f_\rho \mu)\tau(f_\sigma \lambda) &= \alpha(\mu)(\alpha(\lambda))^\rho f_\rho^{-1} f_\sigma^{-1} \\
&= \alpha(\mu)(\alpha(\lambda))^\rho (f_\sigma f_\rho)^{-1} \\
&= \alpha(\mu)(\alpha(\lambda))^\rho (f_{\sigma\rho} \xi_{\sigma,\rho})^{-1} \\
&= \alpha(\mu)(\alpha(\lambda))^\rho \xi_{\sigma,\rho}^{-1} f_{\sigma\rho}^{-1}.
\end{aligned}$$

Since α commutes with the elements of G and $\alpha(\xi_{\sigma,\rho}) = \xi_{\sigma,\rho}^{-1}$ by assumption, we get the desired equality. It remains to prove that $\tau^2 = \mathrm{Id}_B$. Since τ is an antiautomorphism of rings, τ^2 is an automorphism of rings. Hence to prove that τ^2 is the identity map, it is enough to check that $\tau^2(f_\sigma) = f_\sigma$ for all $\sigma \in G$ and that $\tau^2_{|L} = \mathrm{Id}_L$, which is clear from the definition of τ.

We finally prove the last assertion. We will index the entries of a matrix with coefficients in L with the elements of G. Let $b \in B$. If $M_b = (m_{\sigma,\rho})_{\sigma,\rho \in G}$, we have to check that $M_{\tau(b)} = (\alpha(m_{\rho,\sigma}))_{\sigma,\rho \in G}$. Let us write

$$b = \sum_{\sigma \in G} f_\sigma \lambda_\sigma.$$

By Lemma VI.3.1, we have

$$M_b = (\xi_{\sigma\rho^{-1},\rho} \lambda^\rho_{\sigma\rho^{-1}})_{\sigma,\rho \in G}.$$

Now from the equality $f_\sigma f_{\sigma^{-1}\rho} = f_\rho \xi_{\sigma,\sigma^{-1}\rho}$, we get

$$f_\sigma^{-1} f_\rho = f_{\sigma^{-1}\rho} \xi^{-1}_{\sigma,\sigma^{-1}\rho}.$$

Therefore, we have

$$
\begin{aligned}
\tau(b)f_\rho &= \sum_{\sigma \in G} \alpha(\lambda_\sigma) f_\sigma^{-1} f_\rho \\
&= \sum_{\sigma \in G} \alpha(\lambda_\sigma) f_{\sigma^{-1}\rho} \xi^{-1}_{\sigma,\sigma^{-1}\rho} \\
&= \sum_{\sigma \in G} f_{\sigma^{-1}\rho} (\alpha(\lambda_\sigma))^{\sigma^{-1}\rho} \xi^{-1}_{\sigma,\sigma^{-1}\rho} \\
&= \sum_{\sigma \in G} f_\sigma \alpha(\lambda_{\rho\sigma^{-1}})^\sigma \xi^{-1}_{\rho\sigma^{-1},\sigma},
\end{aligned}
$$

the last equality being obtained by performing the change of variables $\sigma \leftrightarrow \sigma^{-1}\rho$. Using again that α commutes with the elements of G and $\alpha(\xi_{\rho\sigma^{-1},\sigma}) = \xi^{-1}_{\rho\sigma^{-1},\sigma}$, we get

$$\tau(b)f_\rho = \sum_{\sigma \in G} f_\sigma \alpha(\lambda_{\rho\sigma^{-1}} \xi_{\rho\sigma^{-1},\sigma}).$$

Thus we get

$$M_{\tau(b)} = (\alpha(\lambda_{\rho\sigma^{-1}} \xi_{\rho\sigma^{-1},\sigma}))_{\sigma,\rho \in G} = M_b^\sharp,$$

and this concludes the proof.   □

REMARK IX.4.4. The description of the involution τ in the lemma above may be made more explicit. As explained in the proof, we have

$$\tau\left(\sum_{\sigma \in G} f_\sigma \lambda_\sigma\right) = \sum_{\sigma \in G} \alpha(\lambda_\sigma) f_\sigma^{-1} \text{ for all } \lambda_\sigma \in L, \sigma \in G.$$

Now we have $f_\sigma f_{\sigma^{-1}} = \xi_{\sigma,\sigma^{-1}}$, and therefore

$$f_\sigma^{-1} = f_{\sigma^{-1}} \xi^{-1}_{\sigma,\sigma^{-1}} \text{ for all } \sigma \in G.$$

Thus, we get

$$\alpha(\lambda_\sigma) f_\sigma^{-1} = f_{\sigma^{-1}} \alpha(\lambda_\sigma)^{\sigma^{-1}} \xi^{-1}_{\sigma,\sigma^{-1}} \text{ for all } \sigma \in G,$$

and performing the change of variables $\sigma \leftrightarrow \sigma^{-1}$ yields

$$\tau\left(\sum_{\sigma \in G} f_\sigma \lambda_\sigma\right) = \sum_{\sigma \in G} f_\sigma \alpha(\lambda_{\sigma^{-1}})^\sigma \xi^{-1}_{\sigma^{-1},\sigma} \text{ for all } \lambda_\sigma \in L, \sigma \in G.$$

Since α commutes with σ and $\alpha(\xi_{\sigma^{-1},\sigma})\xi_{\sigma^{-1},\sigma} = 1$ by assumption, we finally get that

$$\tau\left(\sum_{\sigma \in G} f_\sigma \lambda_\sigma\right) = \sum_{\sigma \in G} f_\sigma \alpha(\lambda_{\sigma^{-1}}^\sigma \xi_{\sigma^{-1},\sigma}) \text{ for all } \lambda_\sigma \in L, \sigma \in G.$$

$\square$

EXAMPLE IX.4.5. Assume that k_0 is a number field. Let L/k be a finite Galois extension of k with Galois group G, and assume that complex conjugation induces a k_0-automorphism α of L which commutes with elements of $\mathrm{Gal}(L/k)$. This automorphism satisfies the conditions explained at the beginning of this section. In particular, if $\xi : G \times G \longrightarrow L^\times$ is a 2-cocycle satisfying

$$|\xi_{\rho,\rho'}|^2 = 1 \text{ for all } \rho, \rho' \in G,$$

then $B = (\xi, L/k, \sigma)$ carries a unitary involution τ such that τ restricts to complex conjugation on L and $\tau(f_\sigma) = f_\sigma^{-1}$ for all $\sigma \in G$.

For example, if $B = (\gamma, L/k, \sigma)$ is a cyclic k-algebra of degree n such that $|\gamma|^2 = 1$, then the unitary involution τ on B given by the previous lemma is defined by

$$B \longrightarrow B$$

$$\tau : \sum_{i=0}^{n-1} e^i \lambda_i \longmapsto \overline{\lambda}_0 + \sum_{i=0}^{n-1} e^i \overline{\gamma \lambda_{n-i}^{\sigma^i}}, \quad ,$$

as it may easily be seen by direct computations, or by using the remark above.

We recover this way the involution obtained by Oggier and Lequeu [**39**]. $\square$

IX.5. Unitary space-time coding

In the first chapter (Section I.3), we introduced space-time codes built over a division k-algebra A as a subset $\mathcal{C}$ of $\varphi_{A,L}(A)$, where L is a subfield of A and

$$\varphi_{A,L} : A \hookrightarrow \mathrm{M}_m(L) \subset \mathrm{M}_m(\mathbb{C})$$

is an injective k-algebra morphism mapping $a \in A$ onto the matrix of left multiplication by a in a chosen L-basis of A. The first code design criterion to be satisfied, the full diversity property, was shown to be achieved when A is a k-division algebra. Later on, further additional properties were introduced. We discuss here a similar problem but with a different flavour. We add to the above the restriction that codewords must be unitary.

Whether space-time codewords should be unitary or not depends on the type of detection made at the receiver: in the so-called coherent case discussed in Section I.3, the receiver knows the channel matrix $\mathbf{H}$ and thus computes $\mathbf{HX}$ for all possible codewords $\mathbf{X}$, and decides that the decoded vector is the one such that $\mathbf{HX}$ is the closest to the received signal $\mathbf{Y}$. This cannot be done in the *non-coherent case* , where the receiver does not know the channel $\mathbf{H}$. In this case, one approach is to ask the transmitter to use a particular encoding strategy, called *differential modulation* [**21, 23**]: we denote the transmitted signal at time t by $\mathbf{S}_t$ (since the signal will be different from a codeword which we keep on denoting by $\mathbf{X}$). The transmitter sends at time $t = 0$ the signal $\mathbf{S}_0 = I$, after which

$$\mathbf{S}_t = \mathbf{S}_{t-1}\mathbf{X}_t, \ t = 1, 2, \ldots,$$

where $\mathbf{X}_t$ is the codeword used at time t. By construction of the scheme, the matrices $\mathbf{X}$ have to be unitary (so that the product $\mathbf{S}$ does not go to zero or infinity).

Now if we further assume that the channel is roughly constant, we get from the channel equation at time t and the differential modulation equation that

$$
\begin{aligned}
\mathbf{Y}_t &= \mathbf{H}\mathbf{S}_t + \mathbf{V}_t \\
&= \mathbf{H}\mathbf{S}_{t-1}\mathbf{X}_t + \mathbf{V}_t \\
&= (\mathbf{Y}_{t-1} - \mathbf{V}_{t-1})\mathbf{X}_t + \mathbf{V}_t \\
&= \mathbf{Y}_{t-1}\mathbf{X}_t + (\mathbf{V}_t - \mathbf{V}_{t-1}\mathbf{X}_t).
\end{aligned}
$$

Note that the matrix $\mathbf{H}$ does not appear anymore, and instead we have the signal received at time $t - 1$. The noise is also further amplified, but the receiver can now decode without knowledge of the channel. This motivates the following definition.

DEFINITION IX.5.1. We will call *unitary (algebra based) code*, a set $\mathcal{C} \subset \mathrm{M}_n(\mathbb{C})$ of matrices satisfying

$$\mathcal{C} = \{\mathbf{U} = \varphi(b) \mid b \in \mathcal{B}\},$$

where $\mathcal{B}$ is a subset of a central simple algebra B over a number field k, $\varphi : B \hookrightarrow \mathrm{M}_n(\mathbb{C})$ is an injective morphism of k-algebras, and $\mathbf{U}$ is unitary, that is, $\mathbf{U}\mathbf{U}^* = I_n$.

Notice that we do not assume the morphism φ to be induced by left multiplication in B, even if it will be the case in the applications. The main reason is that, contrary to the coherent case, the performance of our codes will not depend on the structure of the matrices used as codewords, but only on the central simple algebra B itself.

The first design criterion, based on the probability of decoding wrongly similarly to the coherent case [21, 23], stays the full diversity, namely

$$\det(\mathbf{U}' - \mathbf{U}'') \neq 0, \text{ for all } \mathbf{U}' \neq \mathbf{U}'' \in \mathcal{C}.$$

We also have an upper bound for the probability error which is totally similar to the bound obtained in the coherent case. Hence, as before, the ultimate goal is to maximize the minimum determinant. Obviously, before going further into these considerations, we need to know how to construct our unitary matrices.

There are two ways to look at the design of unitary space-time codes:

(1) first find unitary matrices, among which select a family which is fully diverse;

(2) first consider a family of matrices which is fully diverse, among which look for unitary matrices.

Since we have developed tools to work with division k-algebras, we will focus on the latter. Note that as before, when working with a k-division algebra, then

$$\det(\mathbf{U}' - \mathbf{U}'') \neq 0$$

but of course, there is no reason why $\mathbf{U}' - \mathbf{U}''$ should be a unitary matrix itself, which is something we are not asking.

We now explain how we are going to proceed. Most of the results presented in this section have been taken from [**7**]. Let k/k_0 be a quadratic field extension of number fields, whose non-trivial automorphism is given by the complex conjugation. Let (B, τ) be a central simple k-algebra with a unitary k/k_0-involution.

Notice that, if L is any subfield of $\mathbb{C}$ containing k and $^-$ denotes the complex conjugation, the map $\tau \otimes ^-$ is a unitary involution on $B \otimes_k L$. Thus the following definition makes sense.

DEFINITION IX.5.2. We say that (B, τ) is **positive definite** if there exists a subfield L of $\mathbb{C}$ such that there exists an isomorphism of L-algebras with involutions

$$\varphi : (B \otimes_k L, \tau \otimes ^-) \xrightarrow{\sim} (\mathrm{M}_n(L), ^*).$$

In other words, τ is positive definite if there exists an isomorphism of L-algebras $\varphi : B \otimes_k L \xrightarrow{\sim} \mathrm{M}_n(L)$ such that

$$\varphi \circ (\tau \otimes ^-) = {}^* \circ \varphi.$$

EXAMPLE IX.5.3. The transpose conjugate involution on $\mathrm{M}_n(\mathbb{C})$ is positive definite.

□

REMARK IX.5.4. Notice that since the elements $b \otimes 1, b \in B$ span $B \otimes_k L$ as an L-vector space, the elements $\varphi(b \otimes 1), b \in B$ span $\mathrm{M}_n(L)$ as an L-vector space. Hence, an isomorphism $\varphi : B \otimes_k L \xrightarrow{\sim} \mathrm{M}_n(L)$ induces an isomorphism

$$\varphi : (B \otimes_k L, \tau \otimes ^-) \xrightarrow{\sim} (\mathrm{M}_n(L), ^*)$$

if and only if

$$\varphi(\tau(b) \otimes 1) = \varphi(b \otimes 1)^* \text{ for all } b \in B.$$

□

In view of this definition, it does not seem to be very easy to check whether or not a given unitary involution is positive definite. In fact, one may show that τ is positive definite if and only if $\mathrm{Trd}_B(\tau(b)b) > 0$ for all $b \in B \setminus \{0\}$. Since we will not need this criterion for our purpose, we leave the proof of this criterion to the reader as an exercise (see Exercises 6 and 7).

Assume that τ is positive definite, and set $\mathbf{U}_b = \varphi(b \otimes 1)$ for all $b \in B$. Then the equality above may be rewritten as

$$\mathbf{U}_b^* = \mathbf{U}_{\tau(b)} \text{ for all } b \in B.$$

We may now prove an easy lemma.

LEMMA IX.5.5. *The map*

$$B \longrightarrow \mathrm{M}_n(\mathbb{C})$$
$$b \longmapsto \mathbf{U}_b$$

is an injective morphism of k-algebras. Moreover, the induced group morphism

$$B^\times \longrightarrow \mathrm{GL}_n(\mathbb{C})$$
$$b \longmapsto \mathbf{U}_b$$

is injective.

Proof. Clearly, $\mathbf{U}_1 = I_n$. Let $b, b' \in B$. Since φ is a morphism of L-algebras, we have

$$\mathbf{U}_b \mathbf{U}_{b'} = \varphi(b \otimes 1)\varphi(b' \otimes 1) = \varphi(bb' \otimes 1) = \mathbf{U}_{bb'}.$$

Similarly, one shows that $\mathbf{U}_b + \mathbf{U}_{b'} = \mathbf{U}_{b+b'}$, and $\lambda \mathbf{U}_b = \mathbf{U}_{\lambda b}$ for all $\lambda \in k$.

Moreover, $\mathbf{U}_b = I_n$ if and only if $b = 1$, since φ and the canonical map $B \longrightarrow B \otimes_k L$ are injective. This concludes the proof. $\qquad\qquad\square$

Let us come back to the previous considerations. For all $b \in B$, we have

$$\mathbf{U}_b \mathbf{U}_b^* = \mathbf{U}_b \mathbf{U}_{\tau(b)} = \mathbf{U}_{b\tau(b)}.$$

In particular, $\mathbf{U}_b$ is unitary if and only if $b\tau(b) = 1$. This motivates the following definition.

DEFINITION IX.5.6. Let k/k_0 be any quadratic field extension, and let (B, τ) be a central simple k-algebra with an arbitrary unitary k/k_0-involution. We say that $b \in B$ is **unitary** (with respect to τ) if we have $b\tau(b) = 1$.

The set of unitary elements is easily seen to be a subgroup of $B^\times$, that we denote by $\mathbf{U}(B, \tau)$.

EXAMPLE IX.5.7. If k is a number field, $B = \mathrm{M}_n(k)$ and τ is the transpose conjugate of matrices, a unitary element with respect to τ is nothing but a unitary matrix. $\quad\square$

The previous results may then be summarized as follows.

LEMMA IX.5.8. *Let k/k_0 be a quadratic extension of number fields, whose non-trivial automorphism is the complex conjugation, and let (B, τ) be a central simple k-algebra with a positive definite unitary k/k_0-involution. The map*

$$B \longrightarrow \mathrm{M}_n(\mathbb{C})$$
$$b \longmapsto \mathbf{U}_b$$

induces an injective group morphism

$$\mathbf{U}(B, \tau) \longrightarrow \mathbf{U}_n(\mathbb{C})$$
$$b \longmapsto \mathbf{U}_b.$$

Let (B, τ) be a central simple k-algebra with a positive definite unitary k/k_0-involution. Keeping the previous notation, for any subgroup $\mathcal{G}$ of $\mathbf{U}(B, \tau)$, we get a unitary space-time code

$$\mathcal{C}_{\mathcal{G}} = \{\mathbf{U}_b = \varphi(b \otimes 1) \mid b \in \mathcal{G}\}.$$

Hence, the main idea here is to take our unitary code $\mathcal{C}$ to be a finite subset of some $\mathcal{C}_{\mathcal{G}}$, where $\mathcal{G}$ is a subgroup of $\mathbf{U}(B, \tau)$. In this case, if B is division, we will have $\delta_{min}(\mathcal{C}) > 0$ (i.e. the code is fully diverse), and

$$\delta_{min}(\mathcal{C}) \geq \delta_{min}(\mathcal{C}_{\mathcal{G}}).$$

Of course, we still need to find a way to estimate $\delta_{min}(\mathcal{C}_{\mathcal{G}})$.

EXAMPLE IX.5.9. Assume that B has a maximal subfield $L \subset \mathbb{C}$, and that τ is positive definite. By Proposition IV.1.6, we have a unique isomorphism of L-algebras $\varphi : B \otimes_k L \xrightarrow{\sim} \mathrm{M}_n(L)$ satisfying

$$\varphi(b \otimes 1) = M_b \quad \text{for all } b \in B,$$

where M_b is the matrix of left multiplication by b with respect to a fixed L-basis of $B \otimes_k L$. In this case, for every $b \in \mathbf{U}(B, \tau)$, we will have $\mathbf{U}_b = M_b$, and thus, for any subgroup $\mathcal{G}$ of $\mathbf{U}(B, \tau)$, we will get

$$\mathcal{C}_\mathcal{G} = \{\mathbf{U}_b = M_b \mid b \in \mathcal{G}\}.$$

$\square$

Thus, the difficulty now is to find examples of division algebras B carrying a positive definite unitary involution τ. Lemma IX.4.3 provides such examples.

EXAMPLE IX.5.10. Let k/k_0 be a quadratic extension of number fields, and L/k be a Galois extension of number fields with Galois group G, such that complex conjugation induces a k_0-automorphism of L which commutes with the elements of G.

Let $B = (\xi, L/k, G)$ be a crossed-product algebra of degree n, where ξ is a 2-cocycle satisfying $|\xi_{\sigma,\rho}|^2 = 1$ for all $\sigma, \rho \in G$.

By Lemma IX.4.3, there exists a unique unitary involution τ on B such that

$$M_{\tau(b)} = M_b^* \quad \text{for all } b \in B,$$

where M_b is the matrix of left multiplication by b in the L-basis $(f_\sigma)_{\sigma \in G}$. By Remark IX.5.4 and the previous example, τ is positive definite.

Hence, for any subgroup $\mathcal{G}$ of $\mathbf{U}(B, \tau)$, we have

$$\mathcal{C}_\mathcal{G} = \{\mathbf{U}_b = M_b \mid b \in \mathcal{G}\}.$$

$\square$

As explained above, we would like to find a good estimation of the minimum determinant of our unitary code $\mathcal{C}_\mathcal{G}$. The first step is, as in the coherent case, to find a more tractable expression of it. This is given by the next lemma.

LEMMA IX.5.11. *Let k be a number field, let (B, τ) be a central simple k-algebra with a positive definite unitary involution, and let $\mathcal{G}$ be a subgroup of $\mathbf{U}(B, \tau)$. Then we have*

$$\delta_{min}(\mathcal{C}_\mathcal{G}) = \inf_{b \in \mathcal{G} \setminus \{1\}} |\mathrm{Nrd}_B(1 - b)|^2.$$

Proof. For all $b, b' \in \mathbf{U}(B, \tau), b \neq b'$, using Lemma IX.5.8, we get

$$\mathbf{U}_b - \mathbf{U}_{b'} = \mathbf{U}_b(I_n - \mathbf{U}_b^{-1} \mathbf{U}_{b'}) = \mathbf{U}_b(I_n - \mathbf{U}_{b^{-1}b'}).$$

Now, if b and b' run through all elements of $\mathcal{G}$, $b^{-1}b'$ runs through all elements of $\mathcal{G} \setminus \{1\}$. Since the determinant of a unitary matrix is a complex number of modulus 1, we finally get that

$$\delta_{min}(\mathcal{C}_\mathcal{G}) = \inf_{b \in \mathcal{G} \setminus \{1\}} |\det(I_n - \mathbf{U}_b)|^2.$$

Now we have

$$I_n - \mathbf{U}_b = I_n - \varphi(b \otimes 1) = \varphi((1 - b) \otimes 1),$$

and therefore

$$\det(I_n - \mathbf{U}_b) = \det(\varphi((1 - b) \otimes 1)) \quad \text{for all } b \in \mathcal{G} \setminus \{1\}.$$

Thus, by Remark IV.2.8, as in the coherent case, this equality may be rewritten as

$$\det(I_n - \mathbf{U}_b) = \mathrm{Nrd}_B(1 - b) \quad \text{for all } b \in \mathcal{G} \setminus \{1\},$$

and therefore

$$\delta_{min}(\mathcal{C}_\mathcal{G}) = \inf_{b \in \mathcal{G} \setminus \{1\}} |\mathrm{Nrd}_B(1 - b)|^2.$$

This concludes the proof.   □

It is about time to show how to find classes of unitary elements in a division algebra with a unitary involution (B, τ) by looking at elements of norm 1 in some subfields of B (see [**39**]).

LEMMA IX.5.12. *Let k be an arbitrary field, and let (B, τ) be a division k-algebra with a k/k_0-involution. Then for every $x \in B$, the following conditions are equivalent:*

(1) *x is unitary with respect to τ;*

(2) *there exists a subfield M of B containing x, such that τ restricts to a non-trivial k_0-automorphism of M and $N_{M/M^{\langle \tau \rangle}}(x) = 1$;*

(3) *there exist a subfield M of B containing x and $u \in M^\times$, such that τ restricts to a non-trivial k_0-automorphism of M and $x = u\tau(u)^{-1}$.*

Proof.

(1) $\Rightarrow$ (2). Let $x \in B$ such that $x\tau(x) = 1$. Since B is a division algebra, the subalgebra of B generated by x is a subfield of B, denoted by $k(x)$, as already pointed out in Chapter I. Moreover, it is stable by τ since $\tau(x) = x^{-1}$. Notice that τ does not restrict to the identity on $k(x)$, since $\tau_{|k}$ is the non-trivial automorphism of k/k_0 , and set $M = k(x)$. Then $M/M^{\langle \tau \rangle}$ is a quadratic extension with Galois group $\mathrm{Gal}(M/M^{\langle \tau \rangle}) = \{\mathrm{Id}_M, \tau_{|M}\}$. The condition $x\tau(x) = 1$ then may be rewritten as

$$N_{M/M^{\langle \tau \rangle}}(x) = 1.$$

(2) $\Rightarrow$ (3). This follows from Hilbert 90, since $M/M^{\langle \tau \rangle}$ is a quadratic extension by assumption on τ.

(3) $\Rightarrow$ (1). Let M and u satisfy the conditions of the lemma. Then $\tau(u) \in M^\times$, and we have

$$x\tau(x) = u\tau(u)^{-1}\tau(u\tau(u)^{-1}) = u\tau(u)^{-1}u^{-1}\tau(u) = 1,$$

since M is commutative. This concludes the proof.   □

We now give a worked out example using a cyclic division algebra satisfying the conditions described in Example IX.4.5 (see [**39**]).

EXAMPLE IX.5.13. Let $k = \mathbb{Q}(j)$ and $L = \mathbb{Q}(j)(\zeta_7 + \zeta_7^{-1})$. We have $\mathrm{Gal}(L/\mathbb{Q}(j)) = \langle \sigma \rangle$, where

$$\sigma: \begin{array}{c} L \longrightarrow L \\ \zeta_7 + \zeta_7^{-1} \longmapsto \zeta_7^2 + \zeta_7^{-2}. \end{array}$$

Consider the cyclic algebra $B = (j, L/\mathbb{Q}(j), \sigma)$, which is a division algebra (see Example VII.5.9). Since $|j|^2 = 1$, by Example IX.4.5, there exists a unitary involution τ on B given by

$$B \longrightarrow B$$
$$\tau : \qquad \lambda_0 + e\lambda_1 + e^2\lambda_2 \longmapsto \overline{\lambda_0} + ej^2\overline{\lambda_2^\sigma} + e^2j^2\overline{\lambda_1^{\sigma^2}}.$$

Example IX.5.10 shows that the left multiplication matrix of any unitary element is a unitary matrix. Following the method explained above, we look for subfields M of B which are stable by τ. The first obvious subfield of B one can think of is L. The restriction of τ on L is the complex conjugation. In this case, unitary elements contained in L are elements of the form $z\overline{z}^{-1}, z \in L^\times$.

Let us consider now the subfield generated by e. Since $1, e, e^2$ are linearly independent over L, they are also linearly independent over k. Therefore $[k(e) : k] \geq 3$, and since $e^3 = \gamma$, we have $[k(e) : k] \leq 3$. Thus $k(e)$ is a subfield of B of degree 3 over k, and the minimal polynomial of e over k is $X^3 - j$. Thus we have an isomorphism

$$k(e) \cong_{\mathbb{Q}} \mathbb{Q}(\zeta_9),$$

where ζ_9 is a primitive 9th-root of 1, this isomorphism mapping e onto ζ_9. Since $\tau(e) = e^{-1}$, the previous isomorphism maps $\tau(e)$ onto $\zeta_9^{-1} = \overline{\zeta_9}$. In other words, we have an isomorphism of k-algebras with involution

$$(k(e), \tau_{|k(e)}) \cong_k (\mathbb{Q}(\zeta_9), \overline{}).$$

It follows that unitary elements in $k(e)$ are mapped onto elements of the form $u\overline{u}^{-1}, u \in \mathbb{Q}(\zeta_9)^\times$ by this isomorphism.

Take for example the element $u = 1+j+\zeta_9+\zeta_9^2 j \in \mathbb{Q}(\zeta_9)$. This element corresponds to the element $y = (1 + j) + e + e^2 j \in k(e)$, and the element $\overline{u}$ should correspond to the element $\tau(y)$. Let us check that it is indeed the case. We have

$$\tau(y) = 1 + j^2 + e^{-1} + e^{-2}j^2 = 1 + j^2 + e^2j^2 + e \in k(e),$$

which corresponds to the element $1 + j^2 + \zeta_9^2 j^2 + \zeta_9 \in \mathbb{Q}(\zeta_9)$, which is nothing but $\overline{u}$, since $\overline{\zeta_9} = \zeta_9^8 = \zeta_9^2 j^2$.

Set $\mathbf{Y} = M_y$. Then we have

$$\mathbf{Y} = \begin{pmatrix} 1 + j & j^2 & j \\ 1 & 1 + j & j^2 \\ j & 1 & 1 + j \end{pmatrix}.$$

Now we also have

$$M_{\tau(y)} = \begin{pmatrix} -j & 1 & j^2 \\ j & -j & 1 \\ j^2 & j & -j \end{pmatrix},$$

which can be checked to be $\mathbf{Y}^*$. Then the element $b = y\tau(y)^{-1}$ is unitary, and its multiplication matrix $\mathbf{U}_b = \mathbf{Y}(\mathbf{Y}^*)^{-1}$ is a unitary matrix, as we may check directly by computation.

One may then take the subgroup $\mathcal{G} = \langle b \rangle$ of $\mathbf{U}(B, \tau)$ generated by b, and consider the unitary code $\mathcal{C}_\mathcal{G}$. We then get an infinite unitary code. One way to see this is as follows: after computations, we get

$$\det(\mathbf{U}_b) = \frac{11}{38} - i\frac{21\sqrt{3}}{38}.$$

Hence, we have $\det(\mathbf{U}_b) = e^{i\theta}$, with $\cos(\theta) = \dfrac{11}{38}$. But one may show by induction that $\cos(2m\theta) \neq 1$ for all $m \geq 1$. In particular, $m\theta$ is never a rational multiple of 2π. It follows that $\mathbf{U}_b^m \neq I_3$ for all $m \geq 1$, which is equivalent to saying that $\mathcal{G}$ is infinite. However, the minimum determinant of such a code is 0, as shown by the next proposition. $\qquad\square$

PROPOSITION IX.5.14. *If $\mathcal{G}$ is a subgroup of $\mathbf{U}(B, \tau)$ containing an element of infinite order, then $\delta_{min}(\mathcal{C}_\mathcal{G}) = 0$.*

Proof. Let $b \in \mathcal{G}$ be an element of infinite order. Since $\mathcal{H} = \langle b \rangle \subset \mathcal{G}$, we have

$$0 \leq \delta_{min}(\mathcal{C}_\mathcal{G}) \leq \delta_{min}(\mathcal{C}_\mathcal{H}).$$

Hence, it is enough to prove that $\delta_{min}(\mathcal{C}_\mathcal{H}) = 0$. Notice that, by assumption on b, the corresponding matrix $\mathbf{U}_b$ has infinite order, since the map

$$\mathbf{U}(B, \tau) \longrightarrow \mathbf{U}_n(\mathbb{C})$$
$$b \longmapsto \mathbf{U}_b$$

is an injective group morphism by Lemma IX.5.8. Since $\mathbf{U}_b$ is unitary, it can be diagonalized and all its eigenvalues have modulus 1.

Let $e^{i\theta_j}, j = 1, \ldots, n$ be the (not necessarily distinct) eigenvalues of $\mathbf{U}_b$. For all $m \in \mathbb{Z}$, the matrix $I_n - \mathbf{U}_b^m$ is similar to the diagonal matrix whose diagonal entries are

$$1 - e^{im\theta_j} = -2i \sin(\frac{m\theta_j}{2})e^{i\frac{m\theta_j}{2}}, j = 1, \ldots, n.$$

It follows easily that

$$\delta_{min}(\mathcal{C}_\mathcal{H}) = 4^n \inf_{m \geq 1} \prod_{j=1}^{n} \sin^2(\frac{m\theta_j}{2}).$$

Now, since $\mathbf{U}_b$ has infinite order, at least one θ_j is not a rational multiple of 2π. For this θ_j, the sequence $(\sin(\frac{m\theta_j}{2}))_{m \geq 1}$ is dense in $[-1, 1]$, so we may find an increasing sequence of integers $(\alpha_m)_{m \geq 1}$ such that $\lim_m \sin(\frac{\alpha_m \theta_j}{2}) = 0$. This implies that $\delta_{min}(\mathcal{C}_\mathcal{H}) = 0$, and this concludes the proof. $\qquad\square$

Let us now combine the techniques discussed so far to give 3×3 code constructions.

Let r, m be two positive integers such that $\gcd(r, m) = 1$ and let n be the order of $r + m\mathbb{Z}$ in $(\mathbb{Z}/m\mathbb{Z})^\times$. Consider the cyclotomic field $L = \mathbb{Q}(\zeta_m)$, where ζ_m is a primitive m-th root of unity. It is of degree $\varphi(m)$ over $\mathbb{Q}$ (φ is the Euler function). Recall that $\mathrm{Gal}(\mathbb{Q}(\zeta_m)/\mathbb{Q})$ is isomorphic to $(\mathbb{Z}/m\mathbb{Z})^\times$. Galois theory shows that the $\mathbb{Q}$-automorphism

$$\sigma : \begin{array}{c} L \longrightarrow L \\ \zeta_m \longmapsto \zeta_m^r \end{array}$$

has order n. Set $K = L^{\langle \sigma \rangle}$, so that L/K is cyclic of degree n. Now set $t = \dfrac{m}{\gcd(r - 1, m)}$. Then $\zeta_m^t \in K$, that is ζ_m^t is fixed by σ. Indeed

$$\sigma(\zeta_m^t) = \zeta_m^t \iff t(r - 1) \equiv 0 \pmod{m},$$

which holds since $t(r - 1) = m(r - 1)/\gcd(r - 1, m)$ is a multiple of m.

We can thus consider the cyclic algebra $B = (\zeta_m^t, \mathbb{Q}(\zeta_m)/K, \sigma)$, and the matrix E of left multiplication by e in B (in the $\mathbb{Q}(\zeta_m)$-basis $(1, e, \ldots, e^{n-1})$) is

$$E = \begin{pmatrix} 0 & 0 & 0 & & \zeta_m^t \\ 1 & 0 & 0 & & 0 \\ 0 & 1 & \ddots & & \vdots \\ 0 & & \ddots & & \\ 0 & & & 1 & 0 \end{pmatrix},$$

while the one of $\zeta_m \in L$ is given by

$$D = \begin{pmatrix} \zeta_m & 0 & & 0 \\ 0 & \zeta_m^{\sigma} & & 0 \\ \vdots & & \ddots & \vdots \\ 0 & 0 & & \zeta_m^{\sigma^{n-1}} \end{pmatrix}.$$

Since $|\zeta_m|^2 = 1$, ζ_m is a unitary element of B. Notice that ζ_m has order m and e has order n in $\mathbf{U}(B, \tau)$. Moreover, since $\zeta_m e = e\zeta_m^{\sigma} = e\zeta_m^{-r}$, easy computations show that the subgroup $\mathcal{G}$ of $\mathbf{U}(B, \tau)$ generated by e and ζ_m has order nm, and that

$$\mathcal{G} = \{e^i \zeta_m^j \mid i = 0, \ldots, n-1, j = 0, \ldots, m-1\}.$$

Thus, the unitary code

$$\mathcal{C}_{\mathcal{G}} = \{E^i D^j \mid i = 0, \ldots, n-1, j = 0, \ldots, m-1\}$$

has nm elements. This construction, as well as the following example, has been presented in [**34**].

EXAMPLE IX.5.15. Take $n = 3$, $r = 4$, $m = 21$ and $t = 7$. We thus have the cyclic algebra $B = (j, \mathbb{Q}(\zeta_{21})/K, \sigma)$, where $\sigma : \zeta_{21} \longmapsto \zeta_{21}^4$. We get the family of 63 unitary matrices $E^i D^j$, where

$$D = \begin{pmatrix} \zeta_{21} & 0 & 0 \\ 0 & \zeta_{21}^{16} & 0 \\ 0 & 0 & \zeta_{21}^4 \end{pmatrix}, \quad E = \begin{pmatrix} 0 & 0 & j \\ 1 & 0 & 0 \\ 0 & 1 & 0 \end{pmatrix}.$$

Let us prove that B is a division algebra in this case. Since we have a group isomorphism

$$(\mathbb{Z}/21\mathbb{Z})^{\times} \simeq \mathbb{Z}/2\mathbb{Z} \times \mathbb{Z}/2\mathbb{Z} \times \mathbb{Z}/3\mathbb{Z},$$

$K/\mathbb{Q}$ is the unique biquadratic subextension of $\mathbb{Q}(\zeta_{21})/\mathbb{Q}$, namely

$$K = \mathbb{Q}(j, \sqrt{-7}).$$

One may check that $7\mathcal{O}_K = \mathfrak{p}^2 \bar{\mathfrak{p}}^2$ for a suitable prime ideal $\mathfrak{p}$ of $\mathcal{O}_K$, so that the residue field $\kappa(\mathfrak{p}) = \mathcal{O}_K/\mathfrak{p}$ is isomorphic to $\mathbb{F}_7$. Moreover, $\mathfrak{p}$ totally ramifies in $\mathbb{Q}(\zeta_{21})$.

Since B has degree 3, it is either division or split. By Lemma VII.4.6, B is split if and only if $\bar{j}^2 = \bar{1} \in \mathbb{F}_7$. If it were true, this would imply that any third root of unity in $\mathbb{F}_7$ is equal to $\bar{1}$, which is not the case, since $\bar{2}$ is such a root of unity. Hence B is a division K-algebra, and the family of unitary matrices above is fully diverse.

Here, $\delta_{min}(\mathcal{C}_\mathcal{G}) \leq |\det(I_3 - D)|^2 \approx 0.21$.   □

It is interesting to notice that these families of unitary matrices are very similar to the ones obtained using fixed point free groups representation in [49]. Notice that the family of unitary matrices defined before Example IX.5.15 may be extended further. Indeed, pick any element $x \in \mathbb{Q}(\zeta_m)$ such that $|x|^2 = 1$. Since $\mathbb{Q}(\zeta_m)/\mathbb{Q}(\zeta_m + \zeta_m^{-1})$ is a quadratic field extension, whose unique non-trivial automorphism is the complex conjugation, the multiplication matrix

$$M_x = \begin{pmatrix} x & 0 & & 0 \\ 0 & x^\sigma & & 0 \\ \vdots & & \ddots & \vdots \\ 0 & 0 & & x^{\sigma^{n-1}} \end{pmatrix}$$

is unitary, a fact which also follows from direct computations.

EXAMPLE IX.5.16. As in Example IX.5.15, let us consider the cyclic algebra $B = (j, L/K, \sigma)$, with $K = \mathbb{Q}(j, \sqrt{-7}), L = \mathbb{Q}(\zeta_{21})$ and

$$\sigma : \begin{array}{c} L \longrightarrow L \\ \zeta_{21} \longmapsto \zeta_{21}^4. \end{array}$$

Take for example the following element x and its conjugates

$$\begin{aligned} x &= \frac{1}{2}(-\zeta_{21}^3 - \zeta_{21}^5 + \zeta_{21}^8 - \zeta_{21}^{10} + \zeta_{21}^{11}) \\ x^\sigma &= \frac{1}{2}(1 + 2\zeta_{21}^3 - \zeta_{21}^5 + \zeta_{21}^7 + 2\zeta_{21}^{10} - \zeta_{21}^{11}) \\ x^{\sigma^2} &= \frac{1}{2}(-1 + 2\zeta_{21}^2 - \zeta_{21}^3 + 2\zeta_{21}^5 - \zeta_{21}^7 + \zeta_{21}^8 - \zeta_{21}^{10} + 2\zeta_{21}^{11}) \end{aligned}$$

whose matrix representation is given by

$$F = \begin{pmatrix} x & 0 & 0 \\ 0 & x^\sigma & 0 \\ 0 & 0 & x^{\sigma^2} \end{pmatrix}.$$

Straightforward computations show that x has modulus 1 and that F is unitary.   □

This simple result allows to construct codebooks of the form

$$\mathcal{C}_i = \{E^k D^\ell F^i \mid \ell = 0, \ldots, m-1, \ k = 0, \ldots, n-1\}$$

where i can be chosen to vary into a given range and F is a matrix of the form

$$F = \begin{pmatrix} x & 0 & & 0 \\ 0 & x^\sigma & & 0 \\ \vdots & & \ddots & \vdots \\ 0 & 0 & & x^{\sigma^{n-1}} \end{pmatrix},$$

where $x \in \mathbb{Q}(\zeta_m)$ satisfies $|x|^2 = 1$.

Notice however that if x has finite multiplicative order, then x is an mth-root of 1, and we do not get any new matrices. Therefore, if we want to extend the codebook, x will necessarily have infinite order. A side effect is that the minimum determinant will tend to decrease dramatically by Proposition IX.5.14.

We now prove a result which will allows us to compute the minimum determinant in terms of norms of cyclotomic extensions.

If $n \geq 1$ is an integer, we denote by ϕ_n the nth cyclotomic polynomial.

PROPOSITION IX.5.17. *Let k be a number field, and let D be an arbitrary central division k-algebra of degree n. If $D^\times$ has an element d of order m, the following properties hold:*

(1) *we have $\mu_{d,\mathbb{Q}} = \phi_m$ and $k(d) \cong_k k(\zeta_m)$, where $\zeta_m \in \mathbb{C}$ is some primitive m^{th}-root of 1;*

(2) *$[k(\zeta_m) : k] \mid n$ and either $\zeta_m \in k$ or $D \otimes_k k(\zeta_m)$ is not a division algebra;*

(3) *$\dfrac{\varphi(m)}{gcd(\varphi(m), [k : \mathbb{Q}])} \mid n$. In particular, $\varphi(m) \mid n[k : \mathbb{Q}]$;*

(4) *we have the equalities*

$$\begin{aligned} \mathrm{Nrd}_D(1-d) &= N_{k(\zeta_m)/k}(1-\zeta_m)^{\frac{n}{[k(\zeta_m):k]}} \\ &= \left(\mu_{\zeta_m,k}(1)\right)^{\frac{n}{[k(\zeta_m):k]}}. \end{aligned}$$

Moreover, if D has prime degree and property (2) holds, then $D^\times$ has an element of order m.

Proof. Let $d \in D^\times$ be an element of order m, so we have $d^m = 1$. Hence $\mu_{d,\mathbb{Q}}$ divides $X^m - 1$, and therefore $\mu_{d,\mathbb{Q}}$ is a cyclotomic polynomial ϕ_r, for some $r \mid m$. Since $\phi_r \mid X^r - 1$, we have $d^r - 1 = 0$, and therefore $m \mid r$. Hence $r = m$ and $\mu_{d,\mathbb{Q}} = \phi_m$.

Now $\mu_{d,k} \mid \mu_{d,\mathbb{Q}}$, so there exists $\zeta_m \in \mathbb{C}$, a primitive m^{th}-root of 1, such that $\mu_{d,k}(\zeta_m) = 0$. Elementary Galois theory then shows that we have an isomorphism of k-algebras

$$k(d) \cong_k k(\zeta_m),$$

which maps d onto ζ_m. This proves (1). Notice for later use that such an isomorphism preserves degrees and norms. Therefore, $k(\zeta_m)$ is isomorphic to a subfield of D. By Lemma IV.1.4, $[k(\zeta_m) : k] \mid n$. If $\zeta_m \notin k$, $k(\zeta_m)/k$ has degree at least 2, and Proposition V.3.2 shows that $D \otimes_k k(\zeta_m)$ is not a division algebra. Now assume that D has prime degree, and that $[k(\zeta_m) : k] \mid n$. If $\zeta_m \in k$, then $\zeta_m \in D^\times$ has order m. If $D \otimes_k k(\zeta_m)$ is not a division algebra, then $k(\zeta_m)/k$ is an extension of degree at least 2 dividing n. By Proposition V.3.2, $k(\zeta_m)$ is isomorphic to a subfield of D. Such an isomorphism maps ζ_m onto an element $d \in D^\times$ of order m. This proves (2) and the last part of the proposition.

Now let $t = gcd(\varphi(m), [k : \mathbb{Q}])$, and write $[k : \mathbb{Q}] = rt$ and $\varphi(m) = st$, with $gcd(r, s) = 1$. We have to prove that $s \mid n$. From the equalities

$$[k(\zeta_m) : \mathbb{Q}] = [k(\zeta_m) : k][k : \mathbb{Q}] = [k(\zeta_m) : \mathbb{Q}(\zeta_m)][\mathbb{Q}(\zeta_m) : \mathbb{Q}],$$

we get that $[k(\zeta_m) : k]r = [k(\zeta_m) : \mathbb{Q}(\zeta_m)]s$. In particular, we have $s \mid [k(\zeta_m) : k]$. Since $[k(\zeta_m) : k] = [k(d) : k]$, and $[k(d) : k] \mid n$ by Lemma IV.1.4, we get (3).

It remains to prove (4). Let M be a maximal subfield of D containing d (which exists by Remark IV.1.10 (4)). Then it contains $1 - d$, and by Chapter IV, Exercise 1, we have

$$\mathrm{Nrd}_D(1-d) = N_{M/k}(1-d) = N_{k(d)/k}(1-d)^{\frac{n}{[k(d):k]}}.$$

Thus, we have

$$\mathrm{Nrd}_D(1 - d) = N_{k(\zeta_m)/k}(1 - \zeta_m)^{\frac{n}{[k(\zeta_m):k]}}.$$

Now notice that $k(\zeta_m) = k(1 - \zeta_m)$, and that

$$\mu_{1-\zeta_m,k} = (-1)^{[k(\zeta_m):k]}\mu_{\zeta_m,k}(1 - X).$$

It follows immediately that $N_{k(\zeta_m)/k}(1 - \zeta_m) = \mu_{\zeta_m,k}(1)$, and this proves (4). This concludes the proof. $\square$

COROLLARY IX.5.18. *Let k be a number field, and let D be a central division k-algebra of degree n. Then any subgroup of $D^\times$ is either finite or has an element of infinite order.*

Proof. Let $\mathcal{G}$ be a subgroup of $D^\times$. Assume that every element of $\mathcal{G}$ has finite order. By the previous proposition, if $g \in \mathcal{G}$ has order m, then $\varphi(m) \mid n[k : \mathbb{Q}]$. This implies that m may take only finitely many values. In particular, the least common multiple of the orders of the elements of $\mathcal{G}$ is finite, that is $\mathcal{G}$ has finite exponent. Now if L is a maximal subfield of D, the injective k-algebra morphism

$$\varphi_{D,L} : D \hookrightarrow \mathrm{M}_n(L)$$

induces an injective group morphism $D^\times \hookrightarrow \mathrm{GL}_n(L)$. It follows that $\mathcal{G}$ is isomorphic to a subgroup of $\mathrm{GL}_n(\mathbb{C})$ of finite exponent. By a celebrated theorem of Burnside, this implies that $\mathcal{G}$ is finite. $\square$

We now summarize our results on the minimum determinant of unitary codes in the following theorem.

THEOREM IX.5.19. *Let $\mathcal{G}$ be a subgroup of $\mathbf{U}(B, \tau)$, and assume that B is a division k-algebra of degree n. Then $\mathcal{G}$ is either finite or has an element of infinite order. Moreover, the following properties hold:*

(1) *If $\mathcal{G}$ has an element of infinite order, then $\delta_{min}(\mathcal{C}_G) = 0$;*

(2) *If $\mathcal{G}$ is finite, we have*

$$
\begin{aligned}
\delta_{min}(\mathcal{C}_G) &= \inf_{b \in \mathcal{G}\setminus\{1\}} |N_{k(\zeta_{m_b})/k}(1 - \zeta_{m_b})|^{\frac{2n}{[k(\zeta_{m_b}):k]}} \\
&= \inf_{b \in \mathcal{G}\setminus\{1\}} |\mu_{\zeta_{m_b},k}(1)|^{\frac{2n}{[k(\zeta_{m_b}):k]}},
\end{aligned}
$$

where m_b is the order of b.

Proof.   This follows from Proposition IX.5.14, Proposition IX.5.17 and Corollary IX.5.18, since a subgroup of $\mathbf{U}(B, \tau)$ is a subgroup of $B^\times$. $\square$

REMARK IX.5.20. If $b \in \mathcal{G}$ has finite order m_b, Proposition IX.5.17 shows that that $\mathrm{Nrd}_B(1 - b)$ only depends on m_b. In particular, $\delta_{min}(\mathcal{C}_G)$ only depends on the orders of the elements of $\mathcal{G}$, and not on the group itself. Therefore, to compute the minimum determinant, one may proceed as follows:

(1) compute the set of values $S = \{m_b \mid b \in \mathcal{G} \setminus \{1\}\}$;

(2) choose a subset $\mathcal{S}$ of $\mathcal{G}$ such that each element of S is obtained by a unique element of $\mathcal{S}$;

(3) the observation above shows that we have

$$
\begin{aligned}
\delta_{min}(\mathcal{C}_\mathcal{G}) &= \inf_{b \in \mathcal{S}} |\mathrm{Nrd}_B(1-b)|^2 \\
&= \inf_{b \in \mathcal{S}} |\det(I_n - \mathbf{U}_b)|^2 \\
&= \inf_{b \in \mathcal{S}} |N_{k(\zeta_{m_b})/k}(1-\zeta_{m_b})|^{\frac{2n}{[k(\zeta_{m_b}):k]}} \\
&= \inf_{b \in \mathcal{S}} |\mu_{\zeta_{m_b},k}(1)|^{\frac{2n}{[k(\zeta_{m_b}):k]}} .
\end{aligned}
$$

$\square$

EXAMPLE IX.5.21. As an example, we compute the minimum determinant of the code presented in Example IX.5.15. This code is built of the cyclic division K-algebra $B = (j, \mathbb{Q}(\zeta_{21})/K, \sigma)$, where $K = \mathbb{Q}(j, \sqrt{-7})$ and the group $\mathcal{G}$ is the group of order 63, generated by e and ζ_{21}. The possible values for the order of an element of $\mathcal{G}$ are $1, 3, 7, 9, 21, 63$. Notice that $\mathcal{G}$ is not abelian, hence not cyclic, so $\mathcal{G}$ has no elements of order 63. We also look only at non-trivial elements of $\mathcal{G}$, so we may also discard 1. One may also check that $\mathcal{G}$ has no element of order 9. By considering $\zeta_{21}^7, \zeta_{21}^3$ and ζ_{21}, we see that the other possible values are obtained.

The remark above shows that it is enough to compute $|\det(I_3 - D^m)|^2$ for $m = 1, 3, 7$. Here, the minimum is obtained for $m = 1$, so

$$
\delta_{min}(\mathcal{C}_G) = |\det(I_3 - D)|^2 \approx 0.21.
$$

Computing $\mu_{\zeta_{21},K}$ shows that the exact value is $\dfrac{5 - \sqrt{21}}{2}$.

Notice that we may extend this code by considering the group

$$
\mathcal{G}' = \langle e, \zeta_{21}, -j \rangle = \langle e, \zeta_{21}, -1 \rangle.
$$

It is easy to check that $\mathcal{G}' \simeq \mathcal{G} \times \{\pm 1\}$, so that

$$
\mathcal{C}_{\mathcal{G}'} = \{\pm \mathbf{U} \mid \mathbf{U} \in \mathcal{C}_\mathcal{G}\}.
$$

Hence the orders of non-trivial elements of $\mathcal{G}$ are now $2, 3, 6, 7, 14, 21, 42$, and -1, $-\zeta_{21}^7, -\zeta_{21}^3$ and $-\zeta_{21}$ are elements of order $2, 6, 14$ and 42 respectively. One may compute that

$$
\delta_{min}(\mathcal{C}_{\mathcal{G}'}) = |\det(I_3 + D^2)|^2 = \frac{23 - 5\sqrt{21}}{2} \approx 0.04.
$$

$\square$

REMARK IX.5.22. Let $\mathcal{G}$ be a finite subgroup of $\mathbf{U}(B, \tau)$. One way to get a group $\mathcal{G}$ whose cardinality is as large as possible is to ensure that G contains all the roots of unity lying in k. However, we will often get a small minimum determinant, as we proceed to show now.

Indeed, Theorem IX.5.19 shows in particular that, if $\zeta_m \in k$, then we have

$$
\delta_{min}(\mathcal{C}_\mathcal{G}) \le |1 - \zeta_m|^{2n},
$$

for any finite subgroup $\mathcal{G}$ of $\mathbf{U}(B, \tau)$ (where n is the degree of B over k), that is

$$
\delta_{min}(\mathcal{C}_\mathcal{G}) \le (2\sin(\frac{\pi}{m}))^{2n}.
$$

Now, if $m \geq 7$ and $n \geq 2$, this shows that

$$\delta_{min}(\mathcal{C}_{\mathcal{G}}) \leq (2\sin(\frac{\pi}{7}))^4 < 0.57.$$

The upper bound above also shows that $\delta_{min}(\mathcal{C}_{\mathcal{G}})$ will tend to be very small if the base field k contains roots of 1 of large order. $\square$

The next lemma, used together with the previous proposition, allows to compute the minimum determinant of a unitary code $\mathcal{C}_{\mathcal{G}}$ when $k/\mathbb{Q}$ is a purely imaginary quadratic extension.

LEMMA IX.5.23. *Let $k/\mathbb{Q}$ be a purely imaginary quadratic extension, and let $m \geq 2$. Then we have*

$$|N_{k(\zeta_m)/k}(1 - \zeta_m)|^2 = \begin{cases} p & \text{if } m = p^r, r \geq 1 \text{ and } k \subset \mathbb{Q}(\zeta_m) \\ p^2 & \text{if } m = p^r, r \geq 1 \text{ and } k \not\subset \mathbb{Q}(\zeta_m) \\ 1 & \text{otherwise} . \end{cases}$$

Proof. Since $k/\mathbb{Q}$ is a purely quadratic imaginary extension, we have

$$|N_{k(\zeta_m)/k}(1 - \zeta_m)|^2 = N_{k(\zeta_m)/\mathbb{Q}}(1 - \zeta_m) = N_{\mathbb{Q}(\zeta_m)/\mathbb{Q}}(1 - \zeta_m)^{[k(\zeta_m):\mathbb{Q}(\zeta_m)]}.$$

Therefore, we have

$$|N_{k(\zeta_m)/k}(1 - \zeta_m)|^2 = \begin{cases} N_{\mathbb{Q}(\zeta_m)/\mathbb{Q}}(1 - \zeta_m) \text{ if } k \subset \mathbb{Q}(\zeta_m) \\ N_{\mathbb{Q}(\zeta_m)/\mathbb{Q}}(1 - \zeta_m)^2 \text{ if } k \not\subset \mathbb{Q}(\zeta_m). \end{cases}$$

Notice now that $\mu_{1-\zeta_m,\mathbb{Q}} = (-1)^{\varphi(m)}\phi_m(1 - X)$. It follows that we have

$$N_{\mathbb{Q}(\zeta_m)/\mathbb{Q}}(1 - \zeta_m) = \phi_m(1).$$

If p is a prime number, we have the well-known relations

$$\phi_m(X^p) = \begin{cases} \phi_{mp} & \text{if } p \mid m \\ \phi_{mp}\phi_m & \text{if } p \nmid m. \end{cases}$$

It follows easily that $\phi_m(1) = p$ if $m = p^r, r \geq 1$ and $\phi_m(1) = 1$ otherwise. This concludes the proof. $\square$

REMARKS IX.5.24. Assume that $k/\mathbb{Q}$ is a purely imaginary quadratic extension. Let $\mathcal{G}$ be a subgroup of $\mathbf{U}(B, \tau)$, and assume that B is a division k-algebra of degree n.

(1) It follows from Theorem IX.5.19 that, if $\mathcal{G}$ contains an element of order m, we have

$$\varphi(m) \mid 2n \text{ if } k \subset \mathbb{Q}(\zeta_m)$$

and

$$\varphi(m) \mid n \text{ if } k \not\subset \mathbb{Q}(\zeta_m).$$

(2) If $\mathcal{G}$ is finite, and contains an element whose order is not a prime power, then we have

$$\delta_{min}(\mathcal{C}_{\mathcal{G}}) = 1.$$

Indeed, this is an immediate consequence of Theorem IX.5.19 and Lemma IX.5.23.

$\square$

If we want to find subgroups $\mathcal{G}$ of $\mathbf{U}(B, \tau)$ such that $\delta_{min}(\mathcal{C}_\mathcal{G}) > 0$, Theorem IX.5.19 says that all elements of $\mathcal{G}$ need to have finite (multiplicative) order. Such elements may be found as follows: choose a subfield M of B which is stable by τ, and look for unitary elements among 'roots of 1 in M', that is elements $b \in M$ such that $\mu_{b,\mathbb{Q}} = \phi_m$ for some $m \geq 1$. Moreover, a list of possible values for m may be found using points (2) and (3) of Proposition IX.5.17.

However, the product of elements of finite order is not necessarily an element of finite order. Hence, once we found several unitary elements of finite order, we are not still ensured that the group they generate only have elements of finite order. The next lemma shows how to avoid this problem.

LEMMA IX.5.25. *Let Λ be a subring of B which is finitely generated as an abelian group. Then $\mathbf{U}(B, \tau) \cap \Lambda^\times$ is finite.*

Proof. Let $n = \deg(B)$. By Lemma IX.5.8, the map

$$\psi: \quad \begin{array}{c} \mathbf{U}(B, \tau) \longrightarrow \mathbf{U}_n(\mathbb{C}) \\ b \longmapsto \mathbf{U}_b \end{array}$$

identifies $\mathbf{U}(B, \tau) \cap \Lambda^\times$ to a subgroup of $\mathbf{U}_n(\mathbb{C})$. Since Λ is a finitely generated group, it is countable, and therefore so is $\psi(\mathbf{U}(B, \tau) \cap \Lambda^\times)$. Since $\mathbf{U}_n(\mathbb{C})$ is compact, any countable subset of $\mathbf{U}_n(\mathbb{C})$ is finite. In particular, $\psi(\mathbf{U}(B, \tau) \cap \Lambda^\times)$ is finite, and thus $\mathbf{U}(B, \tau) \cap \Lambda^\times$ is also finite. This concludes the proof. $\square$

REMARK IX.5.26. Such a subring Λ always exists. One may even assume that Λ contains a k-basis of B. For example, let $e_1, \ldots, e_{n^2}$ be a k-basis of B. For all $1 \leq i, j \leq n^2$, there exists $m_{ij} \in \mathbb{Z}$ such that

$$m_{ij} e_i e_j \in \sum_{i=1}^{n^2} e_i \mathcal{O}_k.$$

Let m be the least common multiple of the $m'_{ij}s$. Then we have

$$m e_i e_j \in \sum_{i=1}^{n^2} e_i \mathcal{O}_k \text{ for } 1 \leq i, j \leq n^2.$$

Let Λ be the $\mathcal{O}_k$-module generated by $1, m e_1, \ldots, m e_{n^2}$. By construction, Λ is a subring of B, which contains a k-basis of B, and which is finitely generated as an abelian group (since it is finitely generated as an $\mathcal{O}_k$-module). $\square$

EXAMPLE IX.5.27. Let $k = \mathbb{Q}(i)$, and consider the central simple k-algebra

$$B = (\zeta_8, \frac{1+2i}{\sqrt{5}}, i, k(\sqrt{2}, \sqrt{5})/k, \sigma, \rho),$$

where σ and ρ are defined in a unique way by

$$\sigma(\sqrt{2}) = \sqrt{2}, \sigma(\sqrt{5}) = -\sqrt{5} \text{ and } \rho(\sqrt{2}) = -\sqrt{2}, \rho(\sqrt{5}) = \sqrt{5}.$$

By Example VIII.2.10 (2), this is a division k-algebra. As explained at the beginning of Section VIII.4, in the case where complex conjugation induces an automorphism of L which commutes with the element of $\mathrm{Gal}(L/k)$, the values of the cocycle corresponding to an algebra $(a, b, u, L/k, \sigma, \rho)$ will have modulus 1 if and only if a, b and u have modulus 1. All these conditions are fulfilled here, so by Lemma IX.4.3, there is an involution τ on B such that $\tau_{|L}$ is the complex conjugation, $\tau(e) = e^{-1}$, and $\tau(f) = f^{-1}$, where e, f are the generators of B.

The elements e and f are unitary and e has finite order. However, f has infinite order. Since $\sqrt{5}$ and f commute, $M = k(f, \sqrt{5})$ is a subfield of B which is stable by τ. Let $\alpha \in \mathbb{C}$ such that $\alpha^2 = \dfrac{1 + 2i}{\sqrt{5}}$. Notice that $(\alpha\overline{\alpha})^2 = 1$, and thus $\alpha\overline{\alpha} = 1$. We then have an isomorphism of k-algebras

$$M \cong_k k(\alpha, \sqrt{5})$$

which maps f onto α and $\sqrt{5}$ onto $\sqrt{5}$. Since $\tau(f) = f^{-1}$ is mapped onto $\alpha^{-1} = \overline{\alpha}$, it easily implies that we have an isomorphism of k-algebras with involution

$$(M, \tau_{|M}) \cong_k (k(\alpha, \sqrt{5}), \bar{}).$$

Set $\theta = \dfrac{1 + \sqrt{5}}{2}$. One may check that the element

$$\zeta = -\frac{\theta}{2} + \alpha\left(\frac{1}{2} + i\frac{1 - \theta}{2}\right)$$

satisfies $\zeta^5 = i$, that is ζ is a primitive 20th-root of 1. In particular, $\zeta\overline{\zeta} = 1$. Using the isomorphism above, this yields an element

$$z = -\frac{\theta}{2} + f\left(\frac{1}{2} + i\frac{1 - \theta}{2}\right) \in B,$$

which is unitary and which has order 20.

Straightforward computations show that

$$e^{16} = 1, z^{20} = 1 \text{ and } ze = ez^{-3}.$$

It follows easily that the ring $\Lambda = \mathcal{O}_k[e, z]$ is finitely generated as an $\mathcal{O}_k$-module, hence as an abelian group. One may show that

$$\mathcal{G} = \mathbf{U}(B, \tau) \cap \Lambda^{\times} = \{e^{\ell} z^m \mid \ell = 0, \ldots, 3, m = 0, \ldots, 19\},$$

is a group of order 80. Therefore, the unitary code $\mathcal{C}_\mathcal{G}$ consists of 80 matrices. If $E = \mathbf{U}_e, Z = \mathbf{U}_z$, we have

$$E = \begin{pmatrix} 0 & \zeta_8 & 0 & 0 \\ 1 & 0 & 0 & 0 \\ 0 & 0 & 0 & -\zeta_8 \\ 0 & 0 & 1 & 0 \end{pmatrix},$$

and

$$Z = \begin{pmatrix} -i\frac{\theta}{2} & 0 & \frac{1}{2} + i\frac{\theta-1}{2} & 0 \\ 0 & i\frac{\theta-1}{2} & 0 & -\frac{\theta}{2} - \frac{i}{2} \\ \frac{1}{2} - i\frac{\theta-1}{2} & 0 & -i\frac{\theta}{2} & 0 \\ 0 & -\frac{\theta}{2} + \frac{i}{2} & 0 & i\frac{\theta-1}{2} \end{pmatrix}.$$

In other words,

$$\mathcal{C}_\mathcal{G} = \{E^{\ell} Z^m \mid \ell = 0, \ldots, 3, m = 0, \ldots, 19\}.$$

By Remark IX.5.24 (2), $\delta_{min}(\mathcal{C}_\mathcal{G}) = 1$. $\square$

Let us give a last example.

EXAMPLE IX.5.28. Let $k = \mathbb{Q}(j)$, and let $L = k(\zeta_7)$. Then L/k is a cyclic extension of degree 6, a generator σ of $\mathrm{Gal}(L/k)$ being given by

$$\sigma : \begin{array}{c} L \longrightarrow L \\ \zeta_7 \longmapsto \zeta_7^3. \end{array}$$

Let $B = (-j, k(\zeta_7)/k, \sigma)$. Let us show that B is a division k-algebra.

Notice that the unique subextensions of degree 2 or 3 of L/k are respectively $L^{(2)} = k(\sqrt{-7})/k$ and $L^{(3)} = k(\zeta_7 + \zeta_7^{-1})/k$. By Proposition VII.3.5, B is a division k-algebra if and only if the cyclic k-algebras

$$B_2 = (-j, L^{(2)}/k, \sigma_{|_{L^{(2)}}}) \quad \text{and} \quad B_3 = (-j, L^{(3)}/k, \sigma_{|_{L^{(3)}}})$$

are not split (or, equivalently, are division algebras). A direct application of Proposition VII.5.7 shows that B_2 is a division k-algebra. By Example VII.5.9, B_3 is a division k-algebra (notice that the generator of the Galois group used here differs from the one used in Example VII.5.9, but it does not matter by Corollary VII.1.11 (2)). This proves that B is a division k-algebra. Since B fulfills all the assumptions of Lemma IX.4.3, we may consider the unitary involution τ described in this lemma.

If e is the canonical generator of B, then e is a unitary element of order 36. Moreover, $z = \zeta_7$ is a unitary element of order 7. It follows from the equality $ze = ez^\sigma = ez^{-2}$ that the subgroup $\mathcal{G}$ of $\mathbf{U}(B, \tau)$ generated by e and z is a finite group of order $36 \cdot 7 = 252$. Theorem IX.5.19 and Lemma IX.5.23 then show that $\delta_{min}(\mathcal{C}_\mathcal{G}) = 1$.

In other words, the unitary code

$$\mathcal{C}_\mathcal{G} = \{E^\ell Z^m \mid \ell = 0, \ldots, 35, m = 0, \ldots, 6\}$$

consists of 252 unitary matrices and satisfies $\delta_{min}(\mathcal{C}_\mathcal{G}) = 1$, where

$$E = \begin{pmatrix} 0 & 0 & 0 & 0 & 0 & -j \\ 1 & 0 & 0 & 0 & 0 & 0 \\ 0 & 1 & 0 & 0 & 0 & 0 \\ 0 & 0 & 1 & 0 & 0 & 0 \\ 0 & 0 & 0 & 1 & 0 & 0 \\ 0 & 0 & 0 & 0 & 1 & 0 \end{pmatrix} \quad \text{and} \quad Z = \begin{pmatrix} \zeta_7^{-2} & 0 & 0 & 0 & 0 & 0 \\ 0 & \zeta_7^4 & 0 & 0 & 0 & 0 \\ 0 & 0 & \zeta_7 & 0 & 0 & 0 \\ 0 & 0 & 0 & \zeta_7^2 & 0 & 0 \\ 0 & 0 & 0 & 0 & \zeta_7^{-4} & 0 \\ 0 & 0 & 0 & 0 & 0 & \zeta_7^{-1} \end{pmatrix}.$$

One may also obtain a code with a better minimum determinant by considering a restricted number of matrices. Indeed, let us consider the subgroup $\mathcal{H}$ of $\mathbf{U}(B, \tau)$ generated by e^4 and z. Then $\mathcal{H}$ is a semidirect product of the cyclic group $\langle e^4 \rangle$ of order 9 and of the cyclic group $\langle z \rangle$ of order 7. Straightforward arguments then show that the orders of non-trivial elements of $\mathcal{H}$ are $3, 7$ or 9. By Lemma IX.5.23, the unitary code

$$\mathcal{C}_\mathcal{H} = \{E^{4\ell} Z^m \mid \ell = 0, \ldots, 8, m = 0, \ldots, 6\}$$

has 63 elements and satisfies $\delta_{min}(\mathcal{C}_\mathcal{H}) = 3$. $\qquad\square$

All the previous constructions reveal one weakness, namely the encoding process of the information symbols, or equivalently the parametrization of unitary matrices. In this case, this is done by putting the data in the powers of the unitary matrices.

A much nicer parametrization can be obtained through the Cayley transform as follows. Let $\mathbf{A}$ be an $n \times n$ hermitian matrix. Its Cayley transform is given by

$$\mathbf{U} = (I_n + i\mathbf{A})^{-1}(I_n - i\mathbf{A}),$$

which maps the hermitian matrix $\mathbf{A}$ to a unitary matrix $\mathbf{U}$. Encoding is done by mapping the information symbols $s_1, \ldots, s_r$ into $\mathbf{A}$, typically by fixing a basis of hermitian matrices $(\mathbf{A}_1, \ldots, \mathbf{A}_r)$, $r \le n^2$, and by writing

$$\mathbf{A} = \sum_{\ell=1}^{r} s_\ell \mathbf{A}_\ell.$$

It was shown in [19] that the family of matrices

$$\mathbf{U}_\ell = (I_n + i\mathbf{A}_\ell)^{-1}(I_n - i\mathbf{A}_\ell), \ell = 1, \ldots, r$$

is fully diverse if and only if the family $\mathbf{A}_\ell$, $\ell = 1, \ldots, r$ of hermitian matrices is. It thus makes sense to start with a family of hermitian matrices inside a division algebra [38]. Notice that $i\mathbf{A}$ is in fact skew-hermitian, and that conversely any skew-hermitian matrix may be written in the form $i\mathbf{A}$, where $\mathbf{A}$ is hermitian. Hence, it is equivalent to find a family of skew-hermitian matrices inside a division algebra.

We now briefly explain how to achieve this using algebras with involution. Recall that, if (B, τ) is a central simple k-algebra with a unitary k/k_0-involution, we denote by $\mathrm{Sym}(B, \tau)$ and $\mathrm{Skew}(B, \tau)$ the set of symmetric and skew-symmetric elements (with respect to τ) respectively. We then have the following result:

LEMMA IX.5.29. *Let k be an arbitrary field, and let (B, τ) be any central simple k-algebra of degree n, endowed with a unitary k/k_0-involution. Then the following properties hold:*

(1) *We have $B = \mathrm{Sym}(B, \tau) \oplus \mathrm{Skew}(B, \tau)$;*
(2) $\dim_{k_0}(\mathrm{Sym}(B, \tau)) = \dim_{k_0}(\mathrm{Skew}(B, \tau)) = n^2$;
(3) *for all $a \in \mathrm{Skew}(B, \tau)$ such that $1 \pm a \in B^\times$, the element*

$$(1 + a)^{-1}(1 - a)$$

is unitary.

Proof. Since k has characteristic not 2, we see immediately that

$$\mathrm{Sym}(B, \tau) \cap \mathrm{Skew}(B, \tau) = \{0\}.$$

Moreover, for all $b \in B$, we have $\dfrac{b + \tau(b)}{2} \in \mathrm{Sym}(B, \tau)$, $\dfrac{b - \tau(b)}{2} \in \mathrm{Skew}(B, \tau)$ and

$$b = \frac{b + \tau(b)}{2} + \frac{b - \tau(b)}{2}.$$

This proves (1). Now notice that since k/k_0 is a quadratic extension, there exists $x \in k^\times$ such that $\tau(x) = -x$, since τ restricts to the non-trivial k_0 automorphism of k/k_0. It is easy to check that if $a \in \mathrm{Skew}(B, \tau)$, then $xa \in \mathrm{Sym}(B, \tau)$, and that the map

$$\mathrm{Skew}(B, \tau) \longrightarrow \mathrm{Sym}(B, \tau)$$

$$a \longmapsto xa$$

is an isomorphism of k_0-vector spaces. In particular,

$$\dim_{k_0}(\mathrm{Sym}(B, \tau)) = \dim_{k_0}(\mathrm{Skew}(B, \tau)).$$

The desired result then follows from (1), since we have

$$\dim_{k_0}(B) = 2\dim_k(B) = 2n^2.$$

Finally, if $a \in \mathrm{Skew}(B, \tau)$ satisfies $1 \pm a \in B^\times$, we have

$$\tau((1+a)^{-1}(1-a)) = \tau(1-a)(\tau(1+a))^{-1} = (1+a)(1-a)^{-1},$$

and therefore

$$((1+a)^{-1}(1-a))\tau((1+a)^{-1}(1-a)) = (1+a)^{-1}(1-a)(1+a)(1-a)^{-1} = 1,$$

since $(1-a)$ and $(1+a)$ commute. Hence $(1+a)^{-1}(1-a)$ is unitary, and this concludes the proof. $\qquad\qquad\qquad\qquad\qquad\qquad\qquad\qquad\qquad\qquad\qquad\square$

Now assume as previously that k is a number field, that (B, τ) is a central simple k-algebra of degree n with a positive definite unitary k/k_0-involution, and that complex conjugation is the non-trivial automorphism of k/k_0. Notice that a skew symmetric element a is never equal to ± 1 (since ± 1 is symmetric). Hence, if B is a division k-algebra, for all $a \in \mathrm{Skew}(B, \tau)$, the element $(1+a)^{-1}(1-a)$ is unitary, and its left multiplication matrix is a unitary matrix. Therefore, to construct a family of unitary elements having a good parametrization, it is enough to fix a k_0-basis $(a_1, \dots, a_{n^2})$ of $\mathrm{Skew}(B, \tau)$, and to consider elements of the form

$$(1 + \sum_{i=1}^{n^2} x_i a_i)^{-1}(1 - \sum_{i=1}^{n^2} x_i a_i), x_i \in k_0.$$

In other words, the codewords will have the form

$$(1 + \sum_{i=1}^{n^2} x_i \mathbf{U}_{a_i})^{-1}(1 - \sum_{i=1}^{n^2} x_i \mathbf{U}_{a_i}), x_i \in k_0.$$

Notice that here, the matrices $\mathbf{U}_{a_i}$ are not unitary anymore, but skew-hermitian.

Now, it is easy to see that

$$\mathrm{Skew}(B, \tau) = \{\tau(b) - b \mid b \in B\}.$$

Therefore, to find a k_0-basis of $\mathrm{Skew}(B, \tau)$, it is enough to find n^2 elements $b_1, \dots, b_{n^2} \in B$ such that the elements $\tau(b_i) - b_i, i = 1, \dots, n^2$ are linearly independent over k_0. This condition is equivalent to saying that $\mathrm{Span}_{k_0}(b_1, \dots, b_{n^2}) \cap \mathrm{Sym}(B, \tau) = \{0\}$.

The parametrization above then takes the form

$$(1 + \sum_{i=1}^{n^2} x_i(\tau(b_i) - b_i))^{-1}(1 - \sum_{i=1}^{n^2} x_i(\tau(b_i) - b_i)), x_i \in k_0,$$

and using the relation $\mathbf{U}_{\tau(b)} = \mathbf{U}_b^*$, we get that the corresponding codewords are

$$(1 + \sum_{i=1}^{n^2} x_i(\mathbf{U}_{b_i}^* - \mathbf{U}_{b_i}))^{-1}(1 - \sum_{i=1}^{n^2} x_i(\mathbf{U}_{b_i}^* - \mathbf{U}_{b_i})), x_i \in k_0.$$

Here, the matrices $\mathbf{U}_{b_i}$ are not unitary nor skew-hermitian, but satisfy $\mathbf{U}_{\tau(b_i)} = \mathbf{U}_{b_i}^*$ by construction. As before, most of these elements will have infinite multiplicative order. Once again, to avoid this problem, it will be necessary to find skew-symmetric elements a such that $1 \pm a \in \Lambda^\times$, where Λ is a subring of B which is finitely generated as an abelian group, but we will only obtain a finite number of matrices, corresponding then to a limited number of possibilities for the vectors $(x_1, \ldots, x_{n^2})$. However, the minimum determinant will be large.

To the contrary, if we allow the x_i's to describe a fixed set $\mathcal{S}$ of information symbols, most likely at least one of the corresponding unitary matrices will have infinite order, and the minimum determinant will tend to get closer from zero when the size of $\mathcal{S}$ increases.

<h2 style="text-align:center">EXERCISES</h2>

1. Let k be a field of characteristic different from 2, and let $Q = (a, b)_k$.

   (a) Show that the map
   $$\gamma : \begin{array}{c} Q \longrightarrow Q \\ x + yi + zj + tij \longmapsto x - yi - zj - tij \end{array}$$
   is an involution of the first kind on Q, called the **canonical involution** of Q.

   (b) Let σ be an involution of the first kind on Q. Show that we have $\sigma \circ \gamma = \gamma \circ \sigma$.

   Hint: Write $\sigma = \mathrm{Int}(u) \circ \gamma$ for some $u \in Q^\times$ satisfying $\gamma(u) = \pm u$.

2. Let k/k_0 be a separable quadratic extension, with $\mathrm{char}(k_0) \neq 2$, and let A be a central simple k_0-algebra. Show that we have
   $$\mathrm{Cor}(A \otimes_{k_0} k) \cong_{k_0} A \otimes_{k_0} A,$$
   and deduce that $\mathrm{Cor}_{k/k_0} \circ \mathrm{Res}_{k/k_0}$ is multiplication by 2 in $\mathrm{Br}(k_0)$.

3. Let k/k_0 be a separable quadratic extension, with $\mathrm{char}(k_0) \neq 2$. Let $a \in k^\times$, and let $b \in k_0^\times$. Show that we have
   $$\mathrm{Cor}_{k/k_0}((a, b)_k) \sim_{k_0} (N_{K/k}(a), b)_{k_0}.$$

   Hint: Use the fact that a quaternion algebra is a particular crossed product.

4. Let k/k_0 be a separable quadratic extension, with $\mathrm{char}(k_0) \neq 2$, and let $\bar{}$ its non-trivial k_0-automorphism.

   (a) Let $Q = (a, b)_k$, with $a, b \in k_0^\times$. Show that the map
   $$\tau_0 : \begin{array}{c} Q \longrightarrow Q \\ x + yi + zj + tij \longmapsto \bar{x} - \bar{y}i - \bar{z}j - \bar{t}ij \end{array}$$
   is a unitary involution, and that we have a canonical isomorphism
   $$((a, b)_{k_0} \otimes_{k_0} k, \gamma \otimes \bar{}) \cong_k (Q, \tau_0),$$
   where γ is the canonical involution of $(a, b)_{k_0}$.

   (b) Let Q be a division quaternion k-algebra carrying a unitary involution τ. Show that there exist $a, b \in k_0^\times$ such that $Q = (a, b)_k$ and that $\tau = \tau_0$.

   Hint: Let $Q = (u, v)_k, u, v \in k^\times$, and let Q^0 be the k-linear subspace spanned by i, j and ij. Recall that if $q \in Q^0 \setminus \{0\}$, then q^2 lies in $k^\times$. Show that there exists a non-zero element $e \in \mathrm{Skew}(Q, \tau) \cap Q^0$. Set $e^2 = a$, and show that

$a \in k_0^\times$. Now pick $f \in Q^0$ such that $fe = -ef$, so that $f^2 = a' \in k^\times$ and $1, e, f, ef$ is a k-basis of Q. Show that $\tau(f)e = -e\tau(f)$, and deduce that

$$\tau(f) = zf + tef, \ z, t \in k.$$

Use the fact that $\tau^2(f) = f$ and that a' is not a square (why?) to show that $\tau(f) = \pm f$. Set $i' = e$ and $j' = f$ or $\sqrt{d}f$ (where $k = k_0(\sqrt{d})$) according to the sign. Show that $\tau(j') = -j'$ and that $b = j'^2 \in k_0^\times$. Conclude.

(c) Show that the previous result remains true if Q is split.

5. Let (B, τ) be a central simple k-algebra with a unitary involution. Show that there is a bijection between the set of skew-symmetric elements such that $1 \pm a \in B^\times$ and the set of unitary elements such that $1 \pm u \in B^\times$. In particular, if B is a division algebra, show that there is a 1-1 correspondence between $\mathrm{Skew}(B, \tau)$ and $\mathbf{U}(B, \tau) \setminus \{\pm 1\}$.

6. Let k/k_0 be a quadratic extension of arbitrary fields, let $\bar{}$ its non-trivial k_0-automorphism, and let (B, τ) be a central simple k-algebra with a unitary k/k_0-involution.

(a) Show that for all $b \in B$, we have $\mathrm{Trd}_B(\tau(b)) = \overline{\mathrm{Trd}_B(b)}$. Deduce that the map

$$T_{(B,\tau)}: \begin{array}{c} B \times B \longrightarrow k_0 \\ (b, b') \longmapsto \mathrm{Trd}_B(\tau(b)b'). \end{array}$$

is a hermitian form on B with respect to $(k, \bar{})$.

Let L/k be a field extension, and let $\alpha : L \longrightarrow L$ be a ring automorphism of L extending $\bar{}$. If $h : V \times V \longrightarrow k$ is a hermitian form on a finite dimensional k-vector space V with respect to $(k, \bar{})$, we denote by $h_{(L,\alpha)}$ the unique hermitian form on $V \otimes_k L$ with respect to (L, α) satisfying

$$h_{(L,\alpha)}(v_1 \otimes \lambda_1, v_2 \otimes \lambda_2) = \alpha(\lambda_1)\lambda_2 h(v_1, v_2)$$

for all $v_1, v_2 \in V, \lambda_1, \lambda_2 \in L$.

(b) Let L_0 the subfield of L fixed by α. Check that the map $\tau \otimes \alpha$ is a unitary L/L_0-involution on the L-algebra $B \otimes_k L$, and that we have

$$T_{(B \otimes_k L, \tau \otimes \alpha)} \cong_L (T_{(B,\tau)})_{(L,\alpha)}.$$

(c) If $(B, \tau) \cong_k (B', \tau')$, show that $T_{(B,\tau)} \cong_k T_{(B',\tau')}$.

(d) Let $B = \mathrm{M}_n(k)$ and let $\tau = \mathrm{Int}(H) \circ *$, for some invertible hermitian matrix $H \in \mathrm{M}_n(k)$. Finally, let h_H the hermitian form on k^n defined by

$$h_H: \begin{array}{c} k^n \times k^n \longrightarrow k \\ (X, Y) \longmapsto X^* H Y \end{array}$$

If $h_H \cong_k \langle \lambda_1, \ldots, \lambda_n \rangle, \lambda_i \in k_0^\times$, show that

$$T_{(B,\tau)} \cong_k \langle 1, \lambda_1 \lambda_2^{-1}, \ldots, \lambda_j \lambda_i^{-1}, \ldots \rangle.$$

Hint: Show that, if $H = P^* D P$, where D is diagonal and P is invertible, then $(\mathrm{M}_n(k), \mathrm{Int}(H) \circ *) \cong_k (\mathrm{M}_n(k), \mathrm{Int}(D) \circ *)$.

7. Assume that k/k_0 is a quadratic extension of number fields, whose non-trivial k_0-automorphism is the complex conjugation. In particular, $k_0 \subset \mathbb{R}$. Let (B, τ) be a central simple k_0-algebra with a unitary k/k_0-involution. Show that τ is positive definite if and only if T_τ is a positive definite hermitian form, that is if and only if

$$\mathrm{Trd}_B(\tau(b)b) > 0 \quad \text{for all } b \in B \setminus \{0\}.$$

Hint: Use the previous exercise.

8. Let Q be a division k-algebra, where $k/\mathbb{Q}$ is a quadratic imaginary extension. What the the possible values for the orders of the elements of $Q^\times$?

APPENDIX A

Tensor products

In the following, k will denote an arbitrary field.

A.1. Tensor product of vector spaces

DEFINITION A.1.1. Let $V_1, \ldots, V_n, V$ be k-vector spaces. A map

$$f : V_1 \times \cdots \times V_n \longrightarrow V$$

is n-*linear* if it is k-linear with respect to each argument. In other words, f is n-linear if for all $1 \le i \le n$, the map

$$f_i : V_1 \times \cdots \times V_{i-1} \times V_{i+1} \times \cdots \times V_n \longrightarrow V$$

defined by

$$f_i(x_1, \ldots, x_{i-1}, x_{i+1}, \ldots, x_n) = f(x_1, \ldots, x_n)$$

is k-linear.

The set of n-linear maps $f : V_1 \times \cdots \times V_n \longrightarrow V$ is a k-vector space, denoted by

$$\mathrm{Ml}_n(V_1, \ldots, V_n; V).$$

DEFINITION A.1.2. Let $V_1, \ldots, V_n$ be k-vector spaces. A **tensor product** of $V_1, \ldots, V_n$ is a pair (T, τ), where T is a k-vector space and $\tau : V_1 \times \cdots \times V_n \longrightarrow T$ is a n-linear map satisfying the following universal property:

for every k-vector space V and all $\varphi \in \mathrm{Ml}_n(V_1, \ldots, V_n; V)$, there exists a **unique** $f \in \mathrm{Hom}_k(T, V)$ such that

$$\varphi = f \circ \tau.$$

In other words, (T, τ) is a tensor product of $V_1, \ldots, V_n$ if for every k-vector space V, the map

$$\Theta : \begin{aligned} \mathrm{Hom}_k(T, V) &\longrightarrow \mathrm{Ml}_n(V_1, \ldots, V_n; V) \\ f &\longmapsto f \circ \tau \end{aligned}$$

is an isomorphism of k-vector spaces.

Two tensor products (T_1, τ_1) and (T_2, τ_2) of $V_1, \ldots, V_n$ are **isomorphic** if there exists an isomorphism of k-vector spaces $f : T_1 \longrightarrow T_2$ such that

$$\tau_2 = f \circ \tau_1.$$

We denote it by

$$(T_1, \tau_1) \simeq (T_2, \tau_2).$$

The following lemma shows that if such a tensor product exists, it is unique.

LEMMA A.1.3. *Let (T_1, τ_1) and (T_2, τ_2) be two tensor products of the vector spaces $V_1, \ldots, V_n$. Then there exists a **unique** isomorphism of tensor products*

$$f : (T_1, \tau_1) \xrightarrow{\sim} (T_2, \tau_2).$$

This isomorphism is the unique k-linear map f satisfying

$$\tau_2 = f \circ \tau_1.$$

Proof. Since (T_1, τ_1) is a tensor product of $V_1, \ldots, V_n$ and since τ_2 lies in $\mathrm{Ml}_n(V_1, \ldots, V_n, T_2)$, there exists a linear map $f : T_1 \longrightarrow T_2$ satisfying

$$\tau_2 = f \circ \tau_1.$$

Similarly, since (T_2, τ_2) is a tensor product of $V_1, \ldots, V_n$ and since τ_1 lies in $\mathrm{Ml}_n(V_1, \ldots, V_n, T_1)$, there exists a linear map $g : T_2 \longrightarrow T_1$ satisfying

$$\tau_1 = g \circ \tau_2.$$

We then have

$$\tau_1 = (g \circ f) \circ \tau_1 \text{ et } \tau_2 = (f \circ g) \circ \tau_2.$$

By definition of a tensor product , Id_{T_1} is the unique k-linear map h satisfying

$$\tau_1 = h \circ \tau_1.$$

Thus we have $g \circ f = \mathrm{Id}_{T_1}$. Similarly, we have $f \circ g = \mathrm{Id}_{T_2}$. It follows that f is an isomorphism of k-vector spaces. Since by definition we have $\tau_2 = f \circ \tau_1$, f is then an isomorphism of tensor products.

Suppose now that f' is another isomorphism of tensor products. Then we have

$$\tau_2 = f \circ \tau_1 = f' \circ \tau_1.$$

By definition of a tensor product, we get $f = f'$. $\square$

We are now going to prove that such a tensor product effectively exists. Let $V_1, \ldots, V_n, V$ be k-vector spaces, and let $\mathcal{F}$ be the vector space of all maps from $V_1 \times \cdots \times V_n$ to V.

For all $(x_1, \ldots, x_n) \in V_1 \times \cdots \times V_n$, we denote by $\{x_1, \ldots, x_n\}$ the element of $\mathcal{F}$ defined by

$$\{x_1, \ldots, x_n\}(x_1', \ldots, x_n') = \begin{cases} 1 & \text{if } (x_1', \ldots, x_n') = (x_1, \ldots, x_n) \\ 0 & \text{otherwise} \end{cases}$$

We denote by $\mathcal{M}$ the linear subspace of $\mathcal{F}$ generated by the elements

$$\{x_1, \ldots, x_n\}, x_i \in V_i.$$

REMARK A.1.4. It is easy to check that the family

$$(\{x_1, \ldots, x_n\})_{(x_1, \ldots, x_n) \in V_1 \times \cdots \times V_n}$$

is a k-basis of $\mathcal{M}$. $\square$

Let $\mathcal{N}$ be the linear subspace of $\mathcal{M}$ generated by the elements

$$\{x_1, \ldots, x_{i-1}, x_i + y_i, x_{i+1}, \ldots, x_n\} - \{x_1, \ldots, x_{i-1}, x_i, x_{i+1}, \ldots, x_n\} - \\ \{x_1, \ldots, x_{i-1}, y_i, x_{i+1}, \ldots, x_n\}$$

and
$$\{x_1, \ldots, x_{i-1}, \lambda_i \cdot x_i, x_{i+1}, \ldots, x_n\} - \lambda_i \cdot \{x_1, \ldots, x_{i-1}, x_i, x_{i+1}, \ldots, x_n\},$$
for all $1 \le i \le n, x_i, y_i \in V_i, \lambda_i \in k$.

Set $V_1 \otimes_k \cdots \otimes_k V_n = \mathcal{M}/\mathcal{N}$. We denote by $x_1 \otimes \cdots \otimes x_n$ the equivalence class of $\{x_1, \ldots, x_n\}$ modulo $\mathcal{N}$.

Since the elements of the form $\{x_1, \ldots, x_n\}$ span $\mathcal{M}$, every element of $V_1 \otimes_k \cdots \otimes_k V_n$ may be written as a linear combination of the elements
$$x_1 \otimes \cdots \otimes x_n, (x_1, \ldots, x_n) \in V_1 \times \cdots \times V_n.$$

DEFINITION A.1.5. The elements of $V_1 \otimes_k \cdots \otimes_k V_n$ of the form
$$x_1 \otimes \cdots \otimes x_n$$
are called **elementary tensors**.

Notice that, by definition of the vector space $V_1 \otimes_k \cdots \otimes_k V_n$, we have
$$\lambda \cdot \overline{\{x_1, \ldots, x_n\}} = \overline{\lambda \cdot \{x_1, \ldots, x_n\}} = \overline{\{x_1, \ldots, x_{i-1}, \lambda \cdot x_i, x_{i+1}, \ldots, x_n\}},$$
for all $1 \le i \le n, x_i, y_i \in V_i, \lambda \in k$. We then have
$$\lambda \cdot x_1 \otimes \cdots \otimes x_n = x_1 \otimes \cdots \otimes x_{i-1} \otimes (\lambda \cdot x_i) \otimes x_{i+1} \otimes \cdots \otimes x_n,$$
for all $1 \le i \le n, x_i, y_i \in V_i, \lambda \in k$.

Hence, every element of the vector space $V_1 \otimes_k \cdots \otimes_k V_n$ is in fact a **sum** of elementary tensors. Moreover, by very definition, the map
$$\tau : V_1 \times \cdots \times V_n \longrightarrow V_1 \otimes_k \cdots \otimes_k V_n \text{ defined by}$$
$$\tau(x_1, \ldots, x_n) = x_1 \otimes \cdots \otimes x_n,$$
for all $(x_1, \ldots, x_n) \in V_1 \times \cdots \times V_n$, is n-linear.

LEMMA A.1.6. *The pair $(V_1 \otimes_k \cdots \otimes_k V_n, \tau)$ is a tensor product of $V_1, \ldots, V_n$. In other words, for every k-vector space V and for every n-linear map*
$$\varphi : V_1 \times \cdots \times V_n \longrightarrow V,$$
there exists a unique linear map $f : V_1 \otimes_k \cdots \otimes_k V_n \longrightarrow V$ satisfying
$$f(x_1 \otimes \cdots \otimes x_n) = \varphi(x_1, \ldots, x_n) \text{ for all } x_i \in V_i.$$

Proof. Let V be k-vector space, and let $\varphi \in \mathrm{Ml}_n(V_1, \ldots, V_n; V)$. Since the family
$$(\{x_1, \ldots, x_n\})_{(x_1, \ldots, x_n) \in V_1 \times \cdots \times V_n}$$
is a basis of $\mathcal{M}$, there exists a k-linear map $\psi : \mathcal{N} \longrightarrow V$ such that
$$\psi(\{x_1, \ldots, x_n\}) = \varphi(x_1, \ldots, x_n) \text{ for all } (x_1, \ldots, x_n) \in V_1 \times \cdots \times V_n.$$

Since φ is n-linear, one can see easily that $\ker(\psi)$ contains $\mathcal{N}$. Therefore, there exists a unique linear map $f \in \mathrm{Hom}_k(V_1 \otimes_k \cdots \otimes_k V_n, N)$ satisfying
$$f(\overline{u}) = \psi(u) \text{ for all } u \in \mathcal{M}.$$

In particular, this equality applied to $u = \{x_1, \ldots, x_n\}$ show that the map f satisfies
$$f(x_1 \otimes \cdots \otimes x_n) = \varphi(x_1, \ldots, x_n) \text{ for all } (x_1, \ldots, x_n) \in V_1 \times \cdots \times V_n.$$

In other words, $\varphi = f \circ \tau$. Suppose now that there exists another linear map

$$g \in \mathrm{Hom}_k(V_1 \otimes_k \cdots \otimes_k V_n, V)$$

such that $\varphi = g \circ \tau$. We then have in particular

$$f(x_1 \otimes \cdots \otimes x_n) = \varphi(x_1, \ldots, x_n) = g(x_1 \otimes \cdots \otimes x_n),$$

for all $(x_1, \ldots, x_n) \in V_1 \times \cdots \times V_n$. Since f and g are k-linear, and since $V_1 \otimes_k \cdots \otimes_k V_n$ is spanned by elementary tensors, we get that $f = g$. This conclude the proof. $\quad\square$

In view of Lemma A.1.3, we say that $(V_1 \otimes_k \cdots \otimes_k V_n, \tau)$ is **the** tensor product of the k-vector spaces $V_1, \ldots, V_n$. We will often omit τ.

REMARK A.1.7. Since τ is n-linear, we have

$$\tau(x_1, \ldots, x_n) = 0$$

if one of the $x_i's$ is zero. Therefore,

$$x_1 \otimes \cdots \otimes x_n = 0$$

if $x_i = 0$ for some i. $\quad\square$

We continue by identifying the tensor product $V \otimes_k k$.

LEMMA A.1.8. *Let V be k-vector space. Then there exists a unique isomorphism of k-vector spaces*

$$f : V \otimes_k k \xrightarrow{\sim} V$$

satisfying

$$f(x \otimes \lambda) = \lambda \cdot x \text{ for all } x \in M, \lambda \in k.$$

The inverse map is

$$f^{-1} : \begin{array}{c} V \xrightarrow{\sim} V \otimes_k k \\ x \longmapsto x \otimes 1. \end{array}$$

Proof. Since the map

$$V \times k \longrightarrow V$$
$$(x, \lambda) \longmapsto \lambda \cdot x$$

is k-bilinear by definition of a k-vector space, there exists a unique linear map $f : V \otimes_k k \longrightarrow V$ satisfying

$$f(x \otimes \lambda) = \lambda \cdot x \text{ for all } x \in V, \lambda \in k.$$

It is easy to check that the linear map

$$g : \begin{array}{c} V \longrightarrow V \otimes_k k \\ x \longmapsto x \otimes 1 \end{array}$$

is the inverse of f. Details are left to the reader. $\quad\square$

A.2. Basic properties of the tensor product

LEMMA A.2.1. *Let $V_1, V_1', \ldots, V_n, V_n'$ be k-vector spaces, and let*

$$f_i : V_i \longrightarrow V_i'$$

be k-linear maps. Then there exists a unique linear map

$$f_1 \otimes \cdots \otimes f_n : V_1 \otimes_k \cdots \otimes_k V_n \longrightarrow V_1' \otimes_k \cdots \otimes_k V_n'$$

satisfying

$$(f_1 \otimes \cdots \otimes f_n)(x_1 \otimes \cdots \otimes x_n) = f_1(x_1) \otimes \cdots \otimes f_n(x_n) \text{ for all } x_i \in V_i.$$

Moreover, if $f_1, \ldots, f_n$ are isomorphisms, so is $f_1 \otimes \cdots \otimes f_n$.

Proof. It is easy to check that the map

$$\varphi : V_1 \times \cdots \times V_n \longrightarrow V_1' \otimes_k \cdots \otimes_k V_n'$$

defined by

$$\varphi(x_1, \ldots, x_n) = f_1(x_1) \otimes \cdots \otimes f_n(x_n) \text{ for all } x_i \in V_i$$

is n-linear. The universal property of the tensor product show the existence and the uniqueness of $f_1 \otimes \cdots f_n$.

Assume now that each f_i is an isomorphism. By definition, we have

$$(f_1 \otimes \cdots \otimes f_n) \circ (f_1^{-1} \otimes \cdots \otimes f_n^{-1})(x_1' \otimes \cdots \otimes x_n') = x_1' \otimes \cdots \otimes x_n'$$

and

$$(f_1^{-1} \otimes \cdots \otimes f_n^{-1}) \circ (f_1 \otimes \cdots \otimes f_n)(x_1 \otimes \cdots \otimes x_n) = x_1 \otimes \cdots \otimes x_n,$$

for all $x_i \in V_i, x_i' \in V_i'$. Since the elementary tensors span the tensor product, we get easily that $f_1 \otimes \cdots \otimes f_n$ and $f_1^{-1} \otimes \cdots \otimes f_n^{-1}$ are mutually inverse. $\square$

We now establish a commutativity property.

LEMMA A.2.2. *Let $V_1, \ldots, V_n$ be k-vector spaces, and let $\sigma \in \mathfrak{S}_n$. Then there exists a unique isomorphism*

$$f_\sigma : V_1 \otimes_k \cdots \otimes_k V_n \xrightarrow{\sim} V_{\sigma^{-1}(1)} \otimes_k \cdots \otimes_k V_{\sigma^{-1}(n)}$$

satisfying

$$f(x_1 \otimes \cdots \otimes x_n) = x_{\sigma^{-1}(1)} \otimes \cdots \otimes x_{\sigma^{-1}(n)} \text{ for all } x_i \in V_i.$$

Proof. The map

$$\varphi_\sigma : V_1 \times \cdots \times V_n \longrightarrow V_{\sigma^{-1}(1)} \otimes_k \cdots \otimes_k V_{\sigma^{-1}(n)}$$

defined by

$$\varphi_\sigma(x_1, \ldots, x_n) = x_{\sigma^{-1}(1)} \otimes \cdots \otimes x_{\sigma^{-1}(n)} \text{ for all } x_i \in V_i$$

being n-linear, the existence and uniqueness of f_σ is given by the universal property of the tensor product. To see that it is an isomorphism, notice that the maps $f_\sigma \circ f_{\sigma^{-1}}$ and $f_{\sigma^{-1}} \circ f_\sigma$ coincide with the identity maps on elementary tensors; the usual argument concludes the proof. $\square$

We now study the associativity of the tensor product. We start with a lemma.

Lemma A.2.3. *Let $V_1, \ldots, V_n$ be k-vector spaces, and let $1 \leq r \leq n$. Then there exists a unique k-bilinear map*

$$\mu : (V_1 \otimes_k \cdots \otimes_k V_r) \times (V_{r+1} \otimes_k \cdots \otimes_k V_n) \longrightarrow V_1 \otimes_k \cdots \otimes_k V_n$$

such that

$$\mu(x_1 \otimes \cdots \otimes x_r, x_{r+1} \otimes \cdots \otimes x_n) = x_1 \otimes \cdots \otimes x_n \text{ for all } x_i \in V_i.$$

Proof. If such a map μ exists, it is unique. This comes from the bilinearity of μ and from the fact that elementary tensors span the tensor product. It remains to prove its existence.

Let $(x_1, \ldots, x_r) \in V_1 \times \cdots \times V_r$. One may check that

$$V_{r+1} \times \cdots \times V_n \longrightarrow V_1 \otimes_k \cdots \otimes_k V_n$$
$$(x_{r+1}, \ldots, x_n) \longmapsto x_1 \otimes \cdots \otimes x_r \otimes x_{r+1} \otimes \cdots \otimes x_n$$

is $(n-r)$-linear. Thus, there exists a unique linear map

$$f_{x_1, \ldots, x_r} : V_{r+1} \otimes_k \cdots \otimes_k V_n \longrightarrow V_1 \otimes_k \cdots \otimes_k V_n$$

such that

$$f_{x_1, \ldots, x_r}(x_{r+1} \otimes \cdots \otimes x_n) = x_1 \otimes \cdots \otimes x_n$$

for all $(x_{r+1}, \ldots, x_n) \in V_{r+1} \times \cdots \times V_n$.

It is easy to check that the map

$$V_1 \times \cdots \times V_r \longrightarrow \mathrm{Hom}_k(V_{r+1} \otimes_k \cdots \otimes_k V_n, V_1 \otimes_k \cdots \otimes_k V_n)$$
$$(x_1, \ldots, x_r) \longmapsto f_{x_1, \ldots, x_r}$$

is r-linear. Thus, there exists a unique linear map

$$\varphi : V_1 \otimes_k \cdots \otimes_k V_r \longrightarrow \mathrm{Hom}_k(V_{r+1} \otimes_k \cdots \otimes_k V_n, V_1 \otimes_k \cdots \otimes_k V_n)$$

satisfying

$$\varphi(x_1 \otimes \cdots \otimes x_r) = f_{x_1, \ldots, x_r}$$

for all $(x_1, \ldots, x_r) \in V_1 \times \cdots \times V_r$. In particular, applying this equality to $x_{r+1} \otimes \cdots \otimes x_n$ yields

$$\varphi(x_1 \otimes \cdots \otimes x_r)(x_{r+1} \otimes \cdots \otimes x_n) = x_1 \otimes \cdots \otimes x_n \text{ for all } x_i \in M_i.$$

The map

$$\mu : \begin{aligned} (V_1 \otimes_k \cdots \otimes_k V_r) \times (V_{r+1} \otimes_k \cdots \otimes_k V_n) &\longrightarrow V_1 \otimes_k \cdots \otimes_k V_n \\ (z, z') &\longmapsto \varphi(z)(z') \end{aligned}$$

has all the desired properties. $\square$

We now may prove the associativity of the tensor product.

Lemma A.2.4. *Let $V_1, \ldots, V_n$ be k-vector spaces, and let $1 \leq r \leq n$. Then there exists a unique isomorphism of k-vector spaces*

$$\rho : V_1 \otimes_k \cdots \otimes_k V_n \overset{\sim}{\longrightarrow} (V_1 \otimes_k \cdots \otimes_k V_r) \otimes_k (V_{r+1} \otimes_k \cdots \otimes_k V_n)$$

satisfying

$$\rho(x_1 \otimes \cdots \otimes x_n) = (x_1 \otimes \cdots \otimes x_r) \otimes (x_{r+1} \otimes \cdots \otimes x_n) \text{ for all } x_i \in V_i.$$

In particular, we have canonical isomorphisms

$$(V_1 \otimes_k V_2) \otimes_k V_3 \simeq V_1 \otimes_k V_2 \otimes_k V_3 \simeq V_1 \otimes_k (V_2 \otimes_k V_3),$$

and the tensor product is associative.

Proof. Let τ' be the canonical bilinear map associated to the tensor product $(V_1 \otimes_k \cdots \otimes_k V_r) \otimes_k (V_{r+1} \otimes_k \cdots \otimes_k V_n)$. We are going to prove the uniqueness of the tensor product up to unique isomorphism. By the previous lemma, there exists a unique bilinear map

$$\mu : (V_1 \otimes_k \cdots \otimes_k V_r) \times (V_{r+1} \otimes_k \cdots \otimes_k V_n) \longrightarrow V_1 \otimes_k \cdots \otimes_k V_n$$

such that

$$\mu(x_1 \otimes \cdots \otimes x_r, x_{r+1} \otimes \cdots \otimes x_n) = x_1 \otimes \cdots \otimes x_n \text{ for all } x_i \in V_i.$$

Let V be a k-vector space, and let

$$b : (V_1 \otimes_k \cdots \otimes_k V_r) \times (V_{r+1} \otimes_k \cdots \otimes_k V_n) \longrightarrow V$$

be a bilinear map. Using the properties of the tensor product and the bilinearity of b, one may verify easily that the map

$$\varphi : \begin{array}{c} V_1 \times \cdots \times V_n \longrightarrow V \\ (x_1, \ldots, x_n) \longmapsto b(x_1 \otimes \cdots \otimes x_r, x_{r+1} \otimes \cdots \otimes x_n) \end{array}$$

is n-linear. Thus, there exists a unique k-linear map

$$f \in \mathrm{Hom}_k(V_1 \otimes_k \cdots \otimes_k V_n, V)$$

such that

$$f(x_1 \otimes \cdots \otimes x_n) = b(x_1 \otimes \cdots \otimes x_r, x_{r+1} \otimes \cdots \otimes x_n),$$

for all $x_i \in V_i$. This may be rewritten as

$$f(\mu(x_1 \otimes \cdots \otimes x_r, x_{r+1} \otimes \cdots \otimes x_n)) = b(x_1 \otimes \cdots \otimes x_r, x_{r+1} \otimes \cdots \otimes x_n),$$

for all $x_i \in V_i$. Using the bilinearity of μ and b, the linearity of f, and the fact that elementary tensors span the tensor product, we deduce that the previous equality is equivalent to the equality

$$f \circ \mu = b.$$

Thus, the pair $(V_1 \otimes_k \cdots \otimes_k V_n, \mu)$ is a tensor product of the k-vector spaces $V_1 \otimes_k \cdots \otimes_k V_r$ and $V_{r+1} \otimes_k \cdots \otimes_k V_n$. But the pair

$$((V_1 \otimes_k \cdots \otimes_k V_r) \otimes_k (V_{r+1} \otimes_k \cdots \otimes_k V_n), \tau')$$

is also a tensor product of $V_1 \otimes_k \cdots \otimes_k V_r$ and $V_{r+1} \otimes_k \cdots \otimes_k V_n$. By Lemma A.1.3, there exists a unique isomorphism

$$\rho : V_1 \otimes_k \cdots \otimes_k V_n \longrightarrow (V_1 \otimes_k \cdots \otimes_k V_r) \otimes_k (V_{r+1} \otimes_k \cdots \otimes_k V_n).$$

This isomorphism is uniquely determined by the relation $\tau' = \rho \circ \mu$. Applying $(x_1 \otimes \cdots \otimes x_r, x_{r+1} \otimes \cdots \otimes x_n)$ to this equality, we obtain

$$\rho(x_1 \otimes \cdots \otimes x_n) = (x_1 \otimes \cdots \otimes x_r) \otimes (x_{r+1} \otimes \cdots \otimes x_n) \text{ for all } x_i \in V_i.$$

The last part of the lemma being clear, this concludes the proof. $\square$

COROLLARY A.2.5. *Let $V_1, \ldots, V_n$ be k-vector spaces. Then there exists a unique isomorphism of vector spaces*

$$\Theta : (V_1 \otimes_k \cdots \otimes_k V_n)^{\otimes 2} \overset{\sim}{\longrightarrow} (V_1 \otimes_k V_1) \otimes_k \cdots \otimes_k (V_n \otimes_k V_n)$$

satisfying

$$\Theta((v_1 \otimes \cdots \otimes v_n) \otimes (v_1' \otimes \cdots \otimes v_n')) = (v_1 \otimes v_1') \otimes \cdots \otimes (v_n \otimes v_n')$$

for all $v_i, v_i' \in V_i$.

Proof.   Uniqueness of Θ follows from the fact that elementary tensors span the tensor product. We now prove the existence of Θ and show that it is an isomorphism. Using the associativity and the commutativity of the tensor product, we get successive isomorphisms

$$
\begin{aligned}
(V_1 \otimes_k \cdots \otimes_k V_n)^{\otimes 2} \;\; &\cong_k \;\; V_1 \otimes_k \cdots \otimes_k V_n \otimes_k V_1 \otimes_k \cdots \otimes_k V_n \\
&\cong_k \;\; V_1 \otimes_k V_1 \otimes_k \cdots \otimes_k V_n \otimes_k V_n \\
&\cong_k \;\; (V_1 \otimes_k V_1) \otimes_k \cdots \otimes_k (V_n \otimes_k V_n)
\end{aligned}
$$

Therefore we get an isomorphism

$$\Theta : (V_1 \otimes_k \cdots \otimes_k V_n)^{\otimes 2} \overset{\sim}{\longrightarrow} (V_1 \otimes_k V_1) \otimes_k \cdots \otimes_k (V_n \otimes_k V_n).$$

By inspection, one can easily see that we have

$$\Theta((v_1 \otimes \cdots \otimes v_n) \otimes (v_1' \otimes \cdots \otimes v_n')) = (v_1 \otimes v_1') \otimes \cdots \otimes (v_n \otimes v_n')$$

for all $v_i, v_i' \in V_i$. $\square$

We are now going to establish the distributivity of the tensor product with respect to direct sums.

LEMMA A.2.6. *Let $V_1, \ldots, V_n$ be k-vector spaces, and assume that we have $V = \bigoplus_{\alpha \in I} V_\alpha$. For all $\alpha \in I$, let $\iota_\alpha : V_\alpha \longrightarrow V$ be the canonical inclusion, and let*

$$\eta_i : V_1 \otimes_k \cdots \otimes_k V_n \otimes_k V_\alpha \longrightarrow V_1 \otimes_k \cdots \otimes_k V_n \otimes_k V$$

be the map $\eta_\alpha = \mathrm{Id}_{V_1} \otimes \cdots \otimes \mathrm{Id}_{V_n} \otimes_k \iota_\alpha$. Then

$$V_1 \otimes_k \cdots \otimes_k V_n \otimes_k V = \bigoplus_{\alpha \in I} \eta_\alpha (V_1 \otimes_k \cdots \otimes_k V_n \otimes_k V_\alpha).$$

In particular, η_α is injective for all α, and $V_1 \otimes_k \cdots \otimes_k V_n \otimes_k V_\alpha$ identifies to a linear subspace of $V_1 \otimes_k \cdots \otimes_k V_n \otimes_k V$.

Proof.   For all $\alpha \in I$, let $\pi_\alpha : V \longrightarrow V_\alpha$ be the canonical projection, and let

$$\rho_\alpha : V_1 \otimes_k \cdots \otimes_k V_n \otimes_k V \longrightarrow V_1 \otimes_k \cdots \otimes_k V_n \otimes_k V_\alpha$$

be the map $\rho_\alpha = \mathrm{Id}_{V_1} \otimes \cdots \otimes \mathrm{Id}_{V_n} \otimes_k \pi_\alpha$.

Let $z \in V_1 \otimes_k \cdots \otimes_k V_n \otimes_k V$. We check first that $\rho_\alpha(z)$ is zero for almost all $\alpha \in I$, and that $z = \sum_{\alpha \in I} \eta_\alpha(\rho_\alpha(z))$. By linearity, it is enough to do it on elementary tensors.

Suppose that $z = x_1 \otimes \cdots \otimes x_n \otimes x, x_j \in V_j, x \in V$. By assumption, we have $x = \sum_{\alpha \in I} y_\alpha$, where $y_\alpha \in V_\alpha$ are almost all zero. We then have $\pi_\alpha(x) = y_\alpha$, and by definition of ρ_α, we get

$$\rho_\alpha(z) = x_1 \otimes \cdots \otimes x_n \otimes y_\alpha \in V_1 \otimes_k \cdots \otimes_k V_n \otimes_k V_\alpha \text{ for all } \alpha \in I.$$

We then have

$$\sum_{\alpha \in I} \eta_\alpha(\rho_\alpha(z)) = \sum_{\alpha \in I} x_1 \otimes \cdots \otimes x_n \otimes y_\alpha = x_1 \otimes \cdots \otimes x_n \otimes \sum_{\alpha \in I} y_\alpha.$$

Therefore $\sum_{\alpha \in I} \eta_\alpha(\rho_\alpha(z)) = z$, which is what we wanted to prove. We then just have shown that

$$V_1 \otimes_k \cdots \otimes_k V_n \otimes_k V = \sum_{\alpha \in I} \eta_\alpha(V_1 \otimes_k \cdots \otimes_k V_n \otimes_k V_\alpha).$$

It remains to prove the uniqueness of the decomposition.

Let $z'_\alpha \in \eta_\alpha(V_1 \otimes_k \cdots \otimes_k V_n \otimes_k V_\alpha)$ almost all zero such that

$$\sum_{\alpha \in I} z'_\alpha = 0.$$

Let $z_\alpha \in V_1 \otimes_k \cdots \otimes_k V_n \otimes_k V_\alpha$ such that $z'_\alpha = \eta_\alpha(z_\alpha)$. Then we have

$$\sum_{\alpha \in I} \eta_\alpha(z_\alpha) = 0.$$

Let $\beta \in I$. We then have

$$0 = \rho_\beta(0) = \sum_{\alpha \in I} \rho_\beta(\eta_\alpha(z_\alpha)).$$

Notice now that $\rho_\beta \circ \eta_\beta = \text{Id}$ and $\rho_\beta \circ \eta_\alpha = 0$ for all $\alpha \neq \beta$. Indeed, it is enough to check it on elementary tensors, which follows from the definition in this case. We then obtain $z_\alpha = 0$ for all $\alpha \in I$, and thus $z'_\alpha = 0$ for all $\alpha \in I$, which concludes the proof. $\square$

REMARK A.2.7. Lemma A.2.2 and the previous lemma show that the tensor product respects direct sums with respect to each factor (and not only the first one). $\square$

COROLLARY A.2.8. *Let $V_1, \ldots, V_n$ be k-vector spaces, with respective bases*

$$(e_{i_1}^{(1)})_{i_1 \in I_1}, \ldots, (e_{i_n}^{(n)})_{i_n \in I_n}.$$

Then $(e_{i_1}^{(1)} \otimes \cdots \otimes e_{i_n}^{(n)})_{(i_1, \ldots, i_n) \in I_1 \times \cdots \times I_n}$ is a k-basis of $V_1 \otimes_k \cdots \otimes V_n$. In particular, $V_1 \otimes_k \cdots \otimes V_n$ is finite dimensional over k if and only if $V_1, \ldots, V_n$ are, and in this case, we have

$$\dim_k(V_1 \otimes_k \cdots \otimes_k V_n) = \dim_k(V_1) \cdots \dim_k(V_n).$$

Proof. By Lemma A.2.6 and the remark which follows it, we have

$$V_1 \otimes_k \cdots \otimes_k V_n = \bigoplus_{(i_1, \ldots, i_n) \in I_1 \times \cdots \times I_n} k e_{i_1}^{(1)} \otimes \cdots \otimes k e_{i_n}^{(n)}.$$

Now by definition of the tensor product, we have

$$k e_{i_1}^{(1)} \otimes \cdots \otimes k e_{i_n}^{(n)} = k \cdot (e_{i_1}^{(1)} \otimes \cdots \otimes e_{i_n}^{(n)}).$$

This proves that the family

$$(e_{i_1}^{(1)} \otimes \cdots \otimes e_{i_n}^{(n)})_{(i_1,\ldots,i_n) \in I_1 \times \cdots \times I_n}$$

span $V_1 \otimes_k \cdots \otimes V_n$. Let us prove that the elements of this family are linearly independent over k. Assume that we have a linear dependence relation

$$\sum_{i_1,\ldots,i_n} \lambda_{i_1,\ldots,i_n} e_{i_1}^{(1)} \otimes \cdots \otimes e_{i_n}^{(n)} = 0.$$

The previous considerations imply that

$$\lambda_{i_1,\ldots,i_n} e_{i_1}^{(1)} \otimes \cdots \otimes e_{i_n}^{(n)} = 0 \text{ for all } i_1,\ldots,i_n.$$

To conclude, it is enough to prove that $e_{i_1}^{(1)} \otimes \cdots \otimes e_{i_n}^{(n)} \neq 0$ for all $i_1,\ldots,i_n$. This will easily imply the desired conclusion.

Let $i_1,\ldots,i_n$, and consider the n-linear map

$$\varphi : \begin{aligned} V_1 \times \cdots \times V_n &\longrightarrow k \\ (v_1,\ldots,v_n) &\longmapsto (e_{i_1}^{(1)})^*(v_1) \cdots (e_{i_n}^{(n)})^*(v_n), \end{aligned}$$

where $(e_{i_j}^{(j)})^*$ is the i_jth coordinate form. This induces a unique k-linear map $f : V_1 \otimes_k \cdots \otimes_k V_n \longrightarrow k$ satisfying

$$f(v_1 \otimes \cdots \otimes v_n) = (e_{i_1}^{(1)})^*(v_1) \cdots (e_{i_n}^{(n)})^*(v_n) \text{ for all } v_i \in V_i.$$

Applying this equality to $v_j = e_{i_j}^{(j)}$ yields

$$f(e_{i_1}^{(1)} \otimes \cdots \otimes e_{i_n}^{(n)}) = 1 \neq 0.$$

Since f is linear, we get in particular that $e_{i_1}^{(1)} \otimes \cdots \otimes e_{i_n}^{(n)} \neq 0$. This concludes the proof. $\square$

REMARK A.2.9. In particular, if V and W are non-zero vector spaces, then $V \otimes_k W$ is not zero either, which was not obvious from the definition. $\square$

LEMMA A.2.10. *Let L/k be any field extension, let V be a k-vector space and let W an L-vector space.*

Then there exists a unique k-bilinear map

$$\varphi : L \times V \otimes_k W \longrightarrow V \otimes_k W$$

satisfying

$$\varphi(\lambda, v \otimes w) = v \otimes \lambda w \text{ for all } v \in V, w \in W, \lambda \in L.$$

This map endows naturally $V \otimes_k W$ with the structure of an L-vector space.

Proof. The uniqueness of this map comes from the fact that elementary tensors span $V \otimes_k W$ as a k-vector space. We now prove the existence of φ. Since W is an L-vector space, it is easy to check that the map

$$\begin{aligned} L \times W &\longrightarrow W \\ (\lambda, w) &\longmapsto \lambda w \end{aligned}$$

is k-bilinear. Therefore it induces a k-linear map

$$\psi : L \otimes_k W \longrightarrow W$$

satisfying
$$\psi(\lambda \otimes w) = \lambda w \text{ for all } \lambda \in L, w \in W.$$
The map $\mathrm{Id}_V \otimes \psi : V \otimes_k (L \otimes_k W) \longrightarrow V \otimes_k W$ then satisfies
$$\mathrm{Id}_V \otimes \psi(v \otimes (\lambda \otimes w)) = v \otimes \lambda w \text{ for all } \lambda \in L, v \in V, w \in W.$$
Composing this map with the isomorphisms
$$L \otimes_k (V \otimes_k W) \cong_k (L \otimes_k V) \otimes_k W \cong_k (V \otimes_k L) \otimes_k W \cong_k V \otimes_k (L \otimes_k W),$$
we get a k-linear map
$$\psi' : L \otimes_k (V \otimes_k W) \longrightarrow V \otimes_k W$$
satisfying

$$\psi'(\lambda \otimes (v \otimes w)) = v \otimes \lambda w \text{ for all } v \in V, w \in W, \lambda \in L.$$
Now compose ψ' with the k-bilinear map
$$L \times V \otimes W \longrightarrow L \otimes_k (V \otimes_k W)$$
to obtain the desired map φ. Straightforward computations (left to the reader) show that $V \otimes_k W$ is then an L-vector space for the external product law given by φ. $\qquad\square$

To end this section, we prove a simple but very useful property of scalar extension.

LEMMA A.2.11. *Let L/k be an arbitrary field extension and let V be a k-vector space. The unique k-bilinear map*
$$\theta : L \times V \otimes_k L \longrightarrow V \otimes_k L$$
satisfying
$$\theta(\lambda, v \otimes \lambda') = v \otimes \lambda\lambda' \text{ for all } v \in V, \lambda, \lambda' \in L$$
endows $V \otimes_k L$ with the structure of an L-vector space. Moreover, if $(e_i)_{i \in I}$ is a k-basis of V, then $(e_i \otimes 1)_{i \in I}$ is a L-basis of $V \otimes_k L$.

In particular, $V \otimes_k L$ has finite dimension over L if and only if V has finite dimension over k, and in the case we have
$$\dim_L(V \otimes_k L) = \dim_k(V).$$

Proof. The first part is a particular case of the previous lemma. To prove the second part, let $(e_i)_{i \in I}$ be a k-basis of V. Since $(e_i)_{i \in I}$ is a k-basis of V, one can see that every element z of $V \otimes_k L$ may be written as
$$z = \sum_{i \in I} c_i e_i \otimes \lambda_i, c_i \in k, \lambda_i \in L$$
where the $c_i's$ are almost all zero. By the properties of the tensor product and the definition of the external product law of the L-vector space $V \otimes_k L$, we get
$$z = \sum_{i \in I} e_i \otimes (c_i \lambda_i) = \sum_{i \in I} (c_i \lambda_i)(e_i \otimes 1)$$
Hence $(e_i \otimes 1)_{i \in I}$ spans $V \otimes_k L$. To prove that this family consists of linear independent vectors over L, assume that we have an equation of the form
$$\sum_{i \in I} \lambda'_i(e_i \otimes 1) = 0$$

where the λ_i''s are almost all zero, that is

$$\sum_{i \in I} e_i \otimes \lambda_i' = 0$$

Pick a k-basis $(e_j')_{j \in J}$ of the k-vector space L, and write

$$\lambda_i' = \sum_{j \in J} c_{ij} e_j', c_{ij} \in k$$

where the c_{ij}'s are almost all zero. Using the properties of the tensor product, we easily get

$$\sum_{(i,j) \in I \times J} c_{ij} e_i \otimes e_j' = 0$$

Since $(e_i \otimes e_j')_{(i,j) \in I \times J}$ is a k-basis of $V \otimes_k L$ by Corollary A.2.8, we have $c_{ij} = 0$ for all $(i,j) \in I \times J$. Therefore, $\lambda_i' = 0$ for all $i \in I$ and we are done. $\qquad \square$

A.3. Tensor product of k-algebras

In the sequel, all k-algebras are unital and associative.

Let us consider k-algebras $A_1, \ldots, A_n$. Since $A_1, \ldots, A_n$ are k-vector spaces, we may form the tensor product $A_1 \otimes_k \cdots \otimes_k A_n$. We would like to define a structure of k-algebra on $A_1 \otimes_k \cdots \otimes_k A_n$.

PROPOSITION A.3.1. *There exists a unique k-linear map*

$$\Theta : (A_1 \otimes_k \cdots \otimes_k A_n)^{\otimes 2} \longrightarrow (A_1 \otimes_k \cdots \otimes_k A_n)$$

satisfying

$$\Theta((a_1 \otimes \cdots \otimes a_n) \otimes (a_1' \otimes \cdots \otimes a_n')) = a_1 a_1' \otimes \cdots \otimes a_n a_n' \text{ for all } a_i, a_i' \in A_i.$$

Proof. The bilinear map $A_i \times A_i \longrightarrow A_i$ given by the product law induce unique k-linear maps $\alpha_i : A_i \otimes_k A_i \longrightarrow A_i$ and satisfying

$$\alpha_i(a_i \otimes a_i') = a_i a_i' \text{ for all } a_i, a_i' \in A_i.$$

We then have a unique k-linear map

$$\alpha_1 \otimes \cdots \otimes \alpha_n : (A_1 \otimes_k A_1) \otimes_k \cdots \otimes_k (A_n \otimes_k A_n) \longrightarrow A_1 \otimes_k \cdots \otimes A_n$$

satisfying

$$(\alpha_1 \otimes \cdots \otimes \alpha_n)(z_1 \otimes \cdots \otimes z_n) = \alpha_1(z_1) \otimes \cdots \otimes \alpha_n(z_n) \text{ for all } z_i \in A_i \otimes_k A_i.$$

In particular, for all $a_i, a_i' \in A_i$ we have

$$\alpha_1 \otimes \cdots \otimes \alpha_n((a_1 \otimes a_1') \otimes \cdots \otimes (a_n \otimes a_n')) = a_1 a_1' \otimes \cdots \otimes a_n a_n'.$$

Let $h : (A_1 \otimes_k \cdots \otimes A_n)^{\otimes 2} \overset{\sim}{\longrightarrow} (A_1 \otimes_k A_1) \otimes_k \cdots \otimes_k (A_n \otimes_k A_n)$ be the canonical isomorphism of k-vector spaces defined in Corollary A.2.5. Then the map $\Theta = \alpha_1 \otimes \cdots \otimes \alpha_n \circ h$ satisfies the required condition. Uniqueness follows from the fact elementary tensors span the k-vector space $(A_1 \otimes_k \cdots \otimes A_n)^{\otimes 2}$. $\qquad \square$

DEFINITION A.3.2. Let $A_1, \ldots, A_n$ be k-algebras, and

$$\mu : A_1 \otimes_k \cdots \otimes A_n \times_k A_1 \otimes_k \cdots \otimes A_n \longrightarrow A_1 \otimes_k \cdots \otimes A_n$$

be the map defined by

$$\mu(z, z') = \Theta(z \otimes z') \text{ for all } z, z' \in A_1 \otimes_k \cdots \otimes A_n.$$

In other words, we have

$$\mu(a_1 \otimes \cdots \otimes a_n, a_1' \otimes \cdots \otimes a_n') = a_1 a_1' \otimes \cdots \otimes a_n a_n',$$

and more generally

$$\mu\Big(\Big(\sum_i a_1^{(i)} \otimes \cdots \otimes a_n^{(i)}\Big)\Big(\sum_j \alpha_1^{(j)} \otimes \cdots \otimes \alpha_n^{(j)}\Big)\Big) = \sum_{i,j} a_1^{(i)} \alpha_1^{(j)} \otimes \cdots \otimes a_n^{(i)} \alpha_n^{(j)}.$$

Then $\mu : A_1 \otimes_k \cdots \otimes A_n \times_k A_1 \otimes_k \cdots \otimes A_n \longrightarrow A_1 \otimes_k \cdots \otimes A_n$ is clearly k-bilinear, and therefore endows the k-vector space $A_1 \otimes_k \cdots \otimes A_n$ with a structure of k-algebra. It follows from the explicit formula above that if $A_1, \ldots, A_n$ are associative (resp. unital, resp. commutative), so is $A_1 \otimes_k \cdots \otimes A_n$.

The pair $(A_1 \otimes_k \cdots \otimes A_n, \mu)$ is called the **tensor product** of the algebras $A_1, \ldots, A_n$.

REMARK A.3.3. Once can check that the isomorphisms of k-vector spaces

$$A_1 \otimes_k \cdots \otimes_k A_n \cong_k A_{\sigma^{-1}(1)} \otimes_k \cdots \otimes_k A_{\sigma^{-1}(n)}$$

and

$$A_1 \otimes_k \cdots \otimes_k A_n \xrightarrow{\sim} (A_1 \otimes_k \cdots \otimes_k A_r) \otimes_k (A_{r+1} \otimes_k \cdots \otimes_k A_n)$$

are isomorphisms of k-algebras. In particular, the tensor product of k-algebras is commutative and associative. $\qquad \square$

LEMMA A.3.4. *If $A_1, \ldots, A_n$ are k-algebras, the maps*

$$A_i \longrightarrow A_1 \otimes_k \cdots \otimes A_n$$
$$a_i \longmapsto 1_{A_1} \otimes \cdots \otimes 1_{A_{i-1}} \otimes a_i \otimes 1_{A_{i+1}} \otimes \cdots \otimes 1_{A_n}$$

are injective k-algebra morphisms.

Proof. The fact these maps are morphisms of k-algebras follows from explicit computations. To prove the injectivity, choose a basis of A_i containing 1_{A_i}. Now write a_i in this basis and use Corollary A.2.8. $\qquad \square$

We continue by the following proposition:

PROPOSITION A.3.5. *Let $\varphi_1 : A_1 \longrightarrow B, \ldots, \varphi_n : A_n \longrightarrow B$ be morphisms of k-algebras satisfying*

$$\varphi_i(a_i)\varphi_j(a_j) = \varphi_j(a_j)\varphi_i(a_i) \text{ for all } i \neq j \text{ and all } a_i \in A_i, a_j \in A_j.$$

There exists a unique morphism $h : A_1 \otimes_k \cdots \otimes_k A_n \longrightarrow B$ of unital k-algebras satisfying

$$h(1_{A_1} \otimes \cdots \otimes 1_{A_{i-1}} \otimes a_i \otimes 1_{A_{i+1}} \otimes \cdots \otimes 1_{A_n}) = \varphi_i(a_i)$$

for all $i = 1, \ldots, n$ and all $a_i \in A_i$.

Proof. The map

$$A_1 \times \cdots \times A_n \longrightarrow B$$
$$(a_1, \ldots, a_n) \longmapsto \varphi_1(a_1) \cdots \varphi_n(a_n)$$

is clearly additive in each variable Since B is unital, $k \subset Z(B)$. It follows that this map is k-linear in each variable. Hence there exists a unique k-linear map $h : A_1 \otimes_k \cdots \otimes_k A_n \longrightarrow B$ satisfying

$$h(a_1 \otimes \cdots \otimes a_n) = \varphi_1(a_1) \cdots \varphi_n(a_n) \text{ for all } a_i \in A_i.$$

Using the definition of the product law on $A_1 \otimes_k \cdots \otimes_k A_n$ and the hypothesis on the φ_i's, one can check that h is a k-algebra morphism. Moreover, h clearly satisfies the required conditions.

Now let h' be another morphism of k-algebras satisfying the same conditions. Since we have

$$a_1 \otimes \cdots \otimes a_n = (a_1 \otimes 1_{A_2} \otimes \cdots \otimes 1_{A_n}) \cdots (1_{A_1} \otimes \cdots \otimes 1_{A_{n-1}} \otimes a_n),$$

and h and h' are k-algebra morphisms, we see that we have

$$h'(a_1 \otimes \cdots \otimes a_n) = \varphi_1(a_1) \cdots \varphi_n(a_n) = h(a_1 \otimes \cdots \otimes a_n) \text{ for all } a_i \in A_i.$$

Since elementary tensors span the tensor product, we get $h' = h$. □

COROLLARY A.3.6. *Let* $f_1 : A_1 \longrightarrow B_1, \ldots, f_n : A_n \longrightarrow B_n$ *be morphisms of k-algebras. Then there exists a unique k-algebra morphism*

$$f_1 \otimes \cdots \otimes f_n : A_1 \otimes_k \cdots \otimes_k A_n \longrightarrow B_1 \otimes_k \cdots \otimes_k B_n$$

satisfying

$$(f_1 \otimes \cdots \otimes f_n)(a_1 \otimes \cdots \otimes a_n) = f_1(a_1) \otimes \cdots \otimes f_n(a_n) \text{ for all } a_i \in A_i.$$

If each f_i is an isomorphism, so is $f_1 \otimes \cdots \otimes f_n$.

Proof. The uniqueness part follows form the usual argument. For the existence part, notice that the maps $\varphi_i : A_i \longrightarrow B_1 \otimes_k \cdots \otimes_k B_n$ defined by

$$\varphi(a_i) = 1_{A_1} \otimes \cdots \otimes 1_{A_{i-1}} \otimes f_i(a_i) \otimes 1_{A_{i+1}} \otimes \cdots \otimes 1_{A_n} \text{ for all } a_i \in A_i$$

are k-algebra morphisms with commuting images. By the previous proposition, there exists a unique morphism

$$h : A_1 \otimes_k \cdots \otimes_k A_n \longrightarrow B_1 \otimes_k \cdots \otimes_k B_n$$

of unital k-algebras satisfying

$$h(1_{A_1} \otimes \cdots \otimes 1_{A_{i-1}} \otimes a_i \otimes 1_{A_{i+1}} \otimes \cdots \otimes 1_{A_n}) = \varphi_i(a_i) \text{ for all } i.$$

Using the fact that h is an morphism of k-algebras, we get

$$\begin{aligned} h(a_1 \otimes \cdots \otimes a_n) &= \varphi_1(a_1) \cdots \varphi_n(a_n) \\ &= f_1(a_1) \otimes \cdots \otimes f_n(a_n) \end{aligned}$$

for all $a_i \in A_i$. The last part is clear since $f_1^{-1} \otimes \cdots \otimes f_n^{-1}$ is then the inverse map of $f_1 \otimes \cdots \otimes f_n$. □

Let L/k be an arbitrary field extension, let A be a k-algebra and let B an L-algebra. Recall from Lemma A.2.10 that $A \otimes_k B$ is endowed with a structure of L-vector space, defined on the elementary tensors by

$$\lambda \cdot (a \otimes b) = a \otimes (\lambda \cdot b), \text{ for all } \lambda \in L, a \in A, b \in B.$$

It is easy to check than the product law on $A \otimes_k B$ is then L-bilinear (using the fact that L lies in the center of B), so $A \otimes_k B$ has a natural structure of L-algebra. In particular, $A \otimes_k L$ is an L-algebra, and so is $(A \otimes_k L) \otimes_L B$. Therefore, the following statement makes sense.

LEMMA A.3.7. *Let L/k be any field extension, let A be a k-algebra and let B an L-algebra. Then there exists a unique isomorphism of L-algebras*

$$\Theta : (A \otimes_k L) \otimes_L B \xrightarrow{\sim} A \otimes_k B$$

satisfying

$$\Theta((a \otimes \lambda) \otimes b) = a \otimes (\lambda \cdot b) \text{ for all } \lambda \in L, a \in A, b \in B.$$

Proof. The uniqueness part follows from the usual argument (used twice). The map

$$\iota : \begin{aligned} L &\longrightarrow B \\ \lambda &\longmapsto \lambda 1_B \end{aligned}$$

is an L-algebra morphism, hence a k-algebra morphism. One can check that the k-algebra morphism

$$\mathrm{Id}_A \otimes \iota : A \otimes_k L \longrightarrow A \otimes_k B$$

is also L-linear. Moreover, the map

$$\iota' : \begin{aligned} B &\longrightarrow A \otimes_k B \\ b &\longmapsto 1_A \otimes b \end{aligned}$$

is easily seen to be an L-algebra morphism. Finally, one may check that $\mathrm{Id}_A \otimes \iota$ and ι' have commuting images, so we have a well-defined L-algebra morphism

$$\Theta : (A \otimes_k L) \otimes_L B \longrightarrow A \otimes_k B$$

satisfying

$$\Theta(z \otimes 1_B) = \mathrm{Id}_A \otimes \iota(z) \text{ and } \Theta(1_A \otimes b) = \iota'(b),$$

for all $\lambda \in L, z \in A \otimes_k L, b \in B$. In particular, we have

$$\Theta((a \otimes \lambda) \otimes b) = \Theta((a \otimes \lambda) \otimes 1_B)\Theta(1_A \otimes b) = (a \otimes \lambda 1_B)(1 \otimes b).$$

Hence $\Theta((a \otimes \lambda) \otimes b) = a \otimes \lambda b$. We now have to verify that Θ is an isomorphism. To construct the inverse map, notice that the maps

$$\begin{aligned} A &\longrightarrow (A \otimes_k L) \otimes_L B \\ a &\longmapsto (a \otimes 1) \otimes 1_B \end{aligned}$$

and

$$\begin{aligned} B &\longrightarrow (A \otimes_k L) \otimes_L B \\ b &\longmapsto (1_A \otimes 1) \otimes b \end{aligned}$$

are k-algebra morphisms with commuting images. Then, they induce a k-algebra morphism

$$h : A \otimes_k B \longrightarrow (A \otimes_k L) \otimes_L B$$

satisfying

$$h(a \otimes 1_B) = (a \otimes 1) \otimes 1_B \text{ and } h(1_A \otimes b) = (1_A \otimes 1) \otimes b \text{ for all } a \in A, b \in B.$$

One can check that h is also L-linear, and that h and Θ are mutually inverse.   □

REMARK A.3.8. Similar arguments show that $B \otimes_k A$ and $L \otimes_k A$ are L-algebras and that we have an isomorphism of L-algebras

$$B \otimes_L (L \otimes A) \cong_L B \otimes_k A.$$

□

LEMMA A.3.9. *Let L/k be any field extension, and let $A_1, \ldots, A_n$ be k-algebras. Then there exists a unique L-algebra isomorphism*

$$\rho_L : (A_1 \otimes_k L) \otimes_L \cdots \otimes_L (A_n \otimes_k L) \xrightarrow{\sim} (A_1 \otimes_k \cdots \otimes_k A_n) \otimes_k L$$

satisfying

$$\rho_L((a_1 \otimes \lambda_1) \otimes \cdots \otimes (a_n \otimes \lambda_n)) = (a_1 \otimes \cdots \otimes a_n) \otimes \lambda_1 \cdots \lambda_n$$

for all $a_i \in A, \lambda_i \in L$.

Proof.   The uniqueness of ρ_L may be proved using the usual argument. The map $f_i : A_i \times L \longrightarrow (A_1 \otimes_k \cdots \otimes_k A_n) \otimes_k L$ defined by

$$f_i(a_i, \lambda_i) = (1_{A_1} \otimes \cdots \otimes 1_{A_{i-1}} \otimes a_i \otimes 1_{A_{i+1}} \otimes \cdots \otimes 1_{A_n}) \otimes \lambda_i$$

is k-bilinear, and therefore there exists a unique k-linear map

$$\varphi_i : A \otimes_k L \longrightarrow (A \otimes_k B) \otimes_k L$$

satisfying

$$\varphi_i(a_i \otimes \lambda_i) = (1_{A_1} \otimes \cdots \otimes 1_{A_{i-1}} \otimes a_i \otimes 1_{A_{i+1}} \otimes \cdots \otimes 1_{A_n}) \otimes \lambda_i$$

for all $a_i \in A_i, \lambda_i \in L$. It is not difficult to see that φ_i is an L-algebra morphism, and that the images of $\varphi_1, \ldots, \varphi_n$ pairwise commute (it suffices to check it on elementary tensors and use the commutativity of L). By Proposition A.3.5, there exists a unique L-algebra morphism

$$\rho_L : (A_1 \otimes_k L) \otimes_L \cdots \otimes_L (A_n \otimes_k L) \xrightarrow{\sim} (A_1 \otimes_k \cdots \otimes_k A_n) \otimes_k L$$

satisfying

$$\rho_L(1_{A_1} \otimes \cdots \otimes 1_{A_{i-1}} \otimes z_i \otimes 1_{A_{i+1}} \otimes \cdots \otimes 1_{A_n}) = \varphi_i(z) \text{ for all } z_i \in A_i \otimes_k L.$$

In particular, we have

$$\rho_L((a_1 \otimes \lambda_1) \otimes \cdots \otimes (a_n \otimes \lambda_n)) = (a_1 \otimes \cdots \otimes a_n) \otimes \lambda_1 \cdots \lambda_n$$

for all $a_i \in A, \lambda_i \in L$.

We now construct an inverse map for ρ_L.

The map

$$\begin{array}{ccc} A_1 \times \cdots \times A_n & \longrightarrow & (A_1 \otimes_k L) \otimes_L \cdots \otimes_L (A_n \otimes_k L) \\ (a_1, \ldots, a_n) & \longmapsto & (a_1 \otimes 1) \otimes \cdots \otimes (a_n \otimes 1) \end{array}$$

is n-linear, so there exists a unique map

$$\varphi' : A_1 \otimes_k \cdots \otimes_k A_n \longrightarrow (A_1 \otimes_k L) \otimes_L \cdots \otimes_L (A_n \otimes_k L)$$

satisfying

$$\varphi'(a_1, \ldots, a_n) = (a_1 \otimes 1) \otimes \cdots \otimes (a_n \otimes 1) \text{ for all } a_i \in A_i.$$

Since Id_L is k-linear, we may consider $\rho'_L = \varphi' \otimes \mathrm{Id}_L$. One may check that ρ'_L is a L-algebra morphism, and that ρ_L and ρ'_L are mutually inverse. □

LEMMA A.3.10. *Let $k \subset K \subset L$ be a tower of field extensions, and let A be a k-algebra. Then there exists a unique L-algebra isomorphism*

$$\phi_L : (A \otimes_k K) \otimes_K L \xrightarrow{\sim} A \otimes_k L$$

satisfying

$$\phi_L((a \otimes \lambda) \otimes \mu) = a \otimes \lambda\mu \text{ for all } a \in A, \lambda \in K, \mu \in L$$

Proof. Uniqueness may be proved using the usual arguments. Since $k \subset K \subset L$, K is a k-linear subspace of L, so we have a k-linear map $\eta : A \otimes_k K \longrightarrow A \otimes_k L$ satisfying

$$\eta(a \otimes \lambda) = a \otimes \lambda \text{ for all } a \in A, \lambda \in K$$

One may check that η is a K-algebra morphism. One may also check that the map

$$\eta' : \begin{array}{c} L \longrightarrow A \otimes_k L \\ \mu \longmapsto 1 \otimes \mu \end{array}$$

is also a K-algebra morphism, whose image commutes with the image of η. Hence there exists a unique K-algebra morphism

$$\phi_L : (A \otimes_k K) \otimes_K L \longrightarrow A \otimes_k L$$

satisfying

$$\phi_L(z \otimes 1) = \eta(z) \text{ and } \phi_L(1 \otimes \mu) = \mu \text{ for all } z \in A \otimes_k K, \mu \in L$$

In particular, we have

$$\phi_L((a \otimes \lambda) \otimes \mu) = \phi_L((a \otimes \lambda) \otimes 1)\phi_L(1 \otimes \mu) = (a \otimes \lambda) \otimes \mu$$

But we have

$$(a \otimes \lambda) \otimes \mu = (\lambda(a \otimes 1)) \otimes \mu = (a \otimes 1) \otimes \lambda\mu$$

since the second tensor product is taken is over K. Hence we get

$$\phi_L((a \otimes \lambda) \otimes \mu) = (a \otimes 1) \otimes \lambda\mu.$$

We now construct an inverse map for ϕ_L. The map

$$\iota : \begin{array}{c} A \longrightarrow A \otimes_k K \\ a \longmapsto a \otimes 1 \end{array}$$

and Id_L are K-linear, so we may consider the K-linear map $\phi'_L = \iota \otimes \mathrm{Id}_L$. One may check that ϕ'_L is in fact a L-algebra morphism, and that ϕ_L and ϕ'_L are mutually inverse. □

APPENDIX B

A glimpse of number theory

In this appendix, we recall without proof some well-known facts on local fields and number fields.

B.1. Absolute values

We define the notion of an absolute value. There are several slightly different definitions of this notion. We will take the point of view of [**31**], but some references cited here choose another one. However, the choice of the definition does not have any influence on the validity of the theorems listed in this appendix. Notice that what we call an absolute value here is called a valuation in [**31**].

DEFINITION B.1.1. Let K be a field. An **absolute value** on K is a map $\upsilon : K \longrightarrow \mathbb{R}^+$ such that:

(1) for all $x \in K$, $\upsilon(x) = 0 \iff x = 0$;

(2) for all $x, y \in K$, $\upsilon(xy) = \upsilon(x)\upsilon(y)$;

(3) for all $x, y \in K$, $\upsilon(x + y) \leq \upsilon(x) + \upsilon(y)$.

We will say that υ is **archimedean** if $\operatorname{char}(K) = 0$ and there exists $m \in \mathbb{Z}$ such that $\upsilon(m \cdot 1_K) > 1$, and **non-archimedean** otherwise.

The **trivial** absolute value is the absolute value υ such that $\upsilon(0) = 0$ and $\upsilon(x) = 1$ for all $x \in K^\times$.

Two absolute values υ_1 and υ_2 are **equivalent** if there exists $\lambda \in \mathbb{R}^{+\times}$ such that $\upsilon_2 = \upsilon_1^\lambda$.

One may show that two absolute values are equivalent if and only if they define the same topology on K (cf. [**31**, Theorem 4.1.1]). A **place** of K is an equivalence class of absolute values.

REMARK B.1.2. If υ is an absolute value on K and $s > 0$, then υ^s is not necessarily an absolute value, as the example of the ordinary absolute value on $\mathbb{R}$ already shows. $\qquad\square$

EXAMPLES B.1.3.

(1) Let υ_∞ be the ordinary absolute value on $\mathbb{Q}$. Then υ_∞ is an archimedean absolute value.

(2) Let p be a prime number. For all $x \in \mathbb{Q} \setminus \{0\}$, we denote by $n_p(x)$ the p-adic valuation of x. We also set $n_p(0) = -\infty$. Then the map

$$v_p \colon \begin{aligned} \mathbb{Q} &\longrightarrow \mathbb{R}^+ \\ x &\longmapsto p^{-n_p(x)} \end{aligned}$$

is a non-archimedean absolute value on $\mathbb{Q}$.

$\square$

REMARK B.1.4. A theorem of Ostrowski shows that v_∞ and the absolute values v_p, p prime, form a complete set of pairwise non-equivalent absolute values on $\mathbb{Q}$. We will generalize this statement later on. $\square$

If v is an absolute value on K, we can construct the completion $\hat{K}$ with respect to this absolute value, in the same way we construct $\mathbb{R}$ from $\mathbb{Q}$. This field contains K as a subfield (or more precisely a subfield canonically isomorphic to K) and is equipped with an absolute value which restricts to v on K (see [**31**], Section 4.3). In view of this, this new absolute value will still be denoted by v. Moreover, $\hat{K}$ is complete for the topology defined by v. Two equivalent absolute values will give rise to the same field $\hat{K}$, and the extension of these absolute values to $\hat{K}$ will be equivalent.

EXAMPLE B.1.5. The completion of $\mathbb{Q}$ with respect to v_∞ is the field of real numbers $\mathbb{R}$, while the completion of $\mathbb{Q}$ with respect to v_p is the field of p-adic numbers $\mathbb{Q}_p$. $\square$

DEFINITION B.1.6. Let v be a non-archimedean absolute value on a field K. The set

$$\mathcal{O}_v = \{x \in K | v(x) \leq 1\}$$

is a commutative ring, called the **valuation ring** of (K, v). It is a local ring, with unique maximal ideal

$$\mathfrak{m}_v = \{x \in K | v(x) < 1\}.$$

Therefore, the units of $\mathcal{O}_K$ are

$$\mathcal{O}_v^\times = \{x \in K | v(x) = 1\}.$$

The **residue field** of (K, v) is the field $\kappa(v) = \mathcal{O}_v / \mathfrak{m}_v$. One can show that v and its extension to K_v have canonically isomorphic residue fields.

DEFINITION B.1.7. We say that v is **discrete** if $v(K^\times)$ is a discrete subgroup of $\mathbb{R}^{+\times}$. Any discrete absolute value is non-archimedean.

EXAMPLE B.1.8. For every prime p, v_p is a discrete absolute value on $\mathbb{Q}$. Indeed, $v_p(\mathbb{Q}^\times)$ is the cyclic group of $\mathbb{R}^{+\times}$ generated by p, which is discrete, since it isomorphic to $\mathbb{Z}$. $\square$

Non-trivial discrete absolute values have some nice properties.

PROPOSITION B.1.9. *Let v be a non-trivial discrete absolute value on a field K. Then $\mathcal{O}_v$ is a principal ideal domain, with field of fractions K, and $\mathfrak{m}_v$ is generated by an element $\pi_v \in \mathfrak{m}_v$.*

Moreover, any $x \in K^\times$ may be written in a unique way as

$$x = u\pi^{n_v(x)}, u \in \mathcal{O}_v^\times, n_v(x) \in \mathbb{Z}.$$

DEFINITION B.1.10. Let v be a non-trivial discrete absolute value on a field K. Any generator $\pi_v \in \mathfrak{m}_v$ is called a **local parameter**.

EXAMPLE B.1.11. Let p be a prime number. Then p is a local parameter for the absolute value v_p on $\mathbb{Q}$. $\qquad\qquad\qquad\qquad\qquad\qquad\qquad\qquad\qquad\qquad\qquad\qquad$ $\square$

REMARK B.1.12. If v is a non-trivial discrete valuation on K, with local parameter π_v, it follows from the last part of Proposition B.1.9 that we have

$$v(K^\times) = \langle v(\pi_v) \rangle.$$

$\square$

DEFINITION B.1.13. Let L/K be a field extension, let v and w be two non-trivial discrete absolute values on K and L respectively.

If $w_{|K} = v$, we say that w **extends** v, or that w is an extension of v , and we denote it by $w|v$. In this case, we have

$$\mathfrak{m}_w = \mathfrak{m}_v \cap K.$$

In particular, $\kappa(w) \supset \kappa(v)$. The degree of $\kappa(w)/\kappa(v)$ is called **the residual degree of w over** v and is denoted by $f_{w|v}$ (it may be infinite).

The index $[w(L^\times) : v(K^\times)]$ is finite and called **the ramification index of w over** v; it is denoted by $e_{w|v}$.

We say that the extension $(L, w)/(K, v)$ is

(1) **unramified** if $e_{w|v} = 1$;

(2) **ramified** if $e_{w|v} > 1$;

(3) **totally ramified** if it is ramified and $\kappa(w) = \kappa(v)$.

REMARK B.1.14. Let $(L, w)/(K, v)$ be an extension of non-trivial discrete absolute values, and let π_v and π_w be local parameters for v and w respectively. By the last part of Proposition B.1.9, we may write

$$\pi_v = u\pi_w^e, u \in \mathcal{O}_w^\times, e \geq 1,$$

taking into account that we cannot have $e \leq 0$, since otherwise, we would have

$$v(\pi_v) = w(\pi_v) = w(\pi_w)^e > 1,$$

contradicting the fact that $\pi_v \in \mathfrak{m}_v$.

Set $a = w(\pi_w)$, so that $v(\pi_v) = a^e$. By Remark B.1.12, $w(L^\times)$ and $v(K^\times)$ are generated respectively by a and a^e. It follows that we have

$$e_{w|v} = [\langle a \rangle : \langle a^e \rangle] = e.$$

In other words, the ramification index of $(L, w)/(K, v)$ is also the unique integer $e_{w|v}$ such that

$$\pi_v = u\pi_w^{e_{w|v}}, u \in \mathcal{O}_w^\times,$$

where π_v and π_w are local parameters of v and w respectively.

In particular, $(L, w)/(K, v)$ is unramified if and only if π_v is a local parameter for w. $\qquad\qquad\qquad\qquad\qquad\qquad\qquad\qquad\qquad\qquad\qquad\qquad\qquad\qquad$ $\square$

EXAMPLE B.1.15. Recall that $\mathbb{Z}[i]$ is a principal ideal domain, with field of fractions $\mathbb{Q}(i)$. For any irreducible element π of $\mathbb{Z}[i]$, we may consider the π-adic valuation $n_\pi(x)$ of an element x (as usual, we set $n_\pi(0) = -\infty$).

(1) Recall that $1 + i$ is a prime element of $\mathbb{Z}[i]$. Since $2 = -i(1+i)^2$, it follows that the map

$$w: \begin{aligned} \mathbb{Q}(i) &\longrightarrow \mathbb{R}^+ \\ x &\longmapsto \sqrt{2}^{-n_{1+i}(x)} \end{aligned}$$

is a non-trivial discrete absolute value, with local parameter $1 + i$, which extends the absolute value v_2 on $\mathbb{Q}$. Moreover, the corresponding ramification index is 2.

(2) The element $1 + 2i$ is a prime element of $\mathbb{Z}[i]$. Since we have the equality $5 = (1 - 2i)(1 + 2i)$ and $1 + 2i$ is not associate to $1 - 2i$, it follows that the map

$$w: \begin{aligned} \mathbb{Q}(i) &\longrightarrow \mathbb{R}^+ \\ x &\longmapsto 5^{-n_{1+2i}(x)} \end{aligned}$$

is a non-trivial discrete absolute value, with local parameter $1 - 2i$, which extends the absolute value v_5 on $\mathbb{Q}$. Moreover, the corresponding ramification index is 1.

We will see a generalization of these examples in a forthcoming section.   □

We now define local fields.

DEFINITION B.1.16. A **local field** is a field K of characteristic zero which is complete for a non-trivial discrete absolute value v, such that $\kappa(v)$ is finite.

THEOREM B.1.17. Let L/K be a field extension of degree n, where (K, v) is a local field. Then v extends in a unique way to a non-trivial discrete absolute value w, for which L is complete. Hence, L is also a local field. Moreover, we have

$$n = e_{w|v} f_{w|v}.$$

In particular, L/K is **totally ramified** if and only if $e_{w|v} = n$.

REMARK B.1.18. This theorem shows in particular that any finite extension of $\mathbb{Q}_p$ is a local field. One may prove that the converse holds as well (see [**31**, Theorem 4.7.1], noticing that we have assumed in this appendix that a local field has characteristic zero). The reader will find a proof of this theorem in [**31**], Section 4.5.   □

We end this section by describing the unramified extensions and the totally ramified extensions of a local field.

PROPOSITION B.1.19. Let (K, v) be a local field and let $\kappa(v) \simeq \mathbb{F}_q$, where $q = p^f$, for some prime p.

Then for any integer $m \geq 1$ prime to p, the extension $K(\zeta_m)/K$ is unramified and cyclic of degree d, where d is the order of $\bar{q}$ in $(\mathbb{Z}/m\mathbb{Z})^\times$. Moreover, there exists a canonical generator φ_m for $\mathrm{Gal}(K(\zeta_m)/K)$, uniquely determined by the condition

$$\varphi_m(x) \equiv x^q \mod \pi_L, \text{ for all } x \in \mathcal{O}_L.$$

Conversely, for a given $n \geq 1$, there is a unique unramified extension L/K of degree n, up to K-isomorphism. This extension is cyclic and isomorphic to $K(\zeta_{q^n-1})/K$.

See [**44**, Chapter IV, §4, Proposition 16] for a proof.

DEFINITION B.1.20. The canonical generator of the unique unramified extension L/K of degree n is called the **Frobenius map**, and is denoted by $\mathrm{Frob}(L/K)$.

PROPOSITION B.1.21. *Let (K, v) be a local field.*

A field extension L/K of degree n is totally ramified if and only if L is generated by an element $\alpha \in L$ such that

$$\mu_{\alpha,K} = X^b + a_{n-1}X^{n-1} + \cdots + a_0, n_v(a_i) \geq 1, n_v(a_0) = 1.$$

Moreover, if $\mathrm{char}(v) \nmid n$, and L/K is a Galois totally ramified extension of degree n, then $\mu_n \subset K$ and there exists and local parameter π_v such that $L = K(\sqrt[n]{\pi_K})$.

See [**11**],Theorem 1, p.23 and Proposition 1, p.32 for a proof.

B.2. Factorization of ideals in number fields

DEFINITION B.2.1. A **number field** is a field extension $K/\mathbb{Q}$ of finite degree. The **ring of integers** of K, denoted by $\mathcal{O}_K$, is the subset of elements $x \in K$ which are roots of monic polynomials of $\mathbb{Z}[X]$. It is a subring of K, and its field of fractions is isomorphic to K (apply the results of [**43**] , Section 5.1 to $A = \mathbb{Z}$ and $R = K$).

Moreover, $\mathcal{O}_K$ is a free $\mathbb{Z}$-module of rank $[K : \mathbb{Q}]$ ([**43**, 6.3, Theorem 2]).

EXAMPLES B.2.2.

(1) Let $K = \mathbb{Q}(\sqrt{d})$, where $d \in \mathbb{Z}$ is a non-zero square-free integer. Then we have:
  (i) $\mathcal{O}_K = \mathbb{Z}[\sqrt{d}]$ if $d \not\equiv 1[4]$;
  (ii) $\mathcal{O}_K = \mathbb{Z}[\frac{1+\sqrt{d}}{2}]$ if $d \equiv 1[4]$.
(2) Let $K = \mathbb{Q}(\zeta_n), n \geq 1$. Then $\mathcal{O}_K = \mathbb{Z}[\zeta_n]$.

See [**43**], Sections 5.1 and 16.2 for more details. □

The following theorem is the starting point of algebraic number theory.

THEOREM B.2.3. *Let K be a number field. Then the following properties hold:*

(1) *for every non-zero ideals $\mathfrak{a}, \mathfrak{b}$ of $\mathcal{O}_K$, we have $\mathfrak{a} \mid \mathfrak{b}$ if and only if $\mathfrak{a} \supset \mathfrak{b}$;*

(2) *every non-zero prime ideal $\mathfrak{p}$ is maximal and $\kappa(\mathfrak{p}) = \mathcal{O}_K/\mathfrak{p}$ is a finite field;*

(3) *every non-zero ideal of $\mathcal{O}_K$ decomposes in a unique way (up to permutation) as a product of prime ideals.*

See [**43**], Section 7.1, Part A. and Theorem 1, and Section 7.2.F, for example.

For any non-zero ideal $\mathfrak{a}$, we may then consider its $\mathfrak{p}$-adic valuation $n_{\mathfrak{p}}(\mathfrak{a})$ for any prime ideal $\mathfrak{p}$. We also set $n_{\mathfrak{p}}(0) = -\infty$.

Let $x \in K$, and write $x = \dfrac{a}{b}, a, b \in \mathcal{O}_k$. Then we set

$$n_{\mathfrak{p}}(x) = n_{\mathfrak{p}}(a\mathcal{O}_K) - n_{\mathfrak{p}}(b\mathcal{O}_K).$$

One may check easily that it does not depend on the choice of a and b.

EXAMPLE B.2.4. Let $\mathfrak{p}$ be a prime ideal of $\mathcal{O}_K$, and let $n \geq 1$. Then for any element $a \in \mathfrak{p}^n \setminus \mathfrak{p}^{n+1}$, we have

$$n_\mathfrak{p}(a) = n.$$

Indeed, by assumption, we have $(a) \subset \mathfrak{p}^n$ and $(a) \not\subset \mathfrak{p}^{n+1}$. The first point of Theorem B.2.3 then shows that $\mathfrak{p}^n \mid (a)$ but $\mathfrak{p}^{n+1} \nmid (a)$, that is $n_\mathfrak{p}(a) = n$. $\qquad\square$

If $\mathcal{O}_K$ is a principal ideal domain, every prime ideal $\mathfrak{p}$ is principal, generated by a prime element $\pi \in \mathcal{O}_K$. In this case, we will write n_π rather than $n_{(\pi)}$. If $x \in K$, the integer $n_\pi(x)$ does not depend on the choice of π.

We continue with the following approximation lemma.

LEMMA B.2.5. *Let K be a number field. Let $\mathfrak{p}_1, \ldots, \mathfrak{p}_r$ be r distinct prime ideals of $\mathcal{O}_K$, and let $n_1, \ldots, n_r \in \mathbb{Z}$. Then there exists $x \in K$ satisfying the following conditions:*

(1) $n_{\mathfrak{p}_i}(x) = n_i$ *for $i = 1, \ldots, r$;*

(2) $n_\mathfrak{p}(x) \geq 0$ *for any prime ideal $\mathfrak{p} \neq \mathfrak{p}_i, i = 1, \ldots, r$.*

Proof.   Assume first that $n_1, \ldots, n_r$ are non-negative integers. Since the prime ideals $\mathfrak{p}_1, \ldots, \mathfrak{p}_r$ are all distinct, the ideals $\mathfrak{p}_1^{n_1}, \ldots, \mathfrak{p}_r^{n_r}$ are pairwise coprime. For each i, pick $a_i \in \mathfrak{p}_i^{n_i} \setminus \mathfrak{p}_i^{n_i+1}$ (such an element a_i exists since $\mathfrak{p}_i^{n_i} \neq \mathfrak{p}_i^{n_i+1}$ by the uniqueness of the decomposition into a product of prime ideals).

By the Chinese Remainder Theorem, there exists $a \in \mathcal{O}_K$ such that $a \equiv a_i \mod \mathfrak{p}_i^{n_i}$ for $i = 1, \ldots, r$. Since $a_i \in \mathfrak{p}_i^{n_i} \setminus \mathfrak{p}_i^{n_i+1}$ and $\mathfrak{p}_i^{n_i+1} \subset \mathfrak{p}_i^{n_i}$, we also have $a \in \mathfrak{p}_i^{n_i} \setminus \mathfrak{p}_i^{n_i+1}$. By Example B.2.4, we get $n_{\mathfrak{p}_i}(a) = n_i$ for $i = 1, \ldots, r$. The second condition is automatically satisfied since $a \in \mathcal{O}_K$.

Let us go back to the general case. Renumbering if necessary, we may assume that $n_1, \ldots, n_s \geq 0$ and $n_{s+1}, \ldots, n_r \leq 0$. By the previous point, there exists $b \in \mathcal{O}_K$ such that:

(i) $n_{\mathfrak{p}_i}(b) = 0$ for $i = 1, \ldots, s$;

(ii) $n_{\mathfrak{p}_i}(b) = -n_i$ for $i = s+1, \ldots, r$.

In particular, we may write

$$(b) = \mathfrak{p}_1^{-n_{s+1}} \cdots \mathfrak{p}_r^{-n_r} \mathfrak{q}_1^{m_1} \cdots \mathfrak{q}_t^{m_t},$$

where $\mathfrak{q}_j \neq \mathfrak{p}_1, \ldots, \mathfrak{p}_r$ for all j, and $m_j > 0$. One may then find $a \in \mathcal{O}_K$ satisfying the following conditions:

(iii) $n_{\mathfrak{p}_i}(a) = n_i$ for $i = 1, \ldots, s$;

(iv) $n_{\mathfrak{p}_i}(a) = 0$ for $i = s+1, \ldots, r$;

(v) $n_{\mathbb{Q}_j}(a) = m_j$ for $j = 1, \ldots, t$.

Set $x = \dfrac{a}{b}$.   We claim that x satisfies the required conditions.   Indeed, for $i = 1, \ldots, s$, we have

$$n_{\mathfrak{p}_i}(x) = n_{\mathfrak{p}_i}(a) - n_{\mathfrak{p}_i}(b) = n_i - 0 = n_i$$

by (i) and (ii), and for $i = s+1, \ldots, r$, we have

$$n_{\mathfrak{p}_i}(x) = 0 - (-n_i) = n_i$$

by (ii) and (iv).

Assume now that $\mathfrak{p} \neq \mathfrak{p}_1, \ldots, \mathfrak{p}_r$. If $\mathfrak{p} \neq \mathfrak{q}_1, \ldots, \mathfrak{q}_t$, we have

$$n_{\mathfrak{p}}(x) = n_{\mathfrak{p}}(a) - n_{\mathfrak{p}}(b) = n_{\mathfrak{p}}(a) \geq 0,$$

since $n_{\mathfrak{p}}(b) = 0$ in view of the decomposition of (b) above. Finally, for $j = 1, \ldots, t$, we have

$$n_{\mathfrak{q}_j}(x) = n_{\mathfrak{q}_j}(a) - n_{\mathfrak{q}_j}(b) = m_j - m_j = 0.$$

This concludes the proof.   □

DEFINITION B.2.6. Let L/K be a finite extension of number fields, and let $\mathfrak{p}$ be a prime ideal of $\mathcal{O}_K$. We say that a prime ideal $\mathfrak{P}$ of $\mathcal{O}_L$ **lies above** $\mathfrak{p}$ if we have

$$\mathfrak{P} \cap \mathcal{O}_K = \mathfrak{p}.$$

We denote it by $\mathfrak{P} \mid \mathfrak{p}$.

In this case, $\kappa(\mathfrak{p})$ identifies canonically to a subfield of $\kappa(\mathfrak{P})$, so that $\kappa(\mathfrak{P})/\kappa(\mathfrak{p})$ is an extension of finite fields.

EXAMPLE B.2.7. Let $K = \mathbb{Q}$ and let $K = \mathbb{Q}(i)$. Then $(1 + i)\mathbb{Z}[i]$ is a prime ideal of $\mathcal{O}_L = \mathbb{Z}[i]$ lying above $2\mathbb{Z}$.   □

PROPOSITION B.2.8. *Let L/K be a finite extension of number fields, and let $\mathfrak{p}$ be a prime ideal of $\mathcal{O}_K$. Then a prime ideal $\mathfrak{P}$ of $\mathcal{O}_L$ lies above $\mathfrak{p}$ if and only if $n_{\mathfrak{P}}(\mathfrak{p}\mathcal{O}_L) \geq 1$, that is $\mathfrak{P}$ appears in the decomposition of $\mathfrak{p}\mathcal{O}_L$ into a product of prime ideals of $\mathfrak{p}\mathcal{O}_L$.*

DEFINITION B.2.9. Let L/K be a finite extension of number fields. Let $\mathfrak{p}$ be a prime ideal of $\mathcal{O}_K$, and write

$$\mathfrak{p}\mathcal{O}_L = \mathfrak{P}_1^{e_1} \cdots \mathfrak{P}_r^{e_r}, e_r \geq 1,$$

where $\mathfrak{P}_1, \ldots, \mathfrak{P}_r$ are pairwise distinct prime ideals of $\mathcal{O}_L$.

If $\mathfrak{P} \mid \mathfrak{p}$, then $\mathfrak{P} = \mathfrak{P}_i$ for some i, and the integer e_i is called the **ramification index** of $\mathfrak{P}$ over $\mathfrak{p}$; it is denoted by $e_{\mathfrak{P}|\mathfrak{p}}$. In other words, for all $\mathfrak{P} \mid \mathfrak{p}$, we have

$$e_{\mathfrak{P}|\mathfrak{p}} = n_{\mathfrak{P}}(\mathfrak{p}\mathcal{O}_L).$$

The degree $f_{\mathfrak{P}|\mathfrak{p}} = [\kappa(\mathfrak{P}) : \kappa(\mathfrak{p})]$ is called the **residual degree** of $\mathfrak{P}$ over $\mathfrak{p}$.

We say that:

(1) $\mathfrak{p}$ **does not ramify** in L, or is **unramified** if $e_{\mathfrak{P}|\mathfrak{p}} = 1$ for all $\mathfrak{P} \mid \mathfrak{p}$;
(2) $\mathfrak{p}$ **is inert** in L if $\mathfrak{p}\mathcal{O}_L$ is a prime ideal of $\mathcal{O}_L$;
(3) $\mathfrak{p}$ **totally splits** in L if $\mathfrak{p}\mathcal{O}_L = \mathfrak{P}_1 \cdots \mathfrak{P}_n$, where $\mathfrak{P}_1, \cdots, \mathfrak{P}_n$ are $n = [L : K]$ distinct prime ideals of $\mathcal{O}_L$;
(4) $\mathfrak{p}$ **ramifies** in L if $e_{\mathfrak{P}|\mathfrak{p}} > 1$ for some $\mathfrak{P} \mid \mathfrak{p}$;
(5) $\mathfrak{p}$ **totally ramifies** in L if $\mathfrak{p}\mathcal{O}_L = \mathfrak{P}^n$, for some prime ideal $\mathfrak{P}$ of $\mathcal{O}_L$, where $n = [L : K]$;
(6) $\mathfrak{p}$ **tamely ramifies** in L if $\mathfrak{p}$ ramifies and $e_{\mathfrak{P}|\mathfrak{p}}$ is prime to $\mathrm{char}(\kappa(\mathfrak{p}))$ for all $\mathfrak{P} \mid \mathfrak{p}$, and **wildly ramifies otherwise.**

If $\mathcal{O}_K$ is a principal ideal domain, the prime ideals are generated by the irreducible elements of $\mathcal{O}_K$. In this case, we will say that an irreducible element π of $\mathcal{O}_K$ ramifies (resp. is inert etc.) instead of (π) ramifies (resp. is inert, etc.).

PROPOSITION B.2.10. *Let L/K be an extension of number fields, and let $\mathfrak{p}$ be an ideal of $\mathcal{O}_K$. Then we have*

$$[L:K] = \sum_{\mathfrak{P}|\mathfrak{p}} e_{\mathfrak{P}|\mathfrak{p}} f_{\mathfrak{P}|\mathfrak{p}}.$$

See for example [**31**, (3.10.2)].

The next proposition shows that factorisation of prime ideals is particularly nice in Galois extensions (see [**31**], Proposition 6.1.1 and equality (3.10.2)).

PROPOSITION B.2.11. *Let L/K be a Galois extension of number fields with Galois group G, and let $\mathfrak{p}$ be a prime ideal of $\mathcal{O}_K$. Then G acts transitively of the set of prime ideals of $\mathcal{O}_L$ lying above $\mathfrak{p}$. In particular, the integers $e_{\mathfrak{p}} = e_{\mathfrak{P}|\mathfrak{p}}$ and $f_{\mathfrak{p}} = f_{\mathfrak{P}|\mathfrak{p}}$ do not depend on the choice of a prime ideal $\mathfrak{P}$ of $\mathcal{O}_L$ lying above $\mathfrak{p}$.*

Moreover, if $g_{\mathfrak{p}}$ denotes the number of prime ideals of $\mathcal{O}_L$ lying above $\mathfrak{p}$, we have

$$[L:K] = e_{\mathfrak{p}} f_{\mathfrak{p}} g_{\mathfrak{p}}.$$

The following theorem is very useful to decompose prime ideals of a number field K, especially when $\mathcal{O}_K = \mathbb{Z}[\theta]$ (see [**31**, Theorem 3.8.2] for a proof).

THEOREM B.2.12 (Dedekind). *Let K be a number field, and let $\theta \in \mathcal{O}_K$ such that $K = \mathbb{Q}(\theta)$. For all $p \nmid [\mathcal{O}_K : \mathbb{Z}[\theta]]$, let $\overline{\mu_{\theta,\mathbb{Q}}} \in \mathbb{F}_p[X]$ be the reduction of the minimal polynomial of θ modulo p, and write*

$$\overline{\mu_{\theta,\mathbb{Q}}} = \overline{g}_1^{e_1} \cdots \overline{g}_r^{e_r},$$

where $g_1, \ldots, g_r \in \mathbb{Z}[X]$ are pairwise distinct monic polynomials which are irreducible modulo p and $e_i \geq 1$.

For $i = 1, \ldots, r$, set $\mathfrak{p}_i = (p, g_i(\theta))$. Then $\mathfrak{p}_1, \ldots, \mathfrak{p}_r$ are pairwise distinct prime ideals of $\mathcal{O}_K$ and we have

$$p\mathcal{O}_K = \mathfrak{p}_1^{e_1} \cdots \mathfrak{p}_r^{e_r}.$$

REMARK B.2.13. If $\mathcal{O}_k = \mathbb{Z}[\theta]$, then the condition $p \nmid [\mathcal{O}_K : \mathbb{Z}[\theta]]$ is always fulfilled, and one may obtain the decomposition of $p\mathcal{O}_K$ for any prime p.   □

As an example, we describe the ramification of prime numbers in a quadratic extension $\mathbb{Q}(\sqrt{d})/\mathbb{Q}$.

EXAMPLE B.2.14. Let $K = \mathbb{Q}(\sqrt{d})$, where $d \in \mathbb{Z}$ be a non-zero square free integer, and let p be a prime number.

Assume first that $d \not\equiv 1[4]$. In this case, $\mathcal{O}_K = \mathbb{Z}[\sqrt{d}]$. Since $X^2 - d$ is the minimal polynomial of $\sqrt{d}$ over $\mathbb{Q}$, we get the following results.

(1) If $p = 2$ or $p \mid d$, then $X^2 - d$ is a square modulo $\mathbb{F}_p$, since it is equal to X^2 or $(X - 1)^2$ in all cases. Hence p totally ramifies.

(2) If p is odd, $p \nmid d$ and $d \notin \mathbb{F}_p^{\times 2}$, then $X^2 - d \in \mathbb{F}_p[X]$ is irreducible, and p is inert in $K/\mathbb{Q}$.

(3) If p is odd, $p \nmid d$ and $d \in \mathbb{F}_p^{\times 2}$, then $X^2 - d \in \mathbb{F}_[X]$ is the product of two distinct polynomials of degree 1, and p totally splits in $K/\mathbb{Q}$.

Assume now that $d \equiv 1[4]$, so that $\mathcal{O}_K = \mathbb{Z}[\frac{1+\sqrt{d}}{2}]$. Since $X^2 - X + \frac{1-d}{4}$ is the minimal polynomial of $\dfrac{1 + \sqrt{d}}{2}$ over $\mathbb{Q}$, we get the following results.

(4) Assume that $p = 2$. If $d \equiv 5[8]$, then $X^2 - X + \frac{1-d}{4}$ is irreducible modulo 2 since it is equal to $X^2 - X - 1 \in \mathbb{F}_2[X]$ in this case, and 2 is inert. If $d \equiv 1[8]$, then $X^2 - X + \frac{1-d}{4}$ is equal to $X(X-1) \in \mathbb{F}_2[X]$ in this case, and 2 totally splits.

(5) If $p \mid d$ (so p is odd), then $X^2 - X + \frac{1-d}{4}$ is a square modulo $\mathbb{F}_p$, since it is equal to $(X - 1/2)^2$. Hence p totally ramifies.

(6) If p is odd, $p \nmid d$ and $d \notin \mathbb{F}_p^{\times 2}$, then $X^2 - X + \frac{1-d}{4} \in \mathbb{F}_p[X]$ is irreducible, and p is inert in $K/\mathbb{Q}$.

(7) If p is odd, $p \nmid d$ and $d \in \mathbb{F}_p^{\times 2}$, then $X^2 - X + \frac{1-d}{4} \in \mathbb{F}_p[X]$ is the product of two distinct polynomials of degree 1, and p totally splits in $K/\mathbb{Q}$.

$\square$

The case of cyclotomic extensions is also well-known (see [**31**], Propositions 6.4.6 and 6.4.8 for example).

THEOREM B.2.15. *Let $m \geq 3$ be an integer. Assume that $4 \mid m$ if m is even, and let $L = \mathbb{Q}(\zeta_m)$. For any p prime number, the ramification index e_p is equal to $\varphi(p^{n_p(m)})$, and the residual degree f_p is the multiplicative order of p modulo $\dfrac{m}{p^{n_p(m)}}$.*

COROLLARY B.2.16. *Let $m \geq 3$ be an integer. Assume that $4 \mid m$ if m is even, and let p be a prime number. Then the following properties hold:*

(1) *p ramifies in $\mathbb{Q}(\zeta_m)/\mathbb{Q}$ if and only if $p \mid m$;*

(2) *p is unramified in $\mathbb{Q}(\zeta_m)/\mathbb{Q}$ if and only if $p \nmid m$;*

(3) *p totally splits in $\mathbb{Q}(\zeta_m)/\mathbb{Q}$ if and only if $p \equiv 1[m]$;*

(4) *p is inert in $\mathbb{Q}(\zeta_m)/\mathbb{Q}$ if and only if $\overline{p}$ is a generator of $(\mathbb{Z}/m\mathbb{Z})^\times$.*

We are now interested in the existence of prime ideals having a prescribed behavior in a Galois extension. We start with the case of unramified primes. In this case, the results of Section 25.1.A of [**43**] and Tchebotarev's Density Theorem (see [**43**], Section 25.3, Theorem 1) imply the following result.

PROPOSITION B.2.17. *Let L/K be a Galois extension of number fields. Then the following properties hold:*

(1) *there are infinitely many prime ideals $\mathfrak{p}$ of $\mathcal{O}_K$ which totally splits in L;*

(2) *there exists a prime ideal $\mathfrak{p}$ of $\mathcal{O}_K$ which stays inert in L if and only if L/K is cyclic. In this case, the number of such prime ideals is infinite.*

We will also need the following result (see [**31**], Propositions 4.9.1 and 4.9.2 for a proof).

LEMMA B.2.18. *Let L_1/K and L_2/K be two extensions of a number field K, and let $\mathfrak{p}$ be a prime ideal of $\mathcal{O}_K$. If $\mathfrak{p}$ is unramified (resp. splits completely) in L_1 and L_2, then $\mathfrak{p}$ is unramified (resp. splits completely) in $L_1 L_2$.*

We are now interested in the existence of ramified ideals. First, we need to define the norm of an ideal. The Chinese Remainder Theorem and Theorem B.2.3 show that for any non-zero ideal $\mathfrak{a}$ of $\mathcal{O}_K$, the quotient ring $\mathcal{O}/\mathfrak{a}$ is finite. Therefore, the following definition makes sense.

DEFINITION B.2.19. Let K be a number field. For any non-zero ideal $\mathfrak{a}$ of $\mathcal{O}_K$, we set
$$N_{K/\mathbb{Q}}(\mathfrak{a}) = |\mathcal{O}_K/\mathfrak{a}|.$$
The integer $N_{K/\mathbb{Q}}(\mathfrak{a})$ is called the **absolute norm** of $\mathfrak{a}$.

PROPOSITION B.2.20. *Let K be a number field, and let $\mathfrak{a}, \mathfrak{b}$ be two ideals of $\mathcal{O}_K$. Then the following properties hold:*

(1) *we have $N_{K/\mathbb{Q}}(\mathfrak{a}\mathfrak{b}) = N_{K/\mathbb{Q}}(\mathfrak{a})N_{K/\mathbb{Q}}(\mathfrak{b})$;*

(2) *if $\mathfrak{a} \supset \mathfrak{b}$, then $N_{K/\mathbb{Q}}(\mathfrak{a}) \mid N_{K/\mathbb{Q}}(\mathfrak{b})$. Moreover, equality holds if and only if $\mathfrak{a} = \mathfrak{b}$;*

(3) *for all $x \in K^\times$, we have $N_{K/\mathbb{Q}}(x\mathcal{O}_K) = |N_{K/\mathbb{Q}}(x)|$.*

We will also need a relative version of the norm.

DEFINITION B.2.21. Let L/K be an extension of number fields. For any ideal $\mathfrak{A}$ of $\mathcal{O}_L$, we denote by $\mathcal{N}_{L/K}(\mathfrak{A})$ the ideal of $\mathcal{O}_K$ generated by the elements $N_{L/K}(x), x \in \mathfrak{A}$, and call it the **relative norm** of $\mathfrak{A}$.

We then have the following properties.

PROPOSITION B.2.22. *Let L/K be an extension of number fields, and let $\mathfrak{A}, \mathfrak{B}$ be two ideals of $\mathcal{O}_L$. Then the following properties hold:*

(1) *if $K = \mathbb{Q}$, we have $\mathcal{N}_{L/\mathbb{Q}}(\mathfrak{A}) = N_{L/\mathbb{Q}}(\mathfrak{A})\mathbb{Z}$;*

(2) *we have $\mathcal{N}_{L/K}(\mathfrak{A}\mathfrak{B}) = \mathcal{N}_{L/K}(\mathfrak{A})\mathcal{N}_{L/K}(\mathfrak{B})$;*

(3) *if $\mathfrak{A} \supset \mathfrak{B}$, then $\mathcal{N}_{L/K}(\mathfrak{A}) \subset \mathcal{N}_{L/K}(\mathfrak{B})$. Moreover, equality holds if and only if $\mathfrak{A} = \mathfrak{B}$;*

(4) *if $\mathfrak{P}$ is a prime ideal of $\mathcal{O}_L$ and $\mathfrak{p} = \mathfrak{P} \cap \mathcal{O}_K$, then $\mathcal{N}_{L/K}(\mathfrak{P}) = \mathfrak{p}^{f_{\mathfrak{P}|\mathfrak{p}}}$;*

(5) *if L/K is a Galois extension with Galois group G, then we have*
$$\mathcal{N}_{L/K}(\mathfrak{A}) = \Big(\prod_{\sigma \in G} \sigma(\mathfrak{A}) \Big) \cap \mathcal{O}_K;$$

(6) *if $K \subset M \subset L$, we have*
$$\mathcal{N}_{L/K}(\mathfrak{A}) = \mathcal{N}_{M/K}(\mathcal{N}_{L/M}(\mathfrak{A})).$$
In particular, if $K = \mathbb{Q}$, we have
$$N_{L/\mathbb{Q}}(\mathfrak{A}) = N_{M/\mathbb{Q}}(\mathcal{N}_{L/M}(\mathfrak{A}));$$

(7) *if $\mathfrak{a}$ is an ideal of $\mathcal{O}_K$, then $\mathcal{N}_{L/K}(\mathfrak{a}) = \mathfrak{a}^{[L:K]}$.*

See [**26**] , [**28**], [**31**] or [**43**, Section 13.1].

REMARK B.2.23. Properties (3) and (4) show in particular that, for any non-zero ideal $\mathfrak{A}$ of $\mathcal{O}_L$, a prime ideal $\mathfrak{P}$ of $\mathcal{O}_L$ dividing $\mathfrak{A}$ necessary lie above a prime ideal $\mathfrak{p}$ dividing $\mathcal{N}_{L/k}(\mathfrak{A})$.   □

We would like now to give a characterization of the prime ideals $\mathfrak{p}$ which ramify in L. First, we need some definitions.

DEFINITION B.2.24. Let L/K be an extension of number fields. Let $\mathbf{w} = (w_1, \cdots, w_n)$ be a K-basis of L contained in $\mathcal{O}_L$ (i.e. $w_i \in \mathcal{O}_L$ for all i). The **discriminant** $D(\mathbf{w})$ of $\mathbf{w}$ is the determinant of the matrix $(\mathrm{Tr}_{L/K}(w_i w_j))$. This is an element of $\mathcal{O}_K$.

The **discriminant ideal** $\mathcal{D}_{L/K}$ is the ideal of $\mathcal{O}_K$ generated by the elements $D(\mathbf{w})$, for all K-bases $\mathbf{w}$ of L contained in $\mathcal{O}_L$.

If $\mathcal{O}_K$ is a principal ideal domain, it is the ideal generated by the discriminant of an $\mathcal{O}_K$-basis of $\mathcal{O}_L$. In particular, if $K = \mathbb{Q}$, the absolute value of the discriminant of a $\mathbb{Z}$-basis of $\mathcal{O}_L$, and does not depend on the choice of this basis. It is denoted by $\mathrm{disc}(L)$, and called the **absolute discriminant** of K. In other words, $\mathrm{disc}(L)$ is the unique positive integer such that

$$\mathcal{D}_{L/\mathbb{Q}} = \mathrm{disc}(L)\mathbb{Z}.$$

The **different ideal** of L/K is the ideal $\mathfrak{d}_{L/K}$ of $\mathcal{O}_L$ generated by the elements $f'(x)$, where $x \in \mathcal{O}_L$ and f is the minimal polynomial of x over K.

REMARK B.2.25. It is important to point out that, in some references used in this book, such as [**31**] for example, $\mathfrak{d}_{L/K}$ denotes the discriminant ideal and $\mathcal{D}_{L/K}$ denotes the different ideal. □

We then have the following result.

THEOREM B.2.26. *Let L/K be an extension of number fields of degree n. Then the following properties hold:*

(1) *A prime ideal $\mathfrak{p}$ of $\mathcal{O}_K$ ramifies in L if and only if $\mathfrak{p}$ divides $\mathcal{D}_{L/K}$.*
    In particular, if $K = \mathbb{Q}$, p ramifies in L if and only if $p \mid \mathrm{disc}(L)$;
(2) *let $\alpha \in \mathcal{O}_K$ such that $L = K(\alpha)$, with minimal polynomial f over K. Then*

$$D(1, \theta, \ldots, \theta^{n-1}) = (-1)^{n(n-1)/2} N_{L/K}(f'(\alpha)).$$

   In particular, $\mathcal{D}_{L/K} \mid (N_{L/K}(f'(\alpha)))$.
(3) *the prime ideals of $\mathcal{O}_L$ which divide $\mathfrak{d}_{L/K}$ are exactly those who lie above ideals of $\mathcal{O}_K$ which ramify in L;*
(4) *$\mathcal{N}_{L/K}(\mathfrak{d}_{L/K}) = \mathcal{D}_{L/K}$. In particular, $N_{L/\mathbb{Q}}(\mathfrak{d}_{L/\mathbb{Q}}) = \mathrm{disc}(L)$.*

See [**43**, Section 13.2] for example.

EXAMPLE B.2.27. Let $L = \mathbb{Q}(\sqrt{d})$, where $d \in \mathbb{Z}$ is a non-zero square-free integer. Then it readily follows from Example B.2.2 (1) that $\mathrm{disc}(L) = d$ if $d \equiv 1[4]$ and $4d$ otherwise. □

The following result is particularly useful to compute the ring of integers in some cases.

PROPOSITION B.2.28. *Let L and M be two number fields. Assume that $L/\mathbb{Q}$ and $M/\mathbb{Q}$ are linearly disjoint, and that $\mathrm{disc}(L)$ and $\mathrm{disc}(M)$ are coprime. Then $\mathcal{O}_{LM}$ is a generated as an abelian group by the elements*

$$xy, \ x \in \mathcal{O}_L, y \in \mathcal{O}_M.$$

Moreover, if $(e_1, \ldots, e_n)$ is a $\mathbb{Z}$-basis of $\mathcal{O}_L$ and $(f_1, \ldots, f_m)$ is a $\mathbb{Z}$-basis of $\mathcal{O}_M$, then the elements

$$e_i f_j, i = 1, \ldots, n, j = 1, \ldots, m$$

form a $\mathbb{Z}$-basis of $\mathcal{O}_{LM}$.

Proof. The first point is just a particular case of [**15**, (2.13)]. To prove the second part, notice that the first point shows that the elements

$$e_i f_j, i = 1, \ldots, n, j = 1, \ldots, m$$

span $\mathcal{O}_{LM}$ as an abelian group. It remains to prove that they are linearly independent over $\mathbb{Z}$ or, which is equivalent, over $\mathbb{Q}$. But this comes from the fact that $L/\mathbb{Q}$ and $M/\mathbb{Q}$ are linearly disjoint, noticing that $(e_1, \ldots, e_n)$ and $(f_1, \ldots, f_m)$ are also $\mathbb{Q}$-bases of L and M respectively. $\square$

We now give a proposition which partially describes the ramification in a Kummer extension.

PROPOSITION B.2.29. *Assume that $\mu_n \subset k$, and let $L = k(\sqrt[n]{d})$ be a cyclic extension of k of degree n, where $d \in \mathcal{O}_k \setminus \{0\}$. Then the following properties hold:*

(1) *the prime ideals of $\mathcal{O}_k$ which eventually ramify are those which divide d or n;*

(2) *let $\mathfrak{p} \mid d$. If $n_{\mathfrak{p}}(d)$ is not a multiple of n, then $\mathfrak{p}$ ramifies. If moreover $n_{\mathfrak{p}}(d)$ and n are coprime, then $\mathfrak{p}$ totally ramifies.*

(3) *assume that $\mathfrak{p} \mid d$ and $\mathfrak{p} \nmid n$. Then $\mathfrak{p}$ ramifies (resp. totally ramifies) if and only if $n_{\mathfrak{p}}(d)$ is a not multiple of n (resp. $n_{\mathfrak{p}}(d)$ and n are coprime);*

(4) *assume that n is prime, and let $\mathfrak{p} \nmid n$. Then $\mathfrak{p}$ ramifies if and only if $n_{\mathfrak{p}}(d)$ is not a multiple of n. In this case, it totally ramifies.*

Proof. Let $\alpha = \sqrt[n]{d}$. The minimal polynomial f of α over k is $X^n - d$, whose discriminant is

$$\mathrm{disc}(f) = \pm N_{k(\alpha)/k}(f'(\alpha)) = \pm n^n d^n.$$

Now if $\mathfrak{p}$ ramifies, it divides the discriminant ideal $\mathcal{D}_{L/k}$ by Theorem B.2.26 (1). But $\mathcal{D}_{L/K} \mid \mathrm{disc}(f)$ by Theorem B.2.26 (2), and we get (1).

Let $\mathfrak{p} \mid d$ and let $r = gcd(n, n_{\mathfrak{p}}(d))$. Write $n = rm, n_{\mathfrak{p}}(d) = rs$, where and m and s are coprime. Notice that we have $r \leq n$ by definition. Let e the ramification index of $\mathfrak{p}$ in L. Since $\alpha^n = d$, comparing valuations with respect to any prime ideal $\mathfrak{P}$ of L above $\mathfrak{p}$, we get that $nn_{\mathfrak{P}}(\alpha) = en_{\mathfrak{p}}(d)$. Thus, we have $mn_{\mathfrak{P}}(\alpha) = es$ and therefore $m \mid e$ since m and s are coprime. Assume that $n_{\mathfrak{p}}(d)$ is not a multiple of n. Then we cannot have $r = n$ and therefore $m > 1$. Since $m > 1$, we have $e > 1$, hence $\mathfrak{p}$ ramifies. Moreover, if $n_{\mathfrak{p}}(d)$ and n are coprime, then $m = n$ and thus $n \mid e$. Since $e < n$, this means that $e = n$ and therefore, $\mathfrak{p}$ totally ramifies in this case, hence (2).

We now proceed to show (3). By Lemma B.2.5, there exists $c \in k$ such that $n_{\mathfrak{p}}(c) = -s$ and $n_{\mathfrak{q}}(c) \geq 0$ for all $\mathfrak{q} \neq \mathfrak{p}$. It follows that $n_{\mathfrak{p}}(dc^r) = n_{\mathfrak{p}}(d) - rs = 0$, and that $n_{\mathfrak{q}}(dc^r) \geq 0$ for all $\mathfrak{q} \neq \mathfrak{p}$. In other words, dc^r is an element of $\mathcal{O}_k$ which is not divisible by $\mathfrak{p}$.

Assume that $\mathfrak{p} \nmid n$. In particular, $\mathfrak{p} \nmid r$. Applying (1) to $k(\sqrt[n]{dc^r})$ shows that $\mathfrak{p}$ does not ramify in $k(\sqrt[n]{d})/k$. Now assume that $n_{\mathfrak{p}}(d)$ is a multiple of n. Hence

$r = gcd(n, n_{\mathfrak{p}}(d)) = n$, and $\mathfrak{p}$ does not ramify in L/k. By (2), if $n_{\mathfrak{p}}(d)$ is a multiple of n, then $\mathfrak{p}$ ramifies in L/k. Therefore, $\mathfrak{p}$ ramifies if and only if $n_{\mathfrak{p}}(d)$ is not a multiple of n. In this case, if $n_{\mathfrak{p}}(d)$ and n are not coprime, then $r \geq 2$ and $k(\sqrt[r]{d})/k$ is a non-trivial subextension of L/k in which $\mathfrak{p}$ does not ramify. Hence, $\mathfrak{p}$ ramifies but does not totally ramify in L/k. This proves (3).

Finally, if n is prime, then $n_{\mathfrak{p}}(d)$ is not a multiple of n if and only if $n_{\mathfrak{p}}(d)$ and n are coprime. Point (4) then follows from the previous points. This concludes the proof. □

We end this section by giving a list of results which allow us to compute the different ideal.

THEOREM B.2.30. *Let $K \subset L_1 \subset L_2$ be a tower of number fields. Then we have*

$$\mathfrak{d}_{L_2/K} = \mathfrak{d}_{L_2/L_1} \mathfrak{d}_{L_1/K}.$$

See [**31**, Theorem 3.12.3] for a proof.

The next result shows that the part of the different ideal corresponding to tame ramification is perfectly understood.

THEOREM B.2.31. *Let L/K be an extension of number fields. Let $\mathfrak{p}$ be an ideal of $\mathcal{O}_K$, and let $\mathfrak{P}$ be a ideal of $\mathcal{O}_L$ lying above $\mathfrak{p}$. Then the following properties hold:*

(1) $n_{\mathfrak{P}}(\mathfrak{d}_{L/K}) = e_{\mathfrak{P}|\mathfrak{p}} - 1$ *if $\mathfrak{p} \nmid e_{\mathfrak{P}|\mathfrak{p}}$;*
(2) $n_{\mathfrak{P}}(\mathfrak{d}_{L/K}) > e_{\mathfrak{P}|\mathfrak{p}} - 1$ *if $\mathfrak{p} \mid e_{\mathfrak{P}|\mathfrak{p}}$.*

See [**31**, Theorem 3.12.9] for a proof.

We now explain how to deal with the case of wild ramification when L/K is a Galois extension of number fields with Galois group G. First, we need to define an appropriate filtration of subgroups of G.

DEFINITION B.2.32. Let L/K be a Galois extension of number fields with Galois group G. For any prime ideal $\mathfrak{P}$ of $\mathcal{O}_L$, the subgroup

$$G_{\mathfrak{P}} = \{\sigma \in G \mid \sigma(\mathfrak{P}) = \mathfrak{P}\}$$

is called the **decomposition group of $\mathfrak{P}$**.

For all $n \geq 0$, the $n-$**th ramification group of $\mathfrak{P}$** is the subgroup of G defined by

$$G_n(\mathfrak{P}) = \{\sigma \in G \mid \sigma(\alpha) - \alpha \in \mathfrak{P}^{n+1} \text{ for all } \alpha \in \mathcal{O}_L\}.$$

This is in fact a subgroup of $G_{\mathfrak{P}}$. Indeed, if $\sigma \in G_n(\mathfrak{P})$, then for all $\alpha \in \mathfrak{P}$, we have

$$\sigma(\alpha) \in \mathfrak{P} + \mathfrak{P}^{n+1} \subset \mathfrak{P}.$$

We then have $\sigma(\mathfrak{P}) \subset \mathfrak{P}$, which is equivalent to $\sigma(\mathfrak{P}) = \mathfrak{P}$ by maximality of $\sigma(\mathfrak{P})$. This means that $\sigma \in G_{\mathfrak{P}}$.

Finally, notice that for all $n \geq 0$, $G_n(\mathfrak{P})$ is a subgroup of $G_{n-1}(\mathfrak{P})$ (with the convention $G_{-1}(\mathfrak{P}) = G_{\mathfrak{P}}$).

The interest of this filtration is given by the following theorem.

THEOREM B.2.33. *Let L/K be a Galois extension of number fields. For any prime ideal $\mathfrak{P}$ of $\mathcal{O}_L$, the groups $G_n(\mathfrak{P})$ are trivial for a sufficiently large n, and we have*

$$n_{\mathfrak{P}}(\mathfrak{d}_{L/K}) = \sum_{n \geq 0} (|G_n(\mathfrak{P})| - 1).$$

We let the reader refer to [**30**, Theorem 1.54] for a proof.

B.3. Absolute values on number fields and completion

We start by describing the set of absolute values on a given number field.

Let K be a number field. Any prime ideal $\mathfrak{p}$ of $\mathcal{O}_K$ lies above a unique prime ideal $p\mathbb{Z}$ of $\mathbb{Z}$. We will say that $\mathfrak{p}$ lies above the prime p, and we will denote by $e_{\mathfrak{p}|p}$ and $f_{\mathfrak{p}|p}$ the corresponding ramification index and residual degree.

We then set

$$v_{\mathfrak{p}} : \begin{array}{c} K \longrightarrow \mathbb{R}^+ \\ x \longmapsto p^{-\frac{n_{\mathfrak{p}}(x)}{e_{\mathfrak{p}|p}}} . \end{array}$$

This is a non-trivial discrete absolute value on K, which extends the p-adic valuation v_p of $\mathbb{Q}$.

REMARKS B.3.1.

(1) If $\mathfrak{p}$ is principal, generated by π, then π is a local parameter for $v_{\mathfrak{p}}$.
(2) One may show that the residue field of $v_{\mathfrak{p}}$ is isomorphic to $\kappa(\mathfrak{p})$.

$\square$

Let $\sigma : K \longrightarrow \mathbb{C}$ be a $\mathbb{Q}$-embedding of K into $\mathbb{C}$. If $\sigma(K) \subset \mathbb{R}$, we call σ a **real** embedding and we set

$$v_{\sigma} : \begin{array}{c} K \longrightarrow \mathbb{R}^+ \\ x \longmapsto |\sigma(x)|, \end{array}$$

where $|\cdot|$ denotes the classical absolute value of $\mathbb{R}$.

If $\sigma(K) \not\subset \mathbb{R}$, we call σ a **complex** embedding. We will say that two complex embeddings σ_1, σ_2 are **conjugate** if $\sigma_2(x) = \overline{\sigma_1(x)}$ for all $x \in K$, where $\bar{}$ denotes the complex conjugation. If σ is a complex embedding, we set

$$v_{\sigma} : \begin{array}{c} K \longrightarrow \mathbb{R}^+ \\ x \longmapsto |\sigma(x)|, \end{array}$$

where $|\cdot|$ denotes the modulus of a complex number.

In both cases, v_{σ} is an archimedean valuation.

DEFINITION B.3.2. The absolute value $v_{\mathfrak{p}}$ on K is called the **$\mathfrak{p}$-adic absolute value**.

A place represented by an absolute value $v_{\mathfrak{p}}$ for some prime ideal $\mathfrak{p}$ is called a **finite** place.

A place represented by an absolute value v_{σ} for some real (resp. complex) embedding is called a **real** (resp. **complex**) place.

We then have the following result.

THEOREM B.3.3. *Let K be a number field. Then the set of places of K is represented by the following absolute absolute values, which are pairwise non-equivalent:*

(1) *the absolute values $v_{\mathfrak{p}}$, where $\mathfrak{p}$ describes the set of prime ideals of $\mathcal{O}_K$;*

(2) *the absolute values v_σ, where σ describes the set of real embeddings of K;*

(3) *the absolute values v_σ, where σ describes a maximal set of pairwise non-conjugate complex embeddings of K.*

See [**51**], Lemmas 16 and 17.

Notice now that if L/K is an extension of number fields, and $\mathfrak{P} \mid \mathfrak{p}$, then the restriction of $v_{\mathfrak{P}}$ to K is $v_{\mathfrak{p}}$.

Indeed, it readily follows from the definitions that we have

$$n_{\mathfrak{P}}(x) = e_{\mathfrak{p}|\mathfrak{P}} n_{\mathfrak{p}}(x) \text{ for all } x \in K,$$

as well as

$$e_{\mathfrak{P}|p} = e_{\mathfrak{p}|p} e_{\mathfrak{P}|\mathfrak{p}}.$$

The claim follows easily. Thus, the following result makes sense.

LEMMA B.3.4. *Let L/K be a extension of number fields. Let $\mathfrak{p}$ be a prime ideal of $\mathcal{O}_K$, and let $\mathfrak{P}$ be a prime ideal of $\mathcal{O}_L$ lying above $\mathfrak{P}$. Then we have*

$$e_{v_{\mathfrak{P}}|v_{\mathfrak{p}}} = e_{\mathfrak{P}|\mathfrak{p}} \quad and \quad f_{v_{\mathfrak{P}}|v_{\mathfrak{p}}} = f_{\mathfrak{P}|\mathfrak{p}}.$$

Proof. If $\pi_{\mathfrak{p}}$ is a local parameter of $v_{\mathfrak{p}}$ and $\pi_{\mathfrak{P}}$ is a local parameter of $v_{\mathfrak{P}}$, we have

$$n_{\mathfrak{P}}(\pi_{\mathfrak{p}}) = e_{\mathfrak{P}|\mathfrak{p}} n_{\mathfrak{p}}(\pi_{\mathfrak{p}}) = e_{\mathfrak{P}|\mathfrak{p}} = n_{\mathfrak{P}}(\pi_{\mathfrak{P}}^{e_{\mathfrak{P}|\mathfrak{p}}}),$$

so that $\pi_{\mathfrak{P}} = u \pi_{\mathfrak{P}}^{e_{\mathfrak{P}|\mathfrak{p}}}, u \in \mathcal{O}_{v_{\mathfrak{P}}}^\times$. By Remark B.1.14, we get the equality

$$e_{v_{\mathfrak{P}}|v_{\mathfrak{p}}} = e_{\mathfrak{P}|\mathfrak{p}}.$$

The second equality comes from the fact that the residue fields of $v_{\mathfrak{p}}$ and $v_{\mathfrak{P}}$ are respectively $\kappa(\mathfrak{p})$ and $\kappa(\mathfrak{P})$. $\qquad\square$

To end this section, we would like to relate number fields and local fields via their completions with respect to a valuation. Let K be a number field. Notice that equivalent absolute values give rise to canonically isomorphic completions. If v is a place of K, we will write K_v for the completion of K with respect to any valuation representing this place.

If v is a real place, then $K_v \simeq \mathbb{R}$, and if v is a complex place, we have $K_v \simeq \mathbb{C}$.

Now assume that v is a finite place, and let $\mathfrak{p}$ the corresponding prime ideal of $\mathcal{O}_K$. We will denote by $K_{\mathfrak{p}}$ the corresponding completion. Since $K_{\mathfrak{p}}$ is complete for a non-trivial discrete absolute value with finite residue field $\kappa(\mathfrak{p})$, this is a local field.

Let L/K be an extension of number fields. Let $\mathfrak{p}$ be an ideal of $\mathcal{O}_K$, and let $\mathfrak{P}$ be an ideal of $\mathcal{O}_L$ lying above $\mathfrak{p}$. Since $v_{\mathfrak{P}}$ extends $v_{\mathfrak{p}}$, we get an extension $(L_{\mathfrak{P}}, v_{\mathfrak{P}})/(K_{\mathfrak{p}}, v_{\mathfrak{p}})$ of local fields.

Since completion preserves the ramification index and the residual degree, this extension has ramification index $e_{\mathfrak{P}|\mathfrak{p}}$ and residual degree $f_{\mathfrak{P}|\mathfrak{p}}$. In particular, we have the following result.

PROPOSITION B.3.5. *Let L/K be an extension of number fields. Let $\mathfrak{p}$ be an ideal of $\mathcal{O}_K$, and let $\mathfrak{P}$ be an ideal of $\mathcal{O}_L$ lying above $\mathfrak{p}$. Then we have*

$$[L_{\mathfrak{P}} : K_{\mathfrak{p}}] = e_{\mathfrak{P}|\mathfrak{p}} f_{\mathfrak{P}|\mathfrak{p}}.$$

To conclude this appendix, we take a look to the Galois case. The only thing which does not derive from the previous result results in the next proposition is the equality $L_{\mathfrak{P}} = LK_{\mathfrak{p}}$, which is proved in [**44**] Chapter II, §3.

PROPOSITION B.3.6. *Let L/K be a Galois extension of number fields. Let $\mathfrak{p}$ be an ideal of $\mathcal{O}_K$, and let $\mathfrak{P}$ be any ideal of $\mathcal{O}_L$ lying above $\mathfrak{p}$. Then $L_{\mathfrak{P}} = LK_{\mathfrak{p}}$ (in the completion of an algebraic closure of $k_{\mathfrak{p}}$). In particular, it does not depend on the choice of $\mathfrak{P}$, $L_{\mathfrak{P}}/K_{\mathfrak{p}}$ is Galois, and its Galois group identifies to a subgroup of $\mathrm{Gal}(L/K)$ of order $e_{\mathfrak{p}} f_{\mathfrak{p}}$.*

Finally, $L_{\mathfrak{P}}/K_{\mathfrak{p}}$ is unramified, resp. ramified, resp. totally ramified if and only if $\mathfrak{p}$ is.

REMARK B.3.7. In particular, if L/K is cyclic, so is $L_{\mathfrak{P}}/K_{\mathfrak{p}}$. □

Complex ideal lattices

In this appendix, we first recall few facts on hermitian lattices. We then study more carefully hermitian ideal lattices on an extension L/k of number fields.

C.1. Generalities on hermitian lattices

In this section, we recall some basic definitions on hermitian lattices and introduce some invariants that we will need later on.

DEFINITION C.1.1. Let $k/\mathbb{Q}$ be a totally imaginary quadratic field extension with non-trivial automorphism $k \longrightarrow k$, $u \longmapsto \bar{u}$ (which is nothing but complex conjugation). A **hermitian $\mathcal{O}_k$-lattice** is a pair (M, h), where M is a free $\mathcal{O}_k$-module and $h : M \times M \longrightarrow \mathcal{O}_k$ is a hermitian form with respect to $\bar{\ }$.

We say that two hermitian $\mathcal{O}_k$-lattices (M, h) and (M', h') are **isomorphic** if there is an isomorphism of $\mathcal{O}_k$-modules $f : M \xrightarrow{\sim} M'$ such that

$$h'(f(x), f(y)) = h(x, y) \text{ for all } x, y \in M.$$

A hermitian lattice (M, h) is **positive definite** if $h(x, x) > 0$ for all $x \in M \setminus \{0\}$. This property only depends on the isomorphism class of (M, h).

EXAMPLE C.1.2. Let $n \geq 1$ be an integer. The **cubic lattice** of rank n is the hermitian $\mathcal{O}_k$-lattice on $\mathcal{O}_k^n$ given by

$$h_0 : \begin{aligned} \mathcal{O}_k^n \times \mathcal{O}_k^n &\longrightarrow \mathcal{O}_k \\ (x, y) &\longmapsto \bar{x}^t y. \end{aligned}$$

Therefore, a hermitian $\mathcal{O}_k$-lattice (M, h) is isomorphic to the cubic lattice if and only if M has an orthonormal basis with respect to the hermitian form h. In this case, it is positive definite. $\qquad \square$

LEMMA C.1.3. *Let (M, h) be a hermitian $\mathcal{O}_k$-lattice, and let $\mathbf{e}$ be an $\mathcal{O}_k$-basis of M. Then $\mathrm{Mat}(h, \mathbf{e}) \in \mathbb{Z}$ and does not depend on the choice of $\mathbf{e}$.*

Proof. The matrix $H = \mathrm{Mat}(h, \mathbf{e})$ is a hermitian matrix, i.e., $H = \bar{H}^t$, which implies

$$\det(H) = \det(\overline{H}) = \overline{\det(H)}.$$

Therefore the determinant of H also lies in $\mathbb{R}$. Hence $\det(H) \in \mathcal{O}_k \cap \mathbb{R} = \mathbb{Z}$, since $k/\mathbb{Q}$ is a totally imaginary quadratic field extension.

Let $\mathbf{e}'$ be another basis of M, and let P denote the corresponding base change matrix. If $H' = \mathrm{Mat}(h, \mathbf{e}')$, then we have $H' = \overline{P}^t H P$, and therefore

$$\det(H') = N_{k/\mathbb{Q}}(\det(P)) \det(H).$$

Since $P \in \mathrm{GL}_n(\mathcal{O}_k)$, $\det(P) \in \mathcal{O}_k^\times$, and thus $N_{k/\mathbb{Q}}(\det(P)) = 1$, since $k/\mathbb{Q}$ is a totally imaginary quadratic field extension. This completes the proof. $\qquad\square$

DEFINITION C.1.4. The **determinant** of the lattice (M, h), denoted by $\det(M, h)$, is the determinant of any representative matrix of h. It only depends on the isomorphism class of (M, h).

EXAMPLE C.1.5. Assume that (M, h) is isomorphic to the cubic lattice. Then $\det(M, h) = 1$, since there exists an orthonormal $\mathcal{O}_k$-basis, that is a basis for which the corresponding representative matrix is the identity matrix. $\qquad\square$

REMARK C.1.6. If (M, h) is positive definite, then $\det(M, h)$ is positive. $\qquad\square$

We now introduce the signature of a hermitian $\mathcal{O}_k$-lattice (M, h). Extending scalars to k gives rise to a hermitian form on $V = M \otimes_{\mathcal{O}_k} k$ over k, that we will still denote by h.

It is well-known that the hermitian form $h : V \times V \longrightarrow k$ may be diagonalised, i.e.,
$$h \cong_k \langle a_1, \ldots, a_n \rangle, a_i \in \mathbb{Q}^\times.$$

DEFINITION C.1.7. The **signature** of (M, h) is defined as
$$\mathrm{sign}(M, h) = \sharp\{i \mid a_i > 0\} - \sharp\{i \mid a_i < 0\},$$
for any diagonalisation
$$h \cong_k \langle a_1, \ldots, a_n \rangle, a_i \in \mathbb{Q}^\times.$$
It only depends on the isomorphism class of (M, h).

REMARK C.1.8. It follows from the definition of the signature that (M, h) is positive definite if and only if $\mathrm{sign}(M, h) = \mathrm{rk}(M)$. $\qquad\square$

If (M, h) is a hermitian $\mathcal{O}_k$-lattice, we have $h(x, x) \in \mathcal{O}_k \cap \mathbb{R} = \mathbb{Z}$ for all $x \in M$. Thus, we may define the minimal distance of a hermitian $\mathcal{O}_k$-lattice (M, h) as follows.

DEFINITION C.1.9. Let (M, h) be a hermitian $\mathcal{O}_k$-lattice. The **minimal distance** of (M, h) is the non-negative integer $d(M, h)$ defined by
$$d(M, h) = \min_{x \in M \setminus \{0\}} |h(x, x)|.$$

Once again, two isomorphic hermitian $\mathcal{O}_k$-lattices will have the same minimal distance.

EXAMPLE C.1.10. If (M, h) is isomorphic to the cubic lattice, then $d(M, h) = 1$. $\quad\square$

C.2. Complex ideal lattices

We now study particular hermitian lattices which appears naturally in the context of algebra-based codes. We start with a lemma.

LEMMA C.2.1. *Let L/k be a finite extension of number fields. Assume that L is closed under complex conjugation, and let $L_0 = L \cap \mathbb{R}$. Then for any ideal I of $\mathcal{O}_L$ and any $\lambda \in L_0^\times$, we have*
$$\mathrm{Tr}_{L/k}(\lambda \bar{x} y) \in \mathcal{O}_k \text{ for all } x, y \in I \iff \lambda \bar{I} I \subset \mathfrak{d}_{L/k}^{-1}.$$

Proof. Assume that $\lambda \bar{I} I \subset \mathfrak{d}_{L/k}^{-1}$. Then for all $x, y \in I$ we have $\lambda \bar{x} y \in \mathfrak{d}_{L/k}^{-1}$ by assumption and therefore

$$\mathrm{Tr}_{L/k}(\lambda \bar{x} y) = \mathrm{Tr}_{L/k}(\lambda \bar{x} y \cdot 1_k) \subset \mathcal{O}_k \text{ for all } x, y \in I.$$

Conversely, assume that $\mathrm{Tr}_{L/k}(\lambda \bar{x} y) \in \mathcal{O}_k$ for all $x, y \in I$. Since $\lambda \bar{I} I$ is generated as an additive group by elements of the form $\lambda \bar{x}_1 x_2, x_i \in I$, it is enough to check that $\mathrm{Tr}_{L/k}(\lambda \bar{x}_1 x_2 y) \in \mathcal{O}_k$ for all $x_1, x_2 \in I, y \in \mathcal{O}_L$, which is clear from the assumption. This concludes the proof. $\qquad\square$

We assume from now on that $\mathcal{O}_k$ is a principal ideal domain, and that L/k is a finite field extension of degree n, which is closed under complex conjugation. We will denote by L_0 the maximal real subfield of L, that is $L_0 = L \cap \mathbb{R}$. In this case, L_0 and k are linearly disjoint over $\mathbb{Q}$, and any k-embedding of L into $\mathbb{C}$ is the canonical extension of a $\mathbb{Q}$-embedding of L_0 into $\mathbb{C}$. In particular, for every $\lambda \in L_0$, we have

$$N_{L/k}(\lambda) = N_{L_0/\mathbb{Q}}(\lambda) \in \mathbb{Q}.$$

Notice also that, since $\mathcal{O}_k$ is a principal ideal domain, any ideal I of $\mathcal{O}_L$ is a free $\mathcal{O}_k$-module of rank n. In particular, the following definition makes sense.

DEFINITION C.2.2. Let L/k be an extension of number fields where L is closed under complex conjugation. A **complex ideal lattice** on L/k is a pair (I, h_λ), where I is an ideal of $\mathcal{O}_L$ with $\lambda \in L_0^\times$ satisfying $\lambda \bar{I} I \subset \mathfrak{d}_{L/k}^{-1}$ and h_λ is the hermitian $\mathcal{O}_k$-lattice

$$h_\lambda: \begin{array}{c} I \times I \longrightarrow \mathcal{O}_k \\ (x, y) \longmapsto \mathrm{Tr}_{L/k}(\lambda \bar{x} y). \end{array}$$

The rest of this paragraph is devoted to the computation of the invariants introduced above for a given complex ideal lattice. We start with a definition.

DEFINITION C.2.3. The **relative discriminant** of L/k, denoted by $d_{L/k}$, is the determinant of $(\mathcal{O}_L, h_0)$. In other words,

$$d_{L/k} = \det(\mathrm{Tr}_{L/k}(\overline{w}_i w_j)) \in \mathbb{Z}$$

for any $\mathcal{O}_k$-basis $(w_1, \ldots, w_n)$ of $\mathcal{O}_L$.

We will assume until the end that complex conjugation commutes with all the k-embeddings of L into $\mathbb{C}$.

We are now ready to state our first result.

PROPOSITION C.2.4. *Let (I, h_λ) be an ideal lattice on L/k. Then*

$$\det(I, h_\lambda) = N_{L/k}(\lambda) N_{L/\mathbb{Q}}(I) d_{L/k}.$$

Moreover, we have

$$\mathrm{sign}(I, h_\lambda) = \sharp\{\sigma \in X(L) \mid \sigma(\lambda) > 0\} - \sharp\{\sigma \in X(L) \mid \sigma(\lambda) < 0\}.$$

In particular, (I, h_λ) is positive definite if and only if $\sigma(\lambda) > 0$ for every k-embedding $\sigma : L \longrightarrow \mathbb{C}$.

Proof. Since $\mathcal{O}_k$ is a principal ideal domain and $\mathcal{O}_L$ is a torsion-free finitely generated $\mathcal{O}_k$-module, there exists an $\mathcal{O}_k$-basis $\mathbf{w} = (w_1, \ldots, w_n)$ of $\mathcal{O}_L$ and elements $q_1, \ldots, q_n \in \mathcal{O}_k$ such that $\mathbf{w}' = (q_1 w_1, \ldots, q_n w_n)$ is an $\mathcal{O}_k$-basis of I.

Let $\sigma_1, \ldots, \sigma_n$ be the n k-embeddings of L into $\mathbb{C}$. Since complex conjugation commutes with $\sigma_1, \ldots, \sigma_n$, we have

$$\mathrm{Mat}(h_\lambda, \mathbf{w}') = \overline{W'}^t \mathcal{L} W',$$

where

$$W' = (q_j w_j^{\sigma_i})_{i,j} \text{ and } \mathcal{L} = \begin{pmatrix} \lambda^{\sigma_1} & & \\ & \ddots & \\ & & \lambda^{\sigma_n} \end{pmatrix}.$$

The assertion on the signature of (I, h_λ) then immediately follows.

Clearly, $\det(\mathcal{L}) = N_{L/k}(\lambda)$ and $\det(W') = q_1 \ldots q_n \det(W)$, where we have set $W = (w_j^{\sigma_i})_{i,j}$. Therefore, we get

$$\det(I, h_\lambda) = N_{L/k}(\lambda) \overline{q_1 \cdots q_n} \cdot q_1 \cdots q_n \det(\overline{W}^t) \det(W).$$

If $I = \mathcal{O}_L$ and $\lambda = 1$, we get $d_{L/k} = \det(\overline{W}^t) \det(W)$, and therefore

$$\det(I, h_\lambda) = N_{L/k}(\lambda) \overline{q_1 \cdots q_n} \cdot q_1 \cdots q_n d_{L/k}.$$

Therefore, it remains to show that $N_{L/\mathbb{Q}}(I) = \overline{q_1 \cdots q_n} \cdot q_1 \cdots q_n$. Recall that $N_{L/\mathbb{Q}}(I)$ is by definition the number of elements of $\mathcal{O}_L / I$. Notice now that we have an isomorphism of $\mathcal{O}_k$-modules

$$\mathcal{O}_L / I \cong_{\mathcal{O}_k} \mathcal{O}_k / q_1 \mathcal{O}_k \times \cdots \times \mathcal{O}_k / q_n \mathcal{O}_k.$$

Thus, it remains to show that for a given $q \in \mathcal{O}_k \setminus \{0\}$, the group $\mathcal{O}_k / q \mathcal{O}_k$ has $\bar{q} q$ elements. But the number of elements of $\mathcal{O}_k / q \mathcal{O}_k$ is by definition $N_{k/\mathbb{Q}}(q \mathcal{O}_k)$, which is nothing but $\bar{q} q$. This completes the proof. $\qquad\square$

REMARK C.2.5. The proof above also shows that $d_{L/k} > 0$.

Indeed, this follows from the equalities

$$d_{L/k} = \det(\overline{W}^t) \det(W) = \overline{\det(W)} \det(W) > 0.$$

$\qquad\square$

We now relate $d_{L/k}$ to the norm of the different ideal.

LEMMA C.2.6. *We have $d_{L/k} = \sqrt{N_{k/\mathbb{Q}}(\mathcal{D}_{L/k})} = \sqrt{N_{L/\mathbb{Q}}(\mathfrak{d}_{L/k})}$.*

Proof. Recall that $\mathcal{D}_{L/k}$ is the ideal generated by the elements $\det(\mathrm{Tr}_{L/k}(x_i x_j))$, where $x_1, \ldots, x_n$ run through the k-bases of L consisting of elements of $\mathcal{O}_L$. Since $\mathcal{O}_k$ is a principal ideal domain, $\mathcal{D}_{L/k}$ is generated by $\det(\mathrm{Tr}_{L/k}(w_i w_j))$, where $(w_1, \ldots, w_n)$ is an $\mathcal{O}_k$-basis of $\mathcal{O}_L$. We then have

$$N_{k/\mathbb{Q}}(\mathcal{D}_{L/k}) = N_{k/\mathbb{Q}}(\det(\mathrm{Tr}_{L/k}(w_i w_j))).$$

Now if $W = (w_j^{\sigma_i})$, we have $\det(\mathrm{Tr}_{L/k}(w_i w_j)) = \det(W^t W) = \det(W)^2$, and therefore

$$N_{k/\mathbb{Q}}(\mathcal{D}_{L/k}) = N_{k/\mathbb{Q}}(\det(W))^2 = (\overline{\det(W)} \det(W))^2 = d_{L/k}^2.$$

Since $d_{L/k} > 0$ by Remark C.2.5, we are done. $\qquad\square$

COROLLARY C.2.7. *Let $k \subset M \subset L$ be a tower of field extensions. Then we have*

$$(d_{L/k})^2 = N_{L/\mathbb{Q}}(\mathfrak{d}_{L/M}) \cdot (d_{M/k})^{2 \cdot [L:M]}$$

In particular, $d_{M/k}^{[L:M]} \mid d_{L/k}$.

Proof.   By the previous lemma, we have $(d_{L/k})^2 = N_{k/\mathbb{Q}}(\mathfrak{d}_{L/k})$.   By Theorem B.2.30, we have

$$\mathfrak{d}_{L/k} = \mathfrak{d}_{M/k}\mathfrak{d}_{L/M}.$$

Applying $N_{L/\mathbb{Q}}$ and using Propositions B.2.20 and B.2.22, we get

$$N_{L/\mathbb{Q}}(\mathfrak{d}_{L/k}) = N_{M/\mathbb{Q}}(\mathfrak{d}_{L/M}) \cdot N_{M/\mathbb{Q}}(\mathfrak{d}_{M/k})^{[L:M]}.$$

Now apply the previous lemma to conclude.   □

COROLLARY C.2.8. *Let (I, h_λ) be a complex ideal lattice on L/k.   Then $\det(I, h_\lambda) = \pm 1$ if and only if $\lambda \overline{I} I = \mathfrak{d}_{L/k}^{-1}$.*

Proof.   Since we have $\lambda \overline{I} I \subset \mathfrak{d}_{L/k}^{-1}$ by definition of a complex ideal lattice on L/k, we will have $\lambda \overline{I} I = \mathfrak{d}_{L/k}^{-1}$ if and only if

$$N_{L/\mathbb{Q}}(\lambda \overline{I} I) = N_{L/\mathbb{Q}}(\mathfrak{d}_{L/k}^{-1}),$$

that is

$$N_{L/\mathbb{Q}}(\lambda \overline{I} I) N_{L/\mathbb{Q}}(\mathfrak{d}_{L/k}) = 1.$$

Since complex conjugation is an automorphism of $L/\mathbb{Q}$, we have

$$N_{L/\mathbb{Q}}(\overline{I}) = N_{L/\mathbb{Q}}(I).$$

Moreover, we have

$$N_{L/\mathbb{Q}}(\lambda) = N_{k/\mathbb{Q}}(N_{L/k}(\lambda)) = N_{L/k}(\lambda)^2,$$

since $N_{L/k}(\lambda) \in \mathbb{Q}$. Using Lemma C.2.6, the condition above may be rewritten as

$$(N_{L/k}(\lambda) N_{L/\mathbb{Q}}(I) d_{L/k})^2 = 1,$$

that is $\det(I, h_\lambda)^2 = 1$ by Proposition C.2.4. This completes the proof.   □

REMARK C.2.9. In particular, if (I, h_λ) is positive definite, we have $\det(I, h_\lambda) = 1$ if and only if $\lambda \overline{I} I = \mathfrak{d}_{L/k}^{-1}$, since $\det(I, h_\lambda) > 0$ in this case.   □

We now give an estimation of the minimal distance of a complex ideal lattice.

PROPOSITION C.2.10. *Let (I, h_λ) be a positive definite complex ideal lattice. Then we have*

$$d(I, h_\lambda) \geq n[N_{L/k}(\lambda) N_{L/\mathbb{Q}}(I)]^{1/n}.$$

In particular, if $\det(I, h_\lambda) = 1$, we have

$$d(I, h_\lambda) \geq n \cdot d_{L/k}^{-1/n}.$$

Proof. By Proposition C.2.4, $\sigma(\lambda)$ is a positive real number for every embedding $\sigma : L \longrightarrow \mathbb{C}$. Since σ commutes with complex conjugation, $\sigma(\lambda\overline{x}x) = \sigma(\lambda)\overline{\sigma(x)}\sigma(x)$ is a positive real number for all $x \in L$. In particular, the inequality between the arithmetic and geometric means implies that

$$\frac{1}{n}\mathrm{Tr}_{L/k}(\lambda\overline{x}x) \geq N_{L/k}(\lambda\overline{x}x)^{1/n} \text{ for all } x \in I.$$

Now for all $x \in I$, we have $\lambda\overline{x}x\mathcal{O}_L \subset \lambda\overline{I}I$, and therefore

$$N_{L/\mathbb{Q}}(\lambda)N_{L/\mathbb{Q}}(I)^2 \mid N_{L/\mathbb{Q}}(\lambda\overline{x}x\mathcal{O}_L) = N_{L/\mathbb{Q}}(\lambda\overline{x}x).$$

In particular, if $x \neq 0$, we get

$$N_{L/\mathbb{Q}}(\lambda\overline{x}x) \geq N_{L/\mathbb{Q}}(\lambda)N_{L/\mathbb{Q}}(I)^2.$$

Since $\lambda\overline{x}x \in L_0$, we have $N_{L/k}(\lambda\overline{x}x) \in \mathbb{Q}$ and thus

$$N_{L/\mathbb{Q}}(\lambda\overline{x}x) = N_{L/k}(\lambda\overline{x}x)^2.$$

For the same reason, $N_{L/\mathbb{Q}}(\lambda) = N_{L/k}(\lambda)^2$, and we get

$$N_{L/k}(\lambda\overline{x}x) \geq N_{L/k}(\lambda)N_{L/\mathbb{Q}}(I),$$

taking into account that $N_{L/k}(\lambda)$ and $N_{L/k}(\lambda\overline{x}x)$ are positive. We finally get

$$\mathrm{Tr}_{L/k}(\lambda\overline{x}x) \geq n[N_{L/k}(\lambda)N_{L/\mathbb{Q}}(I)]^{1/n} \text{ for all } x \in I \setminus \{0\},$$

which proves the first part of the proposition. The second part follows from Proposition C.2.4. $\square$

Since the cubic lattice has a minimal distance equal to 1, we get the following corollary.

COROLLARY C.2.11. *If the cubic lattice is isomorphic to a complex ideal lattice on L/k, then $d_{L/k} \geq n^n$.*

Bibliography

1. S.M. Alamouti, *A simple transmit diversity technique for wireless communications*, IEEE J. Selected Areas Communications **16** (1998), 1451–1458.
2. A.A. Albert, *Structure of Algebras*, A.M.S. Coll.Pub., vol. 24, AMS, 1939.
3. S.A. Amitsur, *On central division algebras*, Israel J. Math. **12** (1972), no. 4, 408–420.
4. E. Bayer-Fluckiger, F. Oggier, and E. Viterbo, *New algebraic constructions of rotated $\mathbb{Z}^n$-lattice constellations for the Rayleigh fading channel*, IEEE Transactions on Information Theory **50** (2004), no. 4.
5. J.-C. Belfiore and G. Rekaya, *Quaternionic lattices for space-time coding*, Proceedings of IEEE Information Theory Workshop (ITW), Paris (2003).
6. J.-C. Belfiore, G. Rekaya, and E. Viterbo, *The Golden code: A 2×2 full rate space-time code with non vanishing determinants*, IEEE Trans. on Inf. Theory **51** (2005), no. 4.
7. G. Berhuy, *Algebraic space-time codes based on division algebras with a unitary involution*, (2012), Preprint. Available from http://www-fourier.ujf-grenoble.fr/~berhuy/fichiers/unitarycodes.pdf.
8. G. Berhuy and F. Oggier, *Space-time codes from crossed product algebras of degree 4*, Proceedings of Applied algebra, algebraic algorithms and error-correcting codes, Lecture Notes in Comput. Sci **4851** (2007), 90–99.
9. ———, *On the existence of perfect space-time codes*, IEEE Transactions on Information Theory **55** (2009), no. 5, 2078–2082.
10. G. Berhuy and R. Slessor, *Optimality of codes based on crossed product algebras*, (2011), Preprint. Available from http://www-fourier.ujf-grenoble.fr/~berhuy/Berhuy-Slessor-2209.pdf.
11. J.W.S. Cassels and A. Frölich, *Algebraic number theory*, Acad. Press, 1967.
12. M. O. Damen, A. Tewfik, and J.-C. Belfiore, *A construction of a space-time code based on number theory*, IEEE Transactions on Information Theory **48** (2002), no. 3.
13. P. Elia, B.A. Sethuraman, and P. V. Kumar, *Perfect space-time codes for any number of antennas*, IEEE Transactions on Information Theory **53** (2007).
14. Jr. G. Forney, R. Gallager, G. Lang, F. Longstaff, and S. Qureshi, *Efficient modulation for band-limited channels*, IEEE Journal on Selected Areas in Communications **2** (1984), no. 5, 632–647.
15. A. Fröhlich and M. Taylor, *Algebraic number theory*, Cambridge University Press, 1991.
16. P. Gille and T. Szamuely, *Central simple algebras and Galois cohomology*, Cambridge Studies in Advanced Mathematics, vol. 101, Cambridge University Press, 2006.
17. D. Haile, *A useful proposition for division algebras of small degree*, Proc. Amer. Math. Soc. **106** (1989), no. 2, 317–319.
18. T. Hanke, *A twisted Laurent series ring that is a noncrossed product*, Israel J. Math. **150** (2005).
19. B. Hassibi and B. Hochwald, *Cayley differential unitary space-time codes*, IEEE Trans. on Information Theory **48** (2002).
20. J. Hiltunen, C. Hollanti, and J. Lahtonen, *Dense full-diversity matrix lattices for four transmit antenna MISO channel*, Proceedings of IEEE International Symposium on Information Theory (ISIT) (2005).
21. B. Hochwald and W. Sweldens, *Differential unitary space time modulation*, IEEE Trans. Commun. **48** (2000).
22. J. G. Huard, B. K. Spearman, and K. S. Williams, *Integral bases for quartic fields with quadratic subfields*, J. Number Theory **51** (1995), no. 1, 87–102. MR1321725 (96a:11115)
23. B. Hughes, *Differential space-time modulation*, IEEE Trans. Inform. Theory **46** (2000).

24. K. Iyanaga, *Class numbers of definite hermitian forms*, J. Math. Soc. Japan **21** (1969).

25. N. Jacobson, *Finite-dimensional division algebras over fields*, Springer-Verlag, Berlin, 1996.

26. G. J. Janusz, *Algebraic number fields*, second ed., Graduate Studies in Mathematics, vol. 7, American Mathematical Society, Providence, RI, 1996.

27. S. Karmakar and B Sundar Rajan, *High-rate, multi-symbol-decodable STBCs from Clifford algebras*, IEEE Transactions on Information Theory **55** (2009), no. 6.

28. A. W. Knapp, *Advanced algebra*, Cornerstones, Birkhäuser Boston Inc., Boston, MA, 2007, Along with a companion volume ıt Basic algebra.

29. M.A. Knus, A. Merkurjev, M. Rost, and J.-P. Tignol, *The book of involutions*, Amer. Math. Soc. Coll. Pub., vol. 44, A.M.S, 1998.

30. H. Koch, *Algebraic number theory*, Springer-Verlag,Berlin, 1997.

31. _____, *Number theory: algebraic numbers and functions*, Graduate Studies in Mathematics, vol. 24, A.M.S, 2000.

32. J. Lahtonen, N. Markin, and G. McGuire, *Construction of space-time codes from division algebras with roots of unity as non-norm elements*, IEEE Trans. on Information Theory **54** (2008), no. 11.

33. F. Oggier, *On the optimality of the Golden code*, IEEE Information Theory Workshop (ITW'06) (2006).

34. _____, *Cyclic algebras for noncoherent differential space-time coding*, IEEE Transactions on Information Theory **53 (9)** (2007), 3053–3065.

35. _____, *A survey of algebraic unitary codes*, International Workshop on Coding and Cryptology **5757** (2009), 171–187.

36. F. Oggier, J.-C. Belfiore, and Viterbo E., *Cyclic division algebras: A tool for space-time coding*, Now Publishers Inc., Hanover, MA, USA, 2007.

37. F. Oggier and B. Hassibi, *"an algebraic family of distributed space-time codes for wireless relay networks*, Proceedings of IEEE International Symposium on Information Theory (ISIT) (2006).

38. _____, *Algebraic cayley differential space-time codes*, IEEE Transactions on Information Theory **53** (2007), no. 5.

39. F. Oggier and L. Lequeu, *Families of unitary matrices achieving full diversity*, International Symposium on Information Theory (2005), 1173–1177.

40. F. Oggier, G. Rekaya, J.-C. Belfiore, and E. Viterbo, *Perfect space-time block codes*, IEEE Trans. Inform. Theory **52** (2006), no. 9.

41. R.S. Pierce, *Associative algebras*, Graduate Texts in Mathematics, vol. 88, Springer-Verlag, New York-Berlin, 1982.

42. S. Pumpluen and T. Unger, *Space-time block codes from nonassociative division algebras*, Advances in Mathematics of Communications **5** (2011), no. 3.

43. P. Ribenboim, *Classical theory of algebraic numbers*, Universitext, Springer, 2001.

44. J.-P. Serre, *Corps locaux (4ème édition)*, Hermann, 1997.

45. B. A. Sethuraman, *Division algebras and wireless communication*, Notices of the AMS **57** (2010), no. 11.

46. B.A. Sethuraman and B. Sundar Rajan, *An algebraic description of orthogonal designs and the uniqueness of the Alamouti code*, Proceedings of GlobCom (2002).

47. _____, *STBC from field extensions of the rational field*, Proceedings of IEEE International Symposium on Information Theory (ISIT), Lausanne (2002).

48. B.A. Sethuraman, B. Sundar Rajan, and V. Shashidhar, *Full-diversity, high-rate space-time block codes from division algebras*, IEEE Transactions on Information Theory **49** (2003).

49. A. Shokrollahi, B. Hassibi, B.M. Hochwald, and W. Sweldens, *Representation theory for high-rate multiple-antenna code design*, IEEE Trans. Information Theory **47** (2001), no. 6.

50. R. Slessor, *Performance of codes based on crossed product algebras*, Ph.D. thesis, School of Mathematics, Univ. of Southampton, May 2011 note = Available from http://eprints.soton.ac.uk/197309/3.hasCoversheetVersion/Thesis_--_Richard_Slessor.pdf, OPTannote = annote, 2011.

51. H. P. F. Swinnerton-Dyer, *A brief guide to algebraic number theory*, Cambridge University Press, 2001.

52. V. Tarokh, N. Seshadri, and A. Calderbank, *Space-time codes for high data rate wireless communication : Performance criterion and code construction*, IEEE Trans. Inform. Theory **44** (1998), 744–765.

53. The PARI Group, Bordeaux, *PARI/GP, version* 2.3.3, 2005, available from `http://pari.math.u-bordeaux.fr/`.

54. J.-P. Tignol, *Produits croisés abéliens*, J.of Algebra **70** (1981).

55. T. Unger and N. Markin, *Quadratic forms and space-time block codes from generalized quaternion and biquaternion algebras*, IEEE Trans. on Information Theory **57** (2011), no. 9.

56. R. Vehkalahti, C. Hollanti, J. Lahtonen, and K. Ranto, *On the densest MIMO lattices from cyclic division algebras*, IEEE Transactions on Information Theory **55** (2009), no. 8.

57. S. Vummintala, B. Sundar Rajan, and B.A. Sethuraman, *Information-lossless space-time block codes from crossed-product algebras*, IEEE Trans. Inform. Theory **52** (2006), no. 9.

58. H. Yao and G.W. Wornell, *Achieving the full MIMO diversity-multiplexing frontier with rotation-based space-time codes*, Proceedings of Allerton Conf. on Communication, Control and Computing (2003).

Index

Selected Published Titles in This Series

For a complete list of titles in this series, visit the
AMS Bookstore at **www.ams.org/bookstore/survseries/**.